wise advice and
warmhearted support

Bill

8·10·95

The Clinical Biology of Sodium

The Physiology and Pathophysiology of
Sodium in Mammals

Titles of related interest from Elsevier

Journals

Comparative Biochemistry and Physiology
>Part A: Physiology
>Part B: Biochemistry and Molecular Biology
>Part C: Pharmacology, Toxicology and Endocrinology

Current Advances in Clinical Chemistry (an official journal of the Association of Clinical Biochemists)

Urology and Nephrology

Current Advances in Endocrinology and Metabolism

American Journal of Hypertension

Books

PUSCHETT & GREENBERG
Diuretics IV: Chemistry, Pharmacology and Clinical Applications (Proceedings of the Fourth International Conference on Diuretics, 1992)

VRZGULA
Metabolic Disorders and Their Prevention in Farm Animals

GANTEN & DE JONG
Experimental and Genetic Models of Hypertension

ROBERTSON
Clinical Hypertension

BÜHLER & LARAGH
The Management of Hypertension

The Clinical Biology of Sodium

The Physiology and Pathophysiology of Sodium in Mammals

by

A. R. MICHELL

Professor of Applied Physiology and Comparative Medicine,
The Royal Veterinary College, University of London,
North Mymms AL9 7TA, U.K.

Pergamon

U.K.   Elsevier Science Ltd, The Boulevard, Langford Lane,
       Kidlington, Oxford OX5 1GB, U.K.

U.S.A. Elsevier Science Inc., 660 White Plains Road, Tarrytown,
        New York 10591-5153, U.S.A.

JAPAN   Elsevier Science Japan, Tsunashima Building Annex, 3-20-12
        Yushima, Bunkyo-ku, Tokyo 113, Japan

Copyright © 1995 Elsevier Science Ltd
All Rights Reserved. No part of this publication may be
reproduced, stored in a retrieval system or transmitted in any
form or by any means: electronic, electrostatic, magnetic
tape, mechanical, photocopying, recording or otherwise,
without permission in writing from the copyright holders.

The right of Alastair R. Michell to be identified as author
of this work has been asserted by him in accordance with the
Copyright, Designs and Patents Act, 1988.

First edition 1995

Library of Congress Cataloging in Publication Data
Michell, A. R.
The clinical biology of sodium: the physiology and pathophysiology
of sodium in mammals / by A.R. Michell.—1st ed.
p.   cm.
Includes bibliographical references and index.
1. Sodium—Physiological effect.   2. Sodium—
Pathophysiology.   3. Sodium—Metabolism.   I. Title.
QP535.N2M53   1995
599'.019214—dc20   95-2407

British Library Cataloguing in Publication Data
A catalogue record for this book is available from the British
Library

ISBN 0-08-040842-7

DISCLAIMER
Whilst every effort is made by the Publishers to see that no inaccurate
or misleading data, opinion or statement appear in this book, they
wish to make it clear that the data and opinions appearing in the
articles herein are the sole responsibility of the contributor concerned.
Accordingly, the Publishers and their employees, officers and agents
accept no responsibility or liability whatsoever for the consequences of
any such inaccurate or misleading data, opinion or statement.

Printed in Great Britain by Biddles Ltd, Guildford and King's Lynn

*To Charles 'Chuck' Kleeman
who set my course with wisdom
and the excitement of the subject,
and, characteristically,
with great kindness.*

Contents

Preface

Perhaps as long as they have existed, mammals, especially herbivores, have sought salt. In prehistoric times, bones accumulated at salt licks as animals risked death in pursuit of it. Still today, elephants in Africa hollow out caves to obtain it. Man has mined it, traded in it, treasured it, sworn covenants upon it, robbed and fought for it, taxed it, named towns for it, adulterated it like a drug. Why do we seem to crave so much? Is it a need or an addiction? How much do we need? What harm comes from excess? What does the biology of sodium in various species tell us about the management of clinical disturbances? What do the adaptations and maladaptations of disease tell us about the physiological mechanisms which regulate body sodium?

This book takes a comparative view of mammalian sodium regulation and its clinical disturbances, with particular attention to the insights gained from animals as well as humans. It emphasises the non-renal mechanisms which usually receive less attention, e.g. behavioural and enteric aspects of sodium regulation. A wide range of clinical topics is considered, including oral rehydration for diarrhoea, the physiological basis of fluid therapy, the pathogenesis of oedema, hypovolaemia, endocrine regulation and disturbances, and nutritional requirement, both in relation to hypertension and the design of experiments. In each instance, the concern is not for exhaustive detail, but to show how that aspect forms part of a logical, coherent whole.

Sodium is the osmotic skeleton of our extracellular fluid, the backbone of our internal environment, the foundation of our circulating blood volume. The book should interest doctors, nurses and veterinarians who have to manage disturbances of this most vital ion, particularly those involved in renal, enteric or cardiovascular medicine. The issues discussed include some of the most important in public health—including the continuing, and partly avoidable human and material costs of hypertension and the most miraculously cost-effective discovery in modern medicine—the ability of sugar and salt to tame the once lethal effects of cholera. Yet so much about sodium remains uncertain even in basic clinical areas, for example, how much is needed in pregnancy, what is its impact on hypertension in pregnancy?

This book should also interest those with a general involvement in mammalian physiology or those concerned with the physiological psychology of salt appetite. In addition, it should help those focused on a single aspect, perhaps the molecular biology of sodium transport, to perceive the broader context to which their detailed study relates.

Sodium is not simply pivotal in our physiology, but ingrained in our culture. This book will therefore find a wider readership, for example, among those interested in the controversy about salt and hypertension; it presents sodium in the breadth of its influence, for better or for worse.

Introduction

This introduction begins in an American airport, awaiting a trans-Atlantic flight. That may seem strange, but such waits allow contemplation and especially of the task to be completed in the next 12 months; the translation of a plan, that has evolved over the last 10 years, into a completed manuscript.

My briefcase contains the abstracts of the American Society of Nephrology meetings which I just attended in Boston—a bigger meeting, many will say, than the International Nephrology Congress a few months earlier in Jerusalem. Now, as then, the physical, let alone the intellectual weight of those abstracts is daunting and a substantial proportion concern topics relevant to the regulation of body sodium. For, as most textbooks would have us believe that homeostatic role is assigned totally to the kidney which, in health, fulfils it swiftly, efficiently and effectively, most of the remaining mysteries concern not what the kidney does but how, exactly, it does it so successfully.

Scarcely a handful of those attending these meetings were veterinarians. Moreover, while a successful meeting of the Renal Association in the U.K. might accommodate some hundred or so members, the plenary sessions in Boston looked more like the audience for a world title fight—indeed, from any distance, the speaker was visible only on the vast video-projection screens. The point is, why should a veterinarian, particularly a British veterinarian, have the temerity to write a book about the regulation of body sodium and—especially—its clinical disturbances? Worse, granted the 10,000 papers cited annually in Index Medicus on the subject of sodium, why should he choose to do it alone?

The answer is that the longer I have worked with sodium, since the early stages of my thesis, the more I have realised that its physiology, pathophysiology and clinical disturbances are studied for a variety of reasons by scientists with very different backgrounds. The overlap between these approaches is mostly less than optimal and frequently it is disastrously absent. Moreover, the more I have become committed to comparative medicine, the more I have realised that not only physiology, but clinical science are the poorer—and in some instances the more destitute—for the assumption that humans (and dogs and experimental rats, in so far as they represent them) are the norm for the biology of mammalian sodium balance. Other species, notably herbivores, are regarded as curious backwaters, rather than sources of valuable and relevant insights; I reject that view. It leads, for example, to inadequate awareness of the potential importance of salt appetite and enteric sodium regulation and to erroneous perceptions of the nutritional requirement for sodium.

Whole books are already available on salt appetite, renal excretion, clinical electrolyte disturbances, Na–K ATPase, enteric secretion, etc. As a single author book, this cannot even intend to be comprehensive. Instead, it offers a coherent and individual view of the interactions between a number of different fields which all focus on the biology and clinical implications of mammalian sodium regulation. It aims to clarify the mosaic into which the numerous tiles fit, and so avoid the limitations of analyses focused on particular regions, important and admirable though these may be. Since the kidney dominates most accounts of body sodium and many excellent texts on renal function are already available, this book aims to redress the balance by emphasising non-renal aspects of sodium balance in health and disease. I have frequently cited my own publications, not because they have particular importance but because they contain relevant literature sources. A book which involves such a variety of contexts in which sodium is involved is in danger of having a bibliography longer than the text. References are chosen because they offer an informed entry point to wider areas of the literature and the use of primary references is mainly where they offer unique or pivotal information or where the ideas are very new. While the book is planned as a logical progression, not every reader necessarily reads every chapter. I have, therefore, deliberately included some partial repetitions, usually with cross-reference to the chapter containing the major discussion of the topic concerned.

There is no such field as 'Natriology' nor should such an awful word ever be allowed more than a single, fleeting appearance. But it is my view that the clinical biology of sodium—from cellular effects and membrane transport, through the hierarchy of renal and non-renal regulatory mechanisms, the interactions with other ions, to the clinical disturbances and pathophysiological and pharmacological rationale underlying their management—offers a research field with a body of knowledge as broad and as deep as many recognised '-ologies'.

Lastly, I would like to explain the dedication. Three publications have seminally influenced me as a scientist. 'Perspectives in Biology and Medicine' has excited me, since I was an undergraduate, through its consistent reaffirmation that while the accelerating advance of science may reflect its logarithmic progression in techniques, it remains fundamentally a subject of ideas. The second, as I was about to graduate as a clinician, was a paper on fluid therapy in calves by John Watt. It opened my eyes to an utterly different, unashamedly physiological approach to therapy and transformed the most boring subject I had encountered as a student—fluids and electrolytes—into one with a mesmeric fascination which has remained the driving force behind all my scientific activities. The third—and most formative—receives the dedication of this book.

As I did my PhD on the behavioural regulation of sodium intake, I became increasingly frustrated by the absolute lack of insight among many of those involved, into the nature of real electrolyte disturbances, as opposed to simplistic or even totally theoretical models. The first edition of Maxwell and Kleeman's 'Clinical Disorders of Fluid and Electrolyte Metabolism' had an unsurpassed blend of practicality, clarity and logical development of clinical principles from physiological insights. It also had the exemplary internal consistency which resulted from a small group of authors familiar with one another's views (something which we cherished in our own 'Veterinary Fluid Therapy' some 20 years later). That was lost in its second edition (when it changed to multiple authorship), but regained subsequently and while it is no longer the only superlative textbook in its field, it has, despite the rapidly expanding choice of excellent competitors, managed to remain pre-eminent.

It was my acquaintance with that book that led to my Fellowship at UCLA and allowed me to experience, first-hand, the wisdom and great personal kindness of Charles 'Chuck' Kleeman. It also began a chain of professional contacts which has been influential throughout the rest of my career, notably through Josi Levi, to Adrian Katz who enabled me to develop my interest in sodium transport and Marshall Lindheimer who brought both discipline and rampant enthusiasm—an unusual blend—to bear on my interest in hypertension and the fluid and electrolyte problems of pregnancy. There are others—notably Carl Pfaffmann who encouraged and refined my ideas on salt appetite—who have almost equal reason to be mentioned, but there are limits to such lists and I know that they know their influence remains a matter for my gratitude. I hope that all of them will take some pleasure in this outcome of the direction in which they pointed me. If nothing else, I have never lost the enthusiasm or the biological fascination for this subject which they fostered, and that, above all, is what I hope to convey to every reader. Last but far from least, I must thank my wife, Pauline, and my secretary, Rosemary Forster, for their patience during the production of this book; their contributions were totally different but without either I could not have completed what I began.

A. R. MICHELL

Acknowledgements

A number of illustrations are based on diagrams originally in *Veterinary Fluid Therapy* (Michell *et al.*, 1989, Blackwell) and are reproduced with permission from the publisher. Thanks are also due to 'Save The Children' for permission to reproduce Fig. 3.2.

1

Body Sodium in Context: Distribution, Functions and Regulation

Introduction: Some Problems in Context

Sodium is so abundantly available, its importance so obvious, that it is widely taken for granted. Analytically, it is simple to measure either by flame photometry or by ion-specific electrodes and the concentrations in blood, cells, bone, enteric fluids, saliva, sweat, tears, urine, faeces and cerebrospinal fluid are easily detectable. It is the dominant ion of extracellular fluid (ECF) both quantitatively and functionally, not least it is the single most important factor which dictates its volume. We have inherited its role as the key constituent of "that immortal sea which brought us hither."

The central concept of physiology is homeostasis, the maintenance of a constant internal environment, as conceived by Charles Robin (Neuman, 1969) and immortalised by Claude Bernard and Walter Cannon. Since ECF, as interstitial fluid, provides the internal environment of the cells and, as plasma, the transport medium which constantly replenishes it, the consequences of sodium deficiency are catastrophic. Depletion of extracellular volume, particularly circulating volume (hypovolaemia), leads to inadequate tissue perfusion, deterioration of plasma composition (lactic acidosis, hyperkalaemia, azotaemia, due to anaerobic metabolism, leakage of potassium from damaged cells and compromised renal function), and ultimately to hypovolaemic shock. Untreated, this is a self-accelerating complex of vicious cycles (positive feedback loops) and death is the almost inevitable outcome *(Chapter 8)*.

Granted the extreme danger which results from sodium depletion, it is remarkable that it is so rare. Primitive urine, formed at the glomeruli, has a daily volume equivalent to several months' final output. At this rate, the entire sodium content of ECF would be lost in 9 h, if it were not for regulated reabsorption along the nephrons. If mammalian sodium requirement implies an intake of 0.6 mmol/kg/day, an error of 1% in this process is equivalent to a week's intake. The result of such an error would be to reduce ECF volume by 2 l in 24 h—an amount equal to two-thirds of adult human plasma volume. Clearly sodium depletion, caused for example by diarrhoea, is a potent hazard.

Sodium is not only the predominant ion of seawater, but one of the most plentiful in the Earth's crust as salt, as carbonates, often in mineral waters, as sulphates and silicates. Dietary depletion would seem extremely improbable, except in the interior of continents, removed from the drift of oceanic spray. Even there, movement of fresh water with only traces of sodium can lead to extensive aggregation of salt as a result of evaporation; this is one of the problems of long-term irrigation and the basis of such spectacular phenomena as the Great Salt Lake and the Dead Sea. The greatest likelihood of a low-salt diet arises in herbivores grazing pasture far removed from oceans or salt deposits, but great caution is needed in equating this with 'depletion'. If a healthy sheep can reduce its daily losses to 1 mmol or less, this can be replaced by herbage containing as little as 0.0023% Na^+ in the dry matter. Thus, while sodium is normally a macronutrient, animals can adapt to its presence as a micronutrient or trace element, i.e. one for which the requirement is of the order of parts per million (mg/kg).

One would anticipate, after half a century when sodium was easily quantifiable, and at the end of one in which the human genome will have been mapped, that only the most obscure or banal questions concerning the regulation of body sodium awaited an answer. On the contrary, some of the most fundamental await even a rudimentary explanation. A few examples will suffice; others will emerge subsequently. Where and how is blood volume monitored? In the atria, yes; in 'the great veins' yes, but which and how? And since plasma is only the minority of ECF (Fig. 1.1), how and where is interstitial fluid (ISF) volume monitored? Why, when the latter is expanded to pathological excess (oedema), whether in cardiac, renal or hepatic disease, do we invoke a hypothetical construct such as 'effective plasma volume', i.e. something which fits our theories when our measurements do not? Since the sodium reaching plasma from the intestine vastly exceeds the dietary sodium, as a result of recycling of alimentary secretions, is this flux monitored—and if so, where? In fact, other than the effects of aldosterone on the large intestine, how exactly is enteric sodium absorption regulated? Is sodium intake regulated? If so, why do animals take so much more than they need—and why is there so much research on the role of salt appetite in the defence against sodium depletion, when the

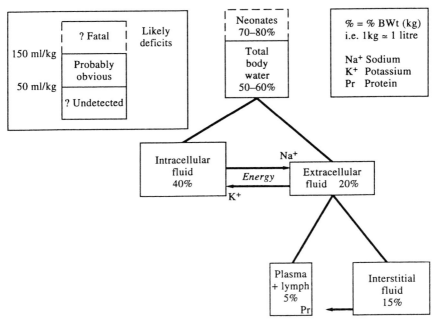

FIG. 1.1. Distribution of body water. For comparison: size of deficits.

main human clinical problem influenced by salt is the exacerbation of hypertension by excess intake? Since there is even more sodium in bone than in ECF, how is that regulated; is it a useful repository for excess, a functional reservoir or an inert deposit? These do not seem like minor details or obscure preoccupations. Indeed, among the major repositories of sodium in the body (Fig. 1.2), plasma (10%), interstitial fluid (30%) and bone (40%), our knowledge is inversely proportional to their magnitude.

Elucidation of the physiology of any element or non-metabolised compound must include consideration of factors regulating its intake, distribution and loss, as well as the availability of internal stores. We begin, therefore, with the distribution and physiological functions of sodium.

Distribution and Concentration of Sodium

Basic Principles

Sodium is present in ECF at about 145 mmol/l (one millimole= molecular weight in mg, i.e. 23 mg Na^+, 58.5 mg NaCl). The concentration is somewhat higher in plasma because of the Gibbs–

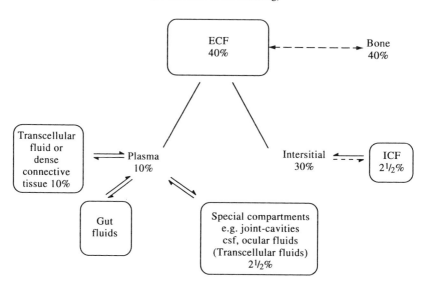

% = approximate % of total body sodium,
(of the order of 60 mmol/kg)*

FIG. 1.2. Distribution of body sodium.

* Figures are approximation because of variation associated with species and sodium intake (e.g. large gut fluid compartment of herbivores) and uncertainty of bone sodium, especially the slowly-exchangeable sodium which is the great majority of bone sodium.

Donnan effect, i.e. the attraction of negative charges on plasma proteins, especially albumin. Unlike, for example, calcium or magnesium, sodium is not bound to protein, hence the total concentration (measured chemically or by flame photometry) is close to that measured by ion-specific electrodes, which detect only the activity of unbound freely dissolved ions. Strong electrolytes are totally dissociated, i.e. NaCl behaves as independent Na^+ and Cl^- ions in aqueous solutions at sufficient dilution; in physiological concentrations this is almost true (less so in the renal medulla where concentrations are far above those in ECF).

The factors determining the balance of distribution between plasma and interstitial fluid, therefore, are essentially those determining fluid exchange across capillaries (Fig. 1.3). Interstitial fluid is replenished from plasma as a result of the outward pressure within capillaries; the greater the arteriolar tone, the further this falls below arterial pressure. Pressure drops across the peripheral resistance (arterioles), just as voltage drops across an electrical resistance. Thus vasoconstriction in response to haemorrhage not only protects arterial

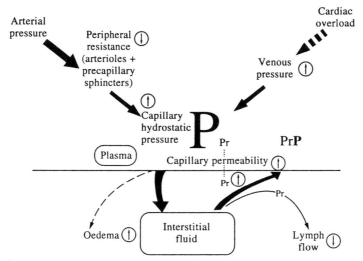

FIG. 1.3. Plasma, interstitial fluid and oedema. Factors affecting normal
distribution. P, hydrostatic pressure; Pr, protein (albumin); ringed arrows
indicate factors contributing to oedema.

pressure (upstream), but also facilitates the uptake of plasma-like fluid
from ISF, downstream. This replenishes circulating volume and may
also promote tissue perfusion by reducing haematocrit (packed cell
volume, PCV) and hence blood viscosity.

Uptake of fluid from ISF is mainly governed by the osmotic
gradient provided by plasma albumin; while sodium is the main
osmotically active ion throughout ECF, albumin provides the main
difference in osmolar concentration between plasma and ISF. The
integrity of this gradient depends on adequate hepatic synthesis of
albumin (and therefore to some extent on dietary protein intake and
utilisation), intact capillary permeability and adequate lymphatic flow
(to return to circulation any albumin leaked from capillaries or cells).
Osmotic gradients depend on the number of particles in solution, not
their identity, i.e. they depend on the number of mmol of solute
rather than the gravimetric concentration. Osmotic concentration
is conveniently measured by machines which detect depression of
freezing point by the solutes and express it in mmol/kg (osmolality) or
mmol/l (osmolarity). Osmolarity is affected by temperature because
this alters the volume of the solution.

Sodium in ISF is not present in a free-flowing pool, except where
there is gross oedema. Indeed, were it so, all animals would almost
inevitably have oedema as their ISF gravitated towards their feet!
Instead, there is much adsorption onto structural components and

the gel matrix in the interstitial spaces; this can be influenced by hormones. The interstitial space is an 'organ of translocation' (Katz, 1980); its contents are mainly present in quantities reflecting a balance between their accumulation from cells or capillaries and their clearance into capillaries or lymphatics.

The ISF permeates the hydrophilic gel formed by the glycosaminoglycan (mucopolysaccharides) produced by fibroblasts. The interstitial space also contains a loose mesh of collagen fibres which partially anchor the gel; the proportions and detailed composition of these constituents vary between tissues (Lassiter, 1990). The key role for these structural components is to prevent gravitational flow of ISF and to maintain favourable diffusion pathways between cells and capillaries. The interstitial proteins in mesentery are unevenly distributed and include possible pathways between the microcirculation and the interstitial matrix (Barber and Nearing, 1990). Structural components of the interstitial space do not contribute to osmotic gradients because of their high molecular weight, thus, despite their major contribution to the dry weight of connective tissue, their 'concentration' in mmol per litre of ISF is small. Temporary mobilisation of free fluid from the gel underlies 'pitting' oedema, whereas excess gel production, as in myxoedema, produces unyielding swelling (Katz, 1980). The extracellular matrix is also anchored to cells by integrins, surface glycoproteins (Fish and Molitoris, 1994).

In contrast with ECF, intracellular fluid (ICF) is rich in potassium and low in sodium, despite the fact that most cells are fairly permeable to sodium. These differences result from the activity of the 'sodium pump', the Na–K ATPase enzyme system. This consumes substantial amounts of ATP in expelling sodium from ICF, supplying it with potassium. By transporting more sodium than potassium, it establishes a transmembrane potential difference (ICF negative). It also prevents cell swelling, which would otherwise result from the osmotic attraction exerted by fixed solutes, notably proteins and other organic compounds, within cells.

Simple Disturbances (Fig. 1.4)

Except for minor movements, e.g. by pinocytosis, water is not primarily transported across membranes, but follows osmotic gradients. By constraining sodium to ECF, the sodium pump allows it to act as the osmotic skeleton of extracellular fluid, dictating its volume. **Changes in sodium balance therefore result predominantly in changes of ECF volume.** It is a common misconception that their main impact is on plasma sodium concentration. This, however, is

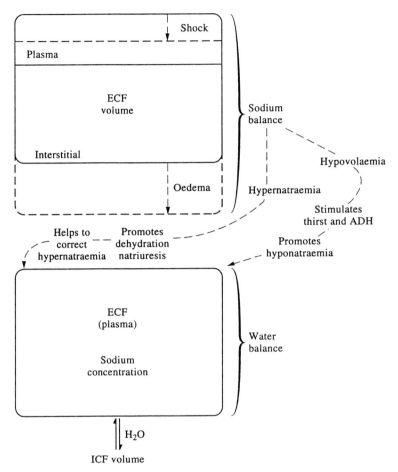

FIG. 1.4. Main relationships between plasma sodium concentration, extracellular fluid (ECF) and intracellular (ICF) volume.

regulated by the mechanisms which determine water balance, i.e. thirst, the secretion of antidiuretic hormone (ADH), and the renal response to it.

The main stimulus to thirst and ADH secretion results from a rise in plasma sodium concentration. This raises the concentration of the interstitial fluid surrounding the osmoreceptors separately responsible for thirst and for ADH release; both groups are within the hypothalamus. ADH is synthesised in other hypothalamic neurons prior to storage in the posterior pituitary gland.

Why then does plasma sodium concentration fall as a result of sodium depletion? It does so belatedly and as an indirect consequence.

The direct effect is hypovolaemia. Initially there is no change in plasma sodium concentration because any fall is corrected by a reduction in ADH secretion and an increase in water excretion (Sutters *et al.*, 1993). Once hypovolaemia is sufficiently severe, the sensitivity of both thirst and ADH secretion to hypovolaemia is enhanced to the extent that it temporarily supersedes osmolality as the dominant stimulus. Water is thus retained in defence of plasma volume, but to the detriment of plasma sodium concentration (*Chapter 8*). If the sodium deficit is replenished, the expanding ECF volume and the reduced plasma sodium concentration (hyponatraemia) facilitate the excretion of the excess water. The thirst associated with hypovolaemia (for example, after severe haemorrhage) is mediated by angiotensin, generated as a result of restricted renal perfusion and the consequent release of renin (*Chapter 2*). The renin–angiotensin system is also a key link in the release of aldosterone from the adrenal cortex in response to sodium depletion (*Chapter 10*); the result is sodium conservation from the distal nephron and the colon.

Water movements in response to deviations of plasma sodium concentration reinforce the importance of the regulation of the osmotic concentration of ECF; on it depends the normality of cell volume and ICF osmolality, although both are also influenced by intracellular mechanisms. In particular, a hypovolaemic patient who also has hyponatraemia, even to a degree which does not seem likely to cause symptoms, still suffers an unnecessary deficit of circulating volume by yielding water from ECF to ICF.

Hyponatraemia, if sufficiently rapid (acute) and severe (a fall of roughly 15 mmol/l), causes symptoms, particularly behavioural disturbance, as a result of redistribution of water into ICF and cell swelling. This is particularly harmful in brain cells because of the constraining cranium. Hypernatraemia, conversely, causes shrinkage. Interestingly, however, hypertonicity due to physiological solutes (e.g. glucose, sodium) elicits, particularly in brain cells, a response seen also in primitive organisms. There is a compensatory synthesis of organic solutes within ICF, thus reducing the osmotic gradient and minimising cell swelling: this response, however, requires time and defends only against gradual changes (*Chapter 8*).

While it is little more than 200 years since William Withering provided the first clear understanding of the cause of 'the dropsy' and, through the cardiac glycosides from foxgloves, the first specific remedy, oedema has been the most visible disturbance of body sodium since humans began to observe themselves or their animals. The possible causes are simple to understand using the principles delineated by Starling in his analysis of exchange across capillary membranes (Fig. 1.3); the actual causes, as we shall see *(Chapter 8)*, remain a matter of considerable mystery.

Oedema

The outward movement of fluid from plasma, via capillary membranes, to ISF depends on the pressure within the capillaries: this is what remains of the pressure generated by the heart once it has been reduced by the variable peripheral resistance, notably the arterioles.

The inward movement, from the interstitial reservoir towards plasma, depends on the osmotic gradient provided by the high concentration of albumin within plasma compared with the lower concentration of albumin and other oncotically (colloid osmotically) active solutes within ISF. This gradient depends on:

1. capillary membranes which are normally only minimally permeable to albumin;
2. lymphatic drainage which swiftly sweeps away any protein accumulating as a result of leaky capillaries or cell turnover.

The influence of leaky capillaries is not, therefore, that they leak water—to which they are necessarily freely permeable in any case—but that they leak protein. It is obvious that a trawler net with a larger mesh still leaks water at essentially the same speed; it is the fish which 'leak'.

The cause of oedema, therefore, is an increased capillary pressure or a reduced transcapillary oncotic gradient.

Expansion of ECF volume physically favours diversion of excess fluid to ISF by diluting plasma albumin. It also favours it physiologically by the reduced peripheral resistance, which corrects any rise in arterial pressure. Arteriolar vasodilation allows greater transmission of pressure to the capillaries; this is also why high environmental temperature favours subcutaneous oedema, and probably why excess alcohol does the same. In a sense, oedema is to ECF what 'weeds' are to a gardener—an abundance of something normal in the wrong place. The reservoir function of ISF buffers plasma against the adverse effects of either excessive contraction or expansion, but the price in the latter case is visible oedema, once the ISF expansion is sufficiently large. Mild oedema, except in brain or lung, is simply cosmetic and even visibly disturbing amounts of oedema are likely to cause discomfort or inconvenience rather than a real threat. Nevertheless, diffusion distances are extended and, once the accumulation is severe, there is mechanical interference with adjacent organs or with movement. Thus ascites (oedema of the peritoneal cavity) impedes the diaphragmatic component of breathing while pulmonary oedema not only impairs diffusion but reduces lung elasticity.

Capillary pressure can also be raised from the venous end, hence the adverse effects of cardiac failure, external pressure leading to venous compression (seat edges on crowded aircraft or inept bandages or plaster casts) or of immobility. The latter reduces the unidirectional venous flow generated by the combination of massage by active skeletal muscle and the venous valves. Since the veins function as the intravascular reservoir or capacitance, extreme expansion which extends them beyond their compliant range will also raise venous pressure, as in extreme fluid retention or cardiac failure.

Disruption of the oncotic gradient can occur at either 'end', i.e. plasma albumin can fall or ISF protein can increase. The latter occurs with lymphatic obstruction or with capillary damage, whether due to physical factors (heat, impact, ischaemia) or pharmaco-physiological mediators, notably those involved in inflammation. Reduced plasma albumin results from defective hepatic synthesis, extreme malnutrition, or leakage from the kidney or gut which exceeds the maximum rate of hepatic synthesis. It can also result from dilution, though this is most likely to be iatrogenic because only colloid-free intravenous infusions are likely to produce a sufficient reduction.

Matters are not so simple. Changes in vascular tone also affect vasomotion, i.e. the balance between short term periods of more or less generous perfusion. Sodium pump (Na–K ATPase) activity is essential for normal vasomotion (Gustafsson, 1993). Accumulation of excessive interstitial fluid is normally prevented by accelerated lymph flow; this returns larger amounts to plasma (Drucker *et al.*, 1981). As a result, the fall of plasma albumin is minimised because, while ISF is normally low in **concentration**, the **rate** of return of albumin to circulation will increase. On the other hand, reduced plasma albumin is far less important as a cause of oedema until a very low concentration is reached; beyond this, ISF accumulates rapidly.

This provides a splendid example of the contrast between the relationship between basic science and clinical science assumed by basic scientists and the one which actually exists. According to conventional wisdom, understanding the basic regulatory mechanism enables the disturbances underlying disease to be predictable. In fact, the ability of physiological knowledge to provide useful explanations of clinical conditions was disappointingly rare until Starling's analyses of both capillary exchange and cardiac function. Starling's Law provided a welcome early example of a physiological concept which could illuminate a clinical problem—heart failure. As a result, there was sustained resistance to the fact that changes in cardiac responsiveness (**between** Starling curves relating cardiac output to myocardial fibre length, represented by derivatives of cardiac load) were at least as important as distension (**along** Starling curves) in the pathogenesis of cardiac failure (Wallace, 1981; Smith, 1989).

In reality, the **discrepancy** between such predictions and what actually happens is as important a source of pressure for the revision of current hypotheses as any data from anaesthetised laboratory animals subjected to surgically contrived imitations of disease. Aldosterone, since it is physiologically responsible for sodium retention, should explain oedema—but it does not *(Chapter 8)*. Excess aldosterone does not cause oedema, hence the need to understand the mechanisms which enable the nephron to resist its salt retaining effects ('renal escape'). Hypoalbuminaemia **ought** to readily cause oedema, but clinically it does not until it is severe.

This anomaly provided much of the spur for Guyton and his colleagues to understand the nature of the 'safety factors' which normally defend against oedema, notably the capacity for accelerated lymphatic return and, more controversially, the role of negative (subatmospheric) pressure in the interstitial space. This results from the combined effect of lymphatic and venous removal of fluid and the adsorption of ISF onto structural components of the interstitial space. The significance of negative interstitial pressure is that it favours lymphatic patency; as soon as fluid accumulates sufficiently to overcome the negative pressure, there is double jeopardy—further increments not only generate more ISF but, by tending to collapse lymphatics, impede its removal.

The role of capillary leakage in oedema provides another example of a stern light cast on physiological theory by clinical fact and where it most matters—in pulmonary oedema, which unlike subcutaneous oedema readily becomes lethal. If experimental animals or patients receive gross amounts of intravenous electrolyte solutions, subcutaneous oedema becomes obvious. It was long assumed, as a result, that such solutions were more likely to cause pulmonary oedema than colloids, which caused less reduction of oncotic gradient.

Clinical evidence suggested, however, that whether patients were treated predominantly with electrolyte solutions (crystalloids) or colloids actually made little difference to the incidence of pulmonary oedema (Michell, 1989b; Geheb *et al.*, 1994). Indeed, patients with ascites due to low plasma albumin do not usually have pulmonary oedema (Murray, 1981). Moreover, it had long been known that pulmonary capillaries are actually rather leaky, hence, apart from anything else, colloids might not actually remain where they were intended to act. Of course, it had been known for even longer, by children taking their biology exams, that capillary pressure, one of the critical factors governing oedema, was actually completely different in the lungs, as a result of the evolution of a double circulation and a four-chambered heart.

If the hydrostatic pressure in pulmonary capillaries is lower, the pulmonary interstitial space would be virtually dry, unless there was a comparable reduction of osmotic gradient. Since plasma albumin cannot temporarily change within the pulmonary artery, the explanation must be an increase in interstitial protein—hence the link to the known high permeability of pulmonary capillaries (Murray, 1981; Michell, 1985c). Otherwise, the latter would appear to predispose the lungs to oedema rather than protecting them from it. Pulmonary interstitial protein is thus more than double the 'average' capillary bed, whereas capillary hydrostatic pressure is slightly less than half (Guyton, 1992a). The basis of protection is 'capillary washdown', i.e. dilution of interstitial protein by any accumulating oedema fluid (Civetta, 1979; Gabel and Drake, 1979; Peters and Hargens, 1981). This therefore re-establishes a more favourable oncotic gradient and provides a 'self-protect' mechanism in addition to accelerated lymphatic drainage (Taylor and Rippe, 1986). It also alerts us to the fact that a raised capillary pressure rather than a reduced oncotic pressure is likely to be the key determinant of iatrogenic pulmonary oedema. Over-rapid infusions, whether of colloids or crystalloids, will thus cause pulmonary oedema, especially if cardiac function is compromised or the lungs damaged, e.g. by inhalation of blood, vomit or smoke, or by direct trauma (Michell, 1989b). In fact, colloids, rate for rate, are probably more likely to cause oedema, since they do not escape into ISF. Clinically this is offset by the fact that they are used in smaller volumes, partly because of expense, partly because of the effects on clotting and because it is known that less is needed if the solution remains intravascular, where it is needed, rather than the majority distributing into ISF as is the case with saline solutions (Michell *et al.*, 1989).

Not everyone accepts that interstitial fluid is at subatmospheric pressure, despite the ingenuity of Guyton and his colleagues in attempting to obtain valid measurements of this fragile and fragmented compartment. Hence, some still view enhanced ISF pressure as a factor limiting the further escape of fluid from capillaries. Moreover, some who accept the importance of 'protein washdown' as a safety factor attribute its greatest power to tissues with tight capillaries, which normally leak least protein (Lassiter, 1990). This is hard to fathom, since any diluting effect of excess fluid transudation must be most where there is a substantial concentration of protein to be diluted. The counterargument is that the additional fluid will contain less protein if the capillaries are tight, but it is flawed because, even in lungs or other leaky capillary beds, the transudate always has less protein than plasma, hence it raises the gradient. Dilution of ISF protein must be a more potent factor in a capillary bed such as that of the lung, where it contributes 50% of the outward force for transudation (Guyton, 1992a) compared with the more usual 15%.

Having considered the active compartments of body sodium, certainly those that exchange most readily with isotopic markers, namely plasma and interstitial fluid, it is worth a thought for the compartment which actually holds more than either—bone. Additional sodium is contained in cartilage and dense connective tissue.

Bone Sodium

Michell (1976b) proposed that bone may be less important as a reservoir for sodium, more important as a 'kidney of accumulation', i.e. a repository for excess. As a result, high salt intake during skeletal growth created a hazard for future hypertensives, while factors causing skeletal demineralisation caused an endogenous sodium load and impaired the ability to sequester excess sodium. One of the factors underlying the protective effect of hard water, therefore, might be the tendency of the associated calcium load to minimise the mobilisation of bone mineral. Conversely, since parathyroid hormone (PTH) mobilises sodium from bone, as well as calcium, the hypothesis was consistent with the hypertensive effect of hyperparathyroidism. It was also consistent with the increasing prevalence of hypertension with age and its associated loss of bone mass. At the time, hyperparathyroidism was not widely perceived as a factor in hypertension, e.g. in discussion by Schrier (1976), Swales (1975) or Liddle and Liddle (1981), although it is now recognised that the majority of patients are hypertensive (Williams and Dluhy, 1994). The hypothesis suggested that one of the potential benefits of potassium was to reduce the amount of sodium stored in bone; the exaggerated natriuresis seen in various forms of hypertension (Williams and Dluhy, 1994) could also reflect an impaired ability to transfer excess sodium to bone.

The underlying importance of bone as a store for excess sodium, as with calcium, is the fact that both ions are actively removed from ICF which cannot, therefore, act as a store in the way that allows transfer of K^+ to ICF to protect against hyperkalaemia. Various studies suggested that, even with due allowance for unmeasured losses, it is often difficult to account for disposition of salt loads and exchangeable sodium increases more than could be explained by increased ECF volume (Michell, 1976b). Recently, Brier and Luft (1994) found that humans distributed perhaps 10% of a salt load into a non-ECF compartment, possibly bone. Michell (1976b) also suggested that endogenous sodium loads associated with mobilisation of bone mineral might contribute to the frequency of oedema in normal pregnancy. The conclusions were that:

> Research into factors regulating skeletal sodium metabolism, including a possible role for calcitonin [was urgently needed and that] until factors regulating skeletal sodium are better understood, animal studies should also focus on the possibility that agents promoting skeletal calcium uptake might prove useful adjuncts to hypertensive therapy.

Some 20 years later, we are little further forward. An increasing range of techniques confirms that bone sodium is largely inert or only very slowly exchangeable. A minority lies in the aqueous compartment of the uniquely regulated bone ISF: the slowly exchanged sodium comprises that on the surface of the bone mineral crystals or within their interior (Michell, 1976b; Rakovic and Pilecka, 1992). The crystal surface area is large and active in physiological exchanges (Posner, 1978). These include substitution of sodium for calcium. The skeletal response to deviations in plasma calcium concentration utilises this large, quiescent surface area rather than sites of active bone remodelling (Green, 1994). Nevertheless, remodelling can cause substantial changes in bone sodium, both gains and losses.

It is important to emphasise that bone exchanges with atypical ISF, rich, for example, in potassium and containing less sodium and calcium (Green, 1994). Mobilisation or deposition of ions in bone mineral can occur via crystal dissolution or crystal growth, respectively; it can also occur via ion exchange. Potassium remains superficial in the crystals and is therefore readily exchangeable, whereas magnesium and sodium are less exchangeable since they lie deeper in the crystals (Green, 1994). Acute potassium loading liberates sufficient bone sodium, both exchangeable and slowly-exchangeable, to offset the impact of the associated natriuresis on ECF volume (Reid *et al.*, 1988). Earlier studies (Mills *et al.*, 1988) showed that potassium loading caused both natriuresis and expanded ECF volume. This is strongly suggestive of natriuresis in response to an endogenous sodium load, rather than the response of a bone reservoir to sodium depletion. It also suggests that the natriuresis often observed with potassium loading does not imply sodium depletion in the normal sense, because it is the relatively inactive reservoir which is depleted while the active compartment (ECF) actually expands.

Bone can act as a reservoir for sodium and responds to sodium depletion (Michell, 1976b; Reid *et al.*, 1988). The effectiveness of the mobilisation is seen in sheep acutely depleted of 420 mmol of sodium; their exchangeable sodium was depressed by only 110 mmol (McDougall *et al.*, 1974). Bone also yields sodium particularly readily in response to metabolic acidosis, thus allowing protons to be buffered by the crystals (Bushinsky, 1994). There is also mobilisation of bicarbonate buffer, together with calcium. Acidosis thus results in dissolution of crystals as well as causing ion exchange; both are likely to liberate bone sodium. If endogenous sodium loads are a factor in hypertension, metabolic acidosis, together with secondary hyperparathyroidism, would heighten their importance in chronic renal failure. Indeed, in **acute** metabolic acidosis, there is evidence that bone buffering liberates sodium rather than calcium (Bushinsky, 1994). Metabolic acidosis has been suggested as a factor in the pathogenesis of essential hypertension *(Chapter 6)*. Both with calcium and sodium, the fact that a high intake of one increases the excretion of the other *(Chapters 6 and 9)* may reflect skeletal as well as renal interactions.

On balance, evidence has increased that bone may act as a source of endogenous sodium loads, and that these could be a factor in many conditions usually viewed in terms of external sodium loads (including hypertension). Nevertheless, after 20 years, it is tantalising that so little attention has been directed towards these intriguing possibilities.

Sodium in Brain and Cerebrospinal Fluid

The extracellular fluid of brain includes the cerebrospinal fluid (CSF) and the small amount of interstitial fluid between brain cells. The brain ISF (Fig. 1.3) is separated from the plasma by the blood–brain barrier (BBB) and from CSF by the ependymal lining of the ventricles of the brain. Transport of solute (and therefore water) across the BBB and also across the choroid plexus into CSF is vital for osmoregulation in the central nervous system (CNS), as are the transport processes across the cell membranes of neurons and glial cells (Strange, 1993). Quite apart from gross damage due to swelling or shrinkage, minor alterations in cell volume can disrupt the normal spatial relationships between brain cells and affect the transmission of impulses. Changes in ionic concentration affect both excitability and levels of neurotransmitter. Maintenance of normal cell volume within the CNS is, therefore, essential to its function.

The swiftest regulatory response is to minimise osmotic gradients by loss or gain of intracellular solute (Strange, 1993). This involves leakage of potassium or inward movement of sodium and chloride, either in exchange for hydrogen ions and chloride or by cotransport with potassium (Na:K:2Cl). Given time, this is reinforced by generation of organic solutes known as idiogenic osmoles *(Chapter 8)*. The initial defence against hyponatraemia, which causes brain swelling, is transfer of excess fluid from cerebral ISF to cerebrospinal fluid; also effective within hours is reduction of intracellular potassium (Sterns, 1990). Loss of extracellular solute, hence reducing the expanded ISF, also contributes to the defence against brain oedema (Gullans and Verbalis, 1993).

Normally, the potassium concentration of brain ECF is below that of plasma and is regulated by Na–K ATPase in the endothelium of brain capillaries; this removes excess K^+ into plasma (Schielke and Betz, 1992). Sodium and chloride concentrations in brain ISF are similar to those of ECF, but differ from plasma. Sodium is slightly higher and chloride considerably higher in CSF, probably because of the absence of protein. Plasma and CSF are in osmotic equilibrium and CSF therefore follows changes in plasma sodium concentration, after some delay. Experimentally it can be shown that the equilibrium concerns osmolality rather than sodium concentration *per se*: if plasma sodium is reduced but osmolality is maintained (e.g. with mannitol), substantial differences can be produced between the sodium concentrations of CSF and plasma. The BBB permeability to sodium in much less than to potassium (Schielke and Betz, 1992).

Ion movements across the BBB are mainly via cellular rather than paracellular pathways and via aqueous channels rather than lipid membranes. Nevertheless, endothelial permeability is low compared with other capillaries (Schielke and Betz, 1992). The luminal surface of the endothelium has a number of sodium cotransport or exchange mechanisms, but their relative importance is uncertain; as in other epithelia, the driving force for movement of sodium is Na–K ATPase present on the cerebral side of the endothelial cells at higher concentration than in other capillaries. Brain capillaries secrete fluid into ISF, which then migrates towards CSF; although the choroid plexuses are the main source of CSF, they are not the sole source. The overall effect of Na–K ATPase in brain capillaries is to remove excess potassium and to generate additional CSF. The capillaries have receptors for hormones such as ADH and ANP (atrial natriuretic peptide) (Schielke and Betz, 1992), but their role in regulating the ionic environment of the neurons remains a matter for speculation and research.

Functions of Sodium

The essence of the physiological need for sodium is *transport*.

1. It enables sodium to act as the **osmotic skeleton** of ECF.

2. It causes the **transmembrane potential differences (PDs)** which underlie excitability in nerve, muscle and other cells.
3. It permits the **movement of water** and thereby influences **cell volume**.
4. It ensures the **supply of K+** to ICF.
5. It permits the **dilution of urine** (hence the excretion of excess water) in the loop of Henlé and the **concentration of urine** (water conservation) in the collecting duct. It also underlies the rise in the concentration of urinary solutes such as urea and creatinine, which results from the contraction of urine volume by isotonic reabsorption of sodium and water in the proximal tubule.
6. It facilitates the **defence against hyperkalaemia** (elevated plasma K+ concentration) by promoting uptake of potassium into cells and thereby facilitating its excretion into urine, under the influence of aldosterone.
7. By **cotransport**, it facilitates both the **enteric uptake of solutes** such as glucose and amino acids, also their renal reabsorption. It also facilitates parallel movement of anions such as bicarbonate and chloride. In response to hypertonic surroundings, cells may increase their intracellular solute by cotransport of sodium with potassium.
8. By **countertransport**, it influences the **concentration of other cations**, e.g. Ca^{2+}, in ICF and may therefore influence, for example, the response of arterioles to vasoconstrictor tone (*see Chapter 6*). Hydrogen ion concentration is influenced by Na^+–H^+ countertransport, a transport mechanism which is artificially engaged by Li^+ (not normally present in animals) in 'Na^+–Li^+ countertransport'. This amiloride sensitive 'antiporter' is essential in the regulation of ICF pH, volume, as well as being involved in cell division and differentiation (Burckhardt and Friedrich, 1988).
9. Because Na–K ATPase is a major consumer of ATP (Milligan and McBride, 1985), **rates of sodium transport potentially affect a variety of metabolic pathways** and influence thermogenesis, e.g. in response to thyroid hormone.

If the essence of the need for sodium is transport, we should not be surprised that there are mechanisms to regulate this process which go beyond the changes within individual cells. This was clear before the discovery of Na–K ATPase:

> *We need not be astonished . . . if we find digitalis-like substances in any cell . . . The digitalis cardiac glycosides, if I may say so, are no drugs at all: they are substitutes for a missing screw in our machinery, which has a cardinal role in one of the most basic physiological regulations.*
>
> (Szent Gyorgyi, 1953, cited by Blaustein, 1993)

Sodium Transport: Na–K ATPase

Active sodium transport, as we have come to understand it, represents a synthesis of observations on intact systems, such as red cells or leukocytes, where ion movement can occur and on various preparations, mostly of membrane fragments or vesicles from broken cells, in which the properties of the enzyme Na–K ATPase have been studied. As already indicated, Na–K ATPase is not the only mechanism available for sodium transport. Various mechanisms permit cotransport or countertransport of sodium with other molecules or ions. They depend, however, on the existence of an underlying sodium gradient ('secondary active transport') and the unique role of Na–K ATPase is to establish that gradient. In doing so, it provides cells with potassium and with the membrane potential underlying their excitability. The latter depends on the asymmetrical movement of sodium and potassium, with the larger outward flux of sodium creating a positive charge outside the cell and also allowing osmotic removal of water from cells. The pump, therefore, fulfils a role analogous to the contractile vacuole of protozoa, as well as establishing ionic and electrical gradients. Secondary active transport is discussed in the context of renal and enteric function and also its possible importance in clinical conditions including hypertension. In ion-transporting epithelia the spatial relationship between sites of passive entry and sites of active extrusion is vital, e.g. Na–H exchange and glucose/amino acid cotransport on the urinary side of proximal tubule cells, and Na–K ATPase on the basolateral surface (Fish and Molitoris, 1994).

The sodium–potassium pump can operate in a number of modes *in vitro*, not least it can be run to generate ATP rather than consume it. Such experiments have greatly expanded our knowledge of the workings of Na–K ATPase, but this account is mainly concerned with its normal function. Since entire volumes are justifiably dedicated to Na–K ATPase, the degree of selectivity involved in an account of this length is almost impertinent. On the other hand, a growing proportion of recent work in this area, as in many others, e.g. mineralocorticoids, has focused on an increasingly detailed molecular description of structures rather than a refined understanding of function.

Isoforms

Na–K ATPase is an ancient molecule, present even in coelenterates and inhibited there, as in vertebrates, by ouabain. It is a heterodimer consisting of an α subunit which appears to account for most of the known properties, e.g. ATP catalysis, ion transport and cardiac glycoside binding (Doris, 1994). The β subunit may affect the conformation of the α subunit or its incorporation in cell membranes; its function remains uncertain. The existence of an additional γ subunit has been suggested but remains controversial (Rose and Vaides, 1994). The molecular structure of Na–K ATPase is very similar to that of Ca-ATPase or H,K-ATPase, though its specificity is very different (Lingrel *et al.*, 1994).

Recently, there has been intense interest in the existence of at least three isoforms of each subunit. Each isoform is more similar between species than the isoforms are to one another, within species (Doris, 1994). The preponderance of these isoforms is responsive to the ionic environment, thus hypokalaemia depresses the expression of α_2 but not α_1 (McDonough *et al.*, 1992). This will also contribute to the tissue specificity of responses, since α_2 is the predominant isoform in skeletal but not cardiac muscle. In the kidney, the ability to enhance potassium reabsorption in response to hypokalaemia depends on increased expression of K-ATPase and Na–K ATPase in the outer medullary collecting duct; the change in Na–K ATPase involves both α_1 and β_1 subunits (McDonough *et al.*, 1992). Inhibition of α_1 and β_1 subunits in the proximal tubule may contribute to the natriuretic response to salt loading. The α_1 isoform, present in most cells, is the predominant renal isoform, whereas in the brain, depending on the region, it is α_2 or α_3 which predominates; β_2 predominates in the brain rather than β_1, which is found in all tissues (Levenson, 1994; Lingrel *et al.*, 1994). There are species differences in these distributions (DeCollogne *et al.*, 1993). The functional significance of this variety of isoforms, particularly within single cells, remains to be explained (Levenson, 1994).

Factors Affecting Activity

Increased intracellular sodium increases pump activity and also stimulates synthesis of additional pump sites if it persists (McDonough *et al.*, 1990; Doris, 1994). This is the main stimulant, as appropriate for a pump primarily responsible for the maintenance of a low intracellular sodium concentration. Cell swelling also stimulates Na–K ATPase activity (Coutry *et al.*, 1994). Upregulation of Na–K ATPase can involve reduced breakdown as well as increased synthesis (McDonough *et al.*, 1990). Potassium stimulates the external site and lithium stimulates both sites, but to a lesser extent than potassium or sodium *(Chapter 11)*; it behaves like sodium, i.e. it is expelled by the pump (Hoffman, 1986). Inhibition by cardiac glycosides occurs at the external site, whereas inhibition by calcium or vanadate occurs at the internal site (Hoffman, 1986; Haupert, 1988). The effect of vanadate involves changes in the optimum pH *(Chapter 11)*. Pump activity depends on adequate concentrations of ATP and of magnesium, which is vital for the formation of a phosphoprotein intermediate. A phosphorylation–dephosphorylation cycle is believed to produce the essential conformational change which underlies ion translocation. Thus the steps involved are:

1. intracellular binding of ATP and three sodium ions;
2. phosphorylation of the pump by ATP;
3. conformational change and sodium release to the extracellular fluid;
4. external binding of two potassium ions and dephosphorylation of the pump;
5. conformational change and release of potassium to intracellular fluid (Glyn, 1993).

The stimulatory effect of potassium at the external site, though real, has often been thought unimportant because it saturates at 1 mmol/l, well below the ECF concentration (4–5 mmol/l). Recent evidence suggests that this may result from the use of broken cell preparations and that in intact renal tubule fragments, for example, stimulation continues until a maximum is reached at 5 mmol/l (Katz, 1988). Na–K ATPase plays a crucial role in the renal adaptation to potassium loads and in the heightened ability of surviving nephrons to maintain adequate potassium excretion until the advanced stages of chronic renal failure (Stanton, 1987). This adaptation involves an increased density of sites rather than an altered affinity and is potentiated by aldosterone but does not depend on it (Katz, 1988). Na–K ATPase is thus involved in adaptation to both surfeit and deficit of potassium as well as playing a pivotal role in a number of pathways likely to underlie both natriuretic and salt-retaining responses (Aperia *et al.*, 1994).

The main hormonal controls on Na–K ATPase are exerted by adrenal corticosteroids, thyroid hormone, insulin and endogenous digitalis-like inhibitors, possibly including ouabain itself. The mineralocorticoids increase inward leakage of sodium, hence stimulating pump activity and, eventually, the number of pump sites (Lechene, 1988; Marver, 1992). The stimulatory effect of insulin depends on activation of existing pump sites via the α subunit in muscle and adipocytes, but in liver it probably depends on a prior increase in cell sodium (Gick *et al.*, 1988; Williams and Epstein, 1989). Thyroid regulation reflects its role in energy regulation and thermogenesis in particular, rather than control of electrolyte balance.

Since the 1980s, it has been clear that there are at least two major groups of natriuretic hormones; natriuretic peptides [e.g. atrial natriuretic peptide (ANP)], which do not inhibit sodium transport, and active sodium transport inhibitors (ASTI), which may also be the natural or endogenous digitalis-like inhibitors (EDLI) of Na–K ATPase (De Wardener and Clarkson, 1985). Attention has recently focused on the role of the adrenal cortex as the source not only of mineralocorticoids, which increase both the activity of the pump and the number of sites, but possibly also of ouabain (Ruegg, 1992; Blaustein, 1993; Doris, 1994). Previous interest in the origin of endogenous inhibitors of sodium transport had centred on the AV3V area of the brain. The significance of these sources, and the interactions between them, remains a matter of conjecture. There are also some enigmatic data, e.g. the failure of substantial changes in blood volume to alter plasma concentrations of endogenous ouabain in conscious dogs, despite measurable increases of EDLI (Boulanger *et al.*, 1993; Ludens *et al.*, 1993).

A novel approach to the importance or otherwise of endogenous inhibitors is that of Lingrel *et al.* (1994) using the techniques of molecular biology and knowledge of subunit structure to breed mice with an α_2 subunit which makes them resistant to cardiac glycosides; the question will be whether they suffer any functional abnormalities as a result. Meanwhile, critical examination continues of alternative, trivial explanations for ouabain-like activity in mammalian tissues such as cross-reactivity of antibodies or dietary ingestion of cardiac glycosides, bearing in mind their activity at very low concentration. Doris (1994) points out that, while the adrenal cortex produces a range of steroids, the 14β-hydroxylation of the steroid nucleus which is characteristic of ouabain would be unique in mammals, though not in other vertebrates, e.g. toads, which use similar steroids as cutaneous venoms. Again, the constituent sugar of ouabain, rhamnose, is almost unknown in mammals, although it has been identified as a constituent of mucopolysaccharides in brain. Ouabain itself is a very polar compound compared with other steroids and this raises problems for its mechanism of secretion from cells. Nevertheless, the evidence that mammals can produce ouabain, probably from the adrenal cortex, now seems substantial (Blaustein, 1993; Laredo *et al.*, 1994; Woolfson *et al.*, 1994); the question remains 'What role does it play and how?'. In particular, what is the functional relationship between the natriuretic effects of endogenous inhibitors of sodium transport and those of natriuretic peptides *(Chapter 7)*?

The concentrations of endogenous ouabain reported so far are sufficiently high to alter the activity of all three α isoforms to varying degrees, according to the tissue involved (Doris, 1994). An interesting question is raised by the fact that cardiac glycosides bind to the external surface of the pump. Steroid hormones usually bind to intracellular receptors and then alter gene expression; are cardiac glycosides unique in acting at an extracellular site, or is there an additional intracellular receptor or second messenger? Either our concepts of the mode of action of steroid hormones require revision, or ouabain is not such a hormone.

It is important to emphasise that ouabain is not the only candidate endogenous cardiac glycoside-like inhibitor of Na–K ATPase (Doris, 1994) and that the hypothalamus remains another likely source of such inhibitors (Woolfson *et al.*, 1994) or of a non-cardiac glycoside, non-peptide, non-lipid inhibitor (Sancho *et al.*, 1993). Nevertheless, with all of them a cardinal problem is whether the amounts produced can have any lasting effect when the resulting rise in intracellular sodium provides a powerful opposing stimulus to the pump (Doris, 1994). There is also considerable scope for artefacts, particularly in preparations derived from cell fragments, therefore it is important that candidate natriuretic hormones are shown to act in a number of models (Woolfson *et al.*, 1994). For example, fatty acids can inhibit Na–K ATPase, but are unlikely to do so when proteins are available to bind them.

While mineralocorticoids are seen as the classic stimulators of sodium transport, glucocorticoids are also important, e.g. in the gut *(Chapter 3)*, vascular smooth muscle (Sterns *et al.*, 1994) and the ascending thick limb of the loop of Henlé (Dietl *et al.*, 1990). Here they have a very indirect mode of action, involving suppression of the production of a local inhibitor of Na–K ATPase activity derived from membrane lipids and arachidonic acid. Production of the latter is suppressed when glucocorticoids induce synthesis of lipomodulin. Since the latter is also involved in the response to ADH and calcitonin, their effects on the loop are also influenced by adrenal function (Dietl *et al.*, 1990). Recently an enteric peptide has been found which activates Na–K ATPase (Kairane *et al.*, 1994).

Aldosterone stimulates an increase in the number of Na–K ATPase pump sites and enhances the associated cation exchange. Initially, there is a lag period, typical of steroid hormones. A fall in aldosterone levels reduces Na–K ATPase activity (O'Neil, 1990). The increase in pump activity caused by aldosterone involves activation of latent channels initially, followed subsequently by increased synthesis of new pump sites. The initial activation of latent channels allows an influx of sodium and the increased ICF sodium concentration stimulates the pump, although activity increases even before this (Beron and Verrey, 1994). The lag period represents the time required for aldosterone to cross cell membranes, bind to cytoplasmic receptors and for the resulting complex to move into the nucleus, bind to an acceptor site on chromatin, initiate transcription and thereby initiate synthesis of new proteins. The lag period is highly variable, depending on the baseline level of Na–K ATPase activity, and this probably contributes to a number of data conflicts concerning the early response to aldosterone (Katz, 1990). There are also the inevitable differences between different tissues, including different nephron sites. Aldosterone stimulates Na–K ATPase gene expression in vascular smooth muscle (Oguchi *et al.*, 1994).

Thyroid hormone activates Na–K ATPase through the binding of triodothyranine (T_3) to specific nuclear binding sites. As a result, the number of pump sites is increased without alteration of either the affinity of the pump for Na, K or ATP or of the energy of activation of Na–K ATPase (Ismail-Beigi, 1992). Synthesis of both α and β subunits is increased, while their breakdown rate is unchanged. Although T_3 and steroids have greatly different structures, their DNA binding receptors are closely related (McDonough *et al.*, 1990). The increase in active transport of Na and K, and the associated breakdown of ATP to ADP (which in turn enhances ATP resynthesis) causes a sustained enhancement of cellular metabolic rate. This contributes to the thermogenic effect of thyroid hormone, though controversy continues concerning the proportion of the additional heat production attributable to Na–K ATPase in different tissues (Ismail-Beigi, 1992). The key factor may be the different proportion of basal energy turnover committed to active transport; in muscle, for example, this is relatively small and leaves the basis of thyroid thermogenesis unexplained (Mandel, 1986). In vascular endothelium, Na–K ATPase only accounts for 5% of basal metabolism (Schrader *et al.*, 1994).

Since both thyroid hormone and mineralocorticoids produce their effects via interactions with the genome (following formation of a receptor complex in the case of mineralocorticoids), there is obvious scope for a common pathway underlying their mode of action. Evidence suggests, however, that they act via independent routes (Lo and Klein, 1992). Although thyroid hormone facilitates entry of sodium into cells, this effect provides insufficient explanation of its stimulatory effect. Similarly, the renal effects were originally attributed to changes in the reabsorptive load of sodium caused by changes in glomerular filtration rate (GFR) (Katz and Lindheimer, 1973). In both instances, however, changes in Na–K ATPase activity can be shown to precede changes in sodium flux (Ismail-Beigi, 1992; Lo and Klein, 1992). Clinically, hypothyroidism is associated with increased plasma concentrations of endogenous inhibitors of Na–K ATPase, while thyrotoxicosis (Graves disease) suppresses their concentration and also increases the number of pump sites, thus perhaps predisposing to cardiac dysrhythmias (Blaustein, 1993).

Na–K ATPase has also been suggested as the basis of noradrenaline-stimulated thermogenesis in brown fat. The process is certainly ouabain-sensitive, but this seems to be an indirect effect via increased intracellular ATP and its influence on fatty acid mobilisation and uncoupling of respiration from phosphorylation. The latter appears to be the main thermogenic action (Mandel, 1986).

Insulin stimulates Na–K ATPase and this is the probable basis of its therapeutic use to reduce hyperkalaemia. Since insulin secretion also responds to hyperkalaemia it is likely that insulin plays a role, albeit a subsidiary one, in the regulation of plasma potassium concentration. Hyperkalaemia also stimulates Na–K ATPase activity within the physiologically plausible range (Sterns and Spital, 1987; Williams and Epstein, 1989; Perrone and Alexander, 1994).

Catecholamines stimulate Na–K ATPase via β-adrenergic receptors and can thereby cause hypokalaemia (Sterns and Spital, 1987; Williams and Epstein, 1989). As a result, stress or high levels of catecholamines in response to cardiac failure may potentiate the effects of diuretics and increase the risk of ventricular dysrhythmias *(Chapter 11)*.

Recent evidence indicates that, alongside the 'classical' modulators of Na–K ATPase, there is a complex of functional networks involving hormones and autocoids affecting intracellular signal transduction pathways; these vary with the tissue, including different nephron segments (Bertorello and Katz, 1993). They provide short-term regulation which, like the effects of ligands such as Na^+, K^+ or ATP, is independent of changes in the rate of biosynthesis. Knowledge of their importance has depended on the emergence of more refined techniques allowing use of intact cells rather than membrane preparations. Among the mediators are cAMP, various cytokines, protein kinases A and C and various eicosanoids, also inositol triphosphate (IP_3) and diacylglycerol (DAG). Both PTH and dopamine inhibit Na–K ATPase, and the latter functions as a local intrarenal modulator. Inhibition requires occupation of both DA_1 and DA_2 receptors, probably in the proximal tubule (Rouse and Suki, 1994). In contrast, noradrenaline and other α-agonists stimulate Na–K ATPase. At high concentration, β-adrenergic inhibition prevails (Fan *et al.*, 1993). Endothelin, either directly or indirectly, inhibits pump activity in the proximal tubule and inner medullary collecting duct, although at low concentrations it may stimulate (Garcia and Garvin, 1994). The effects of calcium are interesting; the established emphasis is on its inhibitory effect, but recent evidence from intact cells indicates that at concentrations within the likely range in ICF it may inhibit or stimulate, depending on the tissue. More important, the effect may be biphasic, with inhibition at high concentration (Bertorello and Katz, 1993). Since Na–K ATPase is a magnesium-dependent enzyme, its activity can be reduced in hypomagnesaemic subjects (Fischer and Giroux, 1987).

The cytoskeleton may be important in the regulation of Na–K ATPase, allowing 'cross-talk' between Na–K ATPase in the baso-lateral membranes of epithelial cells and sodium channels in the apical membranes (Bertorello and Katz, 1993). Actin stimulates Na–K ATPase and its effects depend on its phosphorylation or poly-merisation state (Bertorello and Katz, 1993). The cytoskeleton may also be important in maintaining Na–K ATPase in the correct spatial orientation; this results initially from cell–cell adhesion mediated by attachment of E-cadherin to the cytoskeleton (Fish and Molitoris, 1994). Ischaemia rapidly disrupts this distribution and impedes proximal tubular reabsorption of sodium by recycling it into the lumen (Molitoris, 1992; Fish and Molitoris, 1994). Clearly, more needs to be defined about the specific role of this plethora of regulatory factors in the various functions of Na–K ATPase, such as transepithelial transport or cell volume regulation. Interactions between the cytoskeleton and pump activity might, for example, serve to relate the latter to changes in cell volume. Cell swelling causes recruitment of new pump sites and activation of existing ones (Coutry *et al.*, 1994). Na–K ATPase is the principal but not the only mechanism regulating cell volume (MacKnight and Leaf, 1986). There is likely to be some separation in transporting epithelia between responses regulating net transport and those with a 'housekeeping' function, i.e. maintaining the intracellular environment. In rat brain synaptosomes, swelling caused by hypotonic media stimulates Na–K ATPase without changes in sodium influx, thus osmoregulatory responses can occur independent of the cytoskeleton or changes in internal sodium (Aksentsev *et al.*, 1994).

Na^+–H^+ Exchange

Sodium–lithium exchange, suggested as a marker for hypertension, may well be an *in vitro* marker of the activity of the Na^+–H^+ exchange system *in vivo* *(Chapter 6)*. This removes excess protons, generated during metabolic activity, from cells by exchanging them for sodium; it is specifically inhibited by amiloride. The Na^+–H^+ exchanger may not only be important in the regulation of intracellular pH (Boron, 1982), but also cell volume, cell growth and proliferation (Carr and Thomas, 1994). It is also involved in transepithelial transport of protons and bicarbonate (Pouyssegur *et al.*, 1988). In neutrophils, macrophages and platelets it is also responsible for chemotactic re-sponses, superoxide formation, activation of phospholipase A_2 and release of arachidonic acid and eicosanoids (Decker and Dieter, 1988).

Among the earliest effects following interaction of mitogens with their surface receptors is an amiloride-sensitive sodium influx, first reported in sea-urchin eggs following fertilisation (Pouyssegur *et al.*, 1988). This results from activation of the Na^+–H^+ antiporter. Intracellular pH becomes more alkaline as a result of export of protons in exchange for the sodium influx. Virtually all growth-promoting agents cause similar activation, although it was initially less clear whether this could be a response to an initial acidification of ICF. Subsequently, a number of studies have indicated that activation of the antiporter and, consequently, alkalinisation of ICF are the primary events (Pouyssegur *et al.*, 1988). Mutant cells without an antiporter cannot regulate ICF pH in response to acid loads, do not show a sodium influx in response to growth factors and fail to divide. Nevertheless, Ganz (1991) has suggested that some of the data regarding the role of the Na^+–H^+ exchanger are distorted by the artificiality of the *in vitro* conditions, in particular the exclusion of the bicarbonate buffer system. His contention is that the antiporter serves to preserve the normality of intracellular pH during mitogenic responses rather than mediating a rise of ICF pH.

A number of classic vasoconstrictors, such as angiotensin II, are mitogenic and also stimulate the Na^+–H^+ antiporter (La Pointe and Battle, 1994). Thus they may sustain hypertension by contributing directly to vascular hypertrophy *(Chapter 6)*.

The importance of Na^+–H^+ exchange in bicarbonate conservation by the proximal tubule is beyond dispute: most is reabsorbed by conversion to carbonic acid and the main source of protons for this is the Na^+–H^+ antiporter (Murer *et al.*, 1994). *In vitro* studies of proximal tubule cell lines suggest that its activity is increased by both metabolic and respiratory acidosis. The latter is consistent with the compensatory increase of plasma bicarbonate which offsets the increased pCO_2 of respiratory acidosis, thus maintaining a relatively stable plasma pH (Michell *et al.*, 1989). Activity of the antiporter is suppressed by PTH, consistent with its ability to increase urinary losses of both sodium and bicarbonate from the proximal tubule (Burckhardt and Burckhardt, 1988). While the antiporter is ubiquitous, there may be subtypes and, for example, the renal form is more resistant to amiloride and more responsive to cAMP (La Pointe and Battle, 1994).

Conclusion

Many of the issues raised in this chapter will concern us again as we explore them further or see how they contribute to clinical disturbances. To say that the regulation of body sodium is the pivot around which the defence of ECF volume and circulatory and renal function are centred is sufficient to emphasise its importance. To say that renal sodium regulation, the subject of the next chapter, is inseparable from successful regulation of intravascular volume, as well as the volume and osmolality of ECF (hence the osmolality of ICF too) and the long-term regulation of arterial pressure, is beyond dispute. Yet neither statement hints at the full dimensions or the ramifications of the factors involved in the clinical biology of mammalian sodium regulation.

2

Renal Sodium Regulation

Introduction

Two processes provide the basis of most of the non-endocrine functions of the kidneys; glomerular filtration and tubular re-absorption of sodium. They are inextricably linked and it could not be otherwise: if a 70 kg animal has a glomerular filtration rate (GFR) of 125 ml/min and a plasma sodium concentration of 145 mmol/l, a 1% mismatch between filtration and reabsorption would take only four days to reduce extracellular fluid (ECF) volume (14 l) by 50%—in fact, the animal would already be dead. The task of regulating external sodium balance also requires daunting precision; if the same 70 kg animal has a sodium intake of 0.65 mmol/kg/day (slightly above maintenance requirement, *Chapter 5*), its entire daily intake appears in primitive urine (glomerular filtrate) in less than 2 min; the daily primitive urine contains over 2 years' supply of sodium.

It is not perhaps surprising, therefore, that the great majority of renal energy consumption is expended on a single function, sodium reabsorption; it is, however, bizarre that such a huge reabsorptive workload should be the price of such a minute excretory outcome. Moreover, to the extent that 50% of GFR can be sacrificed with impunity (and some two-thirds without the onset of clinical symptoms), this is a profligate system; no energy-conscious engineer would design it that way.

Inevitably, sodium regulation provides the central focus of renal physiology and related research. As with many aspects of renal physiology, it is fairly easy to define what the kidney does; most of the controversy concerns how it is accomplished. The detail of such controversies, in particular with sodium the pumps, carriers and membrane potentials operating in different segments of the nephron, tends to obscure the fundamental importance of the appropriate balance of function between segments.

Segmental Distribution of Sodium Reabsorption

In its essentials, renal sodium regulation is as simple as depicted in Fig. 2.1. The **proximal tubule** receives glomerular filtrate in large volumes and reabsorbs perhaps two-thirds. The process therefore

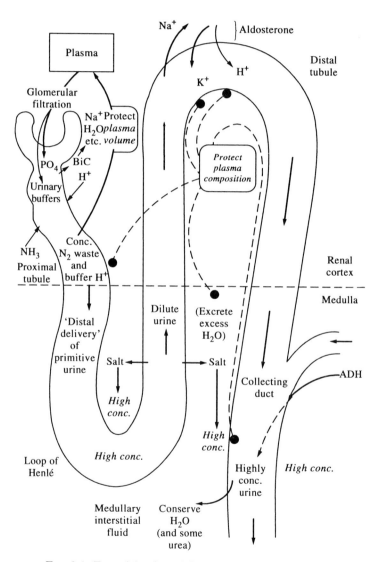

FIG. 2.1. Essentials of renal function in a single nephron.

achieves high volumes of reabsorption rather than high concentration gradients (except for non-reabsorbed solutes which rise in concentration as the volume of tubular fluid falls; these include end-products of protein catabolism such as creatinine). Since the re-absorbed fluid greatly resembles plasma, especially in its sodium and potassium concentration, it is almost ideal for the replenishment of plasma volume, should it be subnormal. The main extra-renal determinant of proximal tubular reabsorption is circulating volume; reabsorption is accelerated during hypovolaemia, and suppressed during volume expansion (e.g. due to excess salt intake or parenteral fluid therapy). While this is appropriate for sodium balance, it means that the excretion of excess sodium will also cause losses of calcium and magnesium, since their reabsorption parallels that of sodium in the proximal tubules. Though incidental, such losses may become clinically significant, e.g. calcium in post-menopausal women on high salt intakes *(Chapter 9)*. While it is usual to emphasise the role of the kidneys in defending **circulating** volume, there is also evidence to suggest that ECF volume as a whole, i.e. **including interstitial fluid (ISF)**, can be the determinant. Thus, saline expansion of ECF volume is more natriuretic than the equivalent volume of blood and the natriuresis persists even in the face of hypotension (Kirchner and Stein, 1994).

Some of the early studies on the effect of volume receptors in the great veins and left atrium can now be reinterpreted in the light of effects on atrial natriuretic peptide (ANP) secretion (see below and *Chapter 7*). Nevertheless, there is substantial evidence for neural responses, including those involving cardiac nerves, e.g. failure of natriuresis following cardiac denervation, despite rises in ANP (Kirchner and Stein, 1994). There could well be species differences, especially between primates and quadrupeds. As well as volume receptors on the venous side, baroreceptors on the arterial side of the circulation influence renal sodium excretion, notably those in the carotid sinus and the afferent arteriole of the kidney (Seldin, 1990). Renal baroreceptors respond to increases in arterial or venous pressure and interstitial pressure (Moss, 1989). Other possible sites for volume receptors include the liver *(Chapter 4)* and brain. Intracarotid or intraventricular hypertonic saline is more natriuretic than systemic infusions and hypotonic saline within the cerebral ventricles reduces sodium excretion. ECF volume depletion reduces the sodium concentration in cerebrospinal fluid (CSF) (Kirchner and Stein, 1994). Reduction of haematocrit, provided ECF volume is not expanded, also reduces sodium excretion; this is true when plasma is used as the diluent, i.e. there is no effect on oncotic pressure.

The other determinant of proximal reabsorption, necessarily, is GFR. **Glomerulo-tubular balance** describes the mechanism whereby changes in GFR, provided they are not too fast or extreme, are virtually matched by parallel changes in tubular reabsorption of sodium and water. **Tubuloglomerular feedback** describes mechanisms whereby 'distal delivery' of sodium is monitored (by the macula densa) and GFR adjusted accordingly to prevent excessive loss of sodium. In both cases, the description is simple but the explanation remains complex and controversial. The combination of these mechanisms ensures that variations in GFR do not prejudice the regulation of sodium balance. This is just as well, since small perturbations can potentially have a drastic impact on sodium excretion. Equally, it implies that the regulation of sodium excretion rests on control of tubular reabsorption rather than GFR. An important stabilising influence on GFR is **autoregulation** of renal blood flow, i.e. over a certain range glomerular perfusion remains stable despite hypovolaemia or hypotension. Tubuloglomerular feedback contributes to this, but is only part of the explanation (Steinhausen *et al.*, 1988).

In **the loop of Henlé**, active removal of salt (perhaps 25% of filtered load) but not water in the ascending thick limb produces the dilute urine necessary for excretion of excess water. It also creates the concentrated interstitial fluid characteristic of the renal medulla; this is the basis of water conservation during dehydration. Antidiuretic hormone (ADH) is released in response to the resulting rise in plasma sodium concentration (or, in more severe dehydration, the fall in circulating volume) and, by making the collecting duct more permeable, it allows the concentrated medullary interstitial fluid to extract water from the final urine and achieve a high concentration and a low volume. The loops are thus the basis of both dilution and concentration of urine (excretion and conservation of water). Inability to secrete appropriate amounts of ADH or to respond to it normally cause diabetes insipidus (central or nephrogenic, respectively). Species particularly adapted to arid environments are characterised by long loops of Henlé. In contrast, some species living in or close to water (e.g. beaver, hippopotamus) have many nephrons with short loops that do not even reach the medulla; carnivores have none and humans are intermediate (Bankir *et al.*, 1989).

The effectiveness of the renal concentrating mechanism depends particularly on:

1. a low blood flow in medullary vessels and the counter-current exchange between their descending and ascending components, thus minimising 'washout' of the concentrated solutes in interstitial fluid;

2. a high concentration of urea in medullary interstitial fluid as well as urine, thus minimising the osmotic gradient which would otherwise result from high concentrations of nitrogenous solutes in final urine, as water is conserved;
3. appropriate release of ADH and normal responsiveness of the collecting duct;
4. adequate delivery of sodium from the proximal nephron;
5. a normal population of loops.

Progressive nephron loss (as in chronic renal failure, CRF) means that there is less likelihood of generating fully concentrated interstitial fluid. Moreover, the surviving nephrons have high flow rates resulting from compensatory increases in their individual GFR (single nephron GFR; SNGFR). Consequently, there is less likelihood of equilibration along solute gradients. Production of maximally concentrated and maximally dilute urine are, therefore, both likely to be impaired. The scope for effective responses to extreme demands (e.g. caused by severe dehydration or over-enthusiastic fluid therapy) is thus restricted; renal function becomes less flexible. Because dogs, unlike humans, tend to drink homeostatically (rather than in response to habit or flavour) their urine is normally more concentrated and polyuria (excess urine output) and compensatory polydipsia (increased fluid intake) are hallmarks of canine CRF. Primary (psychogenic) polydipsia causes a reduction of renal concentrating capacity, partly through medullary washout and partly by chronic effects on the response of the collecting duct to ADH (Morrison and Singer, 1994; Knepper *et al.*, 1994). The response to ADH, therefore, is affected by prior polydipsia, not just by nephrogenic diabetes insipidus *(Chapter 8)*.

In **the distal nephron** less than 10% of filtered sodium remains to be reabsorbed. This fraction may seem trivial but it is functionally of great importance. It determines the final urinary loss of sodium and, in volume depletion, it can fall virtually to zero under the influence of aldosterone. Distal sodium conservation, through its effects on tubular membrane potential, also facilitates the excretion of potassium promoted by aldosterone. Filtered potassium is largely reabsorbed, so potassium balance is vitally dependent on this mechanism. Since, in herbivores, dietary potassium may exceed sodium by 100:1, rather than being closer to 1:1 (as in humans), this role of aldosterone is arguably at least as important as its influence on sodium regulation. Indeed, hyperkalaemia directly stimulates aldosterone secretion from the adrenal cortex, whereas sodium depletion does so indirectly (by release of renin from renal afferent arterioles when perfusion falls as a result of hypovolaemia; renin-induced generation of angiotensin provides the stimulus for aldosterone secretion).

Effects of Hypovolaemia

When aldosterone increases distal reabsorption of sodium, hydrogen ion secretion is also enhanced, particularly if tubular cells are depleted of potassium. Thus, excessive aldosterone secretion or potassium depletion predispose to metabolic alkalosis. On the other hand, inadequate distal delivery of sodium will impede the excretion of hydrogen ions or of potassium, despite elevations of aldosterone secretion.

Mild hypovolaemia, especially if sustained, predisposes to potassium depletion and hypokalaemia; this is reinforced by enhanced colonic exchange of K for Na, also under the influence of aldosterone. In severe volume depletion, however, the patient may be hyperkalaemic, despite cell deficits of potassium, and despite increased secretion of aldosterone. The cause is inadequate delivery of sodium to the distal nephron. Hyperkalaemia is also likely to be exacerbated by any accompanying metabolic acidosis; movement of hydrogen ions into cells is usually accompanied by outward movement of potassium (Fig. 2.2).

Since plasma sodium concentration is mainly regulated by control of water balance, inadequate delivery of tubular fluid to the loop and distal nephron undermines the defence of plasma sodium concentration as well as potassium and pH. Thus, the enhancement of proximal reabsorption in defence of plasma volume undermines the defence of plasma composition by subsequent segments of the nephron. That is why, in fluid therapy, restoration of circulating volume often allows simultaneous correction of plasma composition, by improving renal function, e.g. saline, though neutral rather than acidic, corrects metabolic alkalosis. Hypovolaemia, *per se*, undermines the ability of normal kidneys to resist the development of metabolic alkalosis by increasing bicarbonate excretion.

This general survey leads to two conclusions:

1. regulation of sodium excretion depends not only on the individual mechanisms characterising successive segments of the nephron, but on a normal balance of function between segments;
2. even mild hypovolaemia potently distorts renal function by enhancing proximal reabsorption and reducing distal delivery of sodium.

Redistribution of renal blood flow (RBF) between superficial

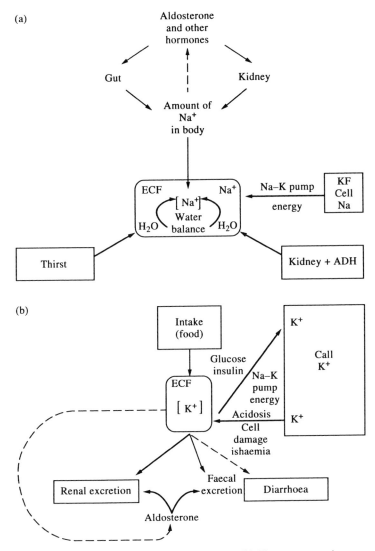

FIG. 2.2. Regulation of plasma: (a) Na^+; (b) K^+ concentration.

and deep nephrons (cortical or juxtamedullary glomeruli) can also influence sodium excretion.

Endocrine Aspects of Renal Sodium Regulation

Three key components of endocrine enhancement of sodium conservation have already been mentioned; renin, angiotensin and aldosterone. Twenty years ago they would have constituted most of the topic. Moreover, angiotensin would have been seen mainly as a link between sodium depletion and aldosterone secretion, also as a stimulus to thirst and increased arterial pressure (through vasoconstriction). Now angiotensin is regarded as an important influence on sodium excretion. The main developments, however, have been:

1. a rapid growth in knowledge of the mechanisms for excretion of excess sodium (as distinct from simply suppressing the mechanisms responsible for sodium conservation);
2. an increasing range of hormones known to affect sodium excretion, whether or not they contribute to regulation of sodium balance;
3. realisation that, with a number of hormones, intrarenal generation rather than systemic concentration is the main influence on water and electrolyte excretion.

These are discussed mainly in *Chapter 7*.

Glomerular Filtration

The basis of normal renal function is an adequate glomerular filtration rate. This rests on autoregulation of renal perfusion and tubulo-glomerular feedback. The key determinants of GFR are:

1. filtration pressure within the glomerular capillaries; this can be increased by constriction of the efferent arteriole (e.g. in response to angiotensin II) or the relaxation of the afferent arteriole;
2. oncotic gradient between plasma and primitive urine (glomerular filtrate);
3. renal blood flow;
4. packed cell volume (PCV) [since it influences renal plasma flow (RPF) at a given RBF];
5. capillary permeability and other factors affecting the ultra-filtration coefficient;
6. factors affecting the tone of mesangial cells.

GFR is the dominant influence on renal function because it dictates the necessary workload (i.e. rate of consumption of ATP and oxygen) associated with sodium reabsorption and because various renal gradients or exchanges are sensitive to flow rate, especially if it is either very low or very high. While GFR does not appear to be a primary factor in the regulation of any excretory rate, it has a strong influence, particularly because glomerulotubular balance is focused on sodium and changes affect other ions as well. In addition, changes in GFR resulting from changes in filtration fraction will also affect fractional reabsorption (the proportion of filtrate reabsorbed) in the proximal tubule: as a result, the delivery of water and solute to the remaining segments of the nephron also changes. There would seem, therefore, to be strong reasons why GFR should be tightly regulated, although some of its 'constancy' does not stand up to scrutiny. Changes of 1% are potentially disastrous (unless offset by glomerulotubular balance), yet they are well beyond the precision of measurements, especially in intact animals. Moreover, given time, the body adjusts excellently to progressive loss of nephrons and the consequent reduction in GFR. The basis of this adaptation, especially the increased SNGFR of surviving intact nephrons in CRF, is discussed in *Chapter 8*.

While GFR may change, for example as a result of redistribution of blood flow between superficial or deep nephrons (which have higher SNGFRs; Navar *et al.*, 1986), most of our understanding of the control of filtration is based on the factors affecting SNGFR in micropuncture studies of anaesthetised experimental animals. Much of it is also based on experiments with unspecified, usually high, salt intakes (*Chapter 5*), despite the fact that these increase ECF volume and GFR (Kirchner and Stein, 1994). GFR is also affected by dietary protein, increasing in response to amino acids. In this sense, therefore, a reduction of dietary protein will 'rest' the kidneys since, like a reduction of sodium intake, it reduces the amount of energy committed to reabsorption of sodium (Katz, 1988). There are considerable species differences, not only in renal morphology, e.g. the multilobed kidney of cattle, but in nephron number and nephron density per kidney and also nephron size (Crispin and Stickland, 1983; Navar *et al.*, 1986; Reece, 1993).

Despite the assumption that the kidneys normally work in parallel, responding independently to the same stimuli, recent evidence suggests that each kidney can influence the function of the other, via reflex pathways (Golin *et al.*, 1987). In disease, interaction between kidneys is well known, for example underperfusion of one kidney, leading to increased production of renin and angiotensin, leads to responses in the other. The influence of renal nerves on kidney function remains poorly understood. It is clear that kidneys work reasonably well without them, as in transplant recipients, but there is also growing awareness of the ability of renal nerves to 'fine-tune' normal renal function. This does not simply reflect the effects on renal vasculature, but direct innervation of nephron segments. For example, changes in α_2-adrenergic receptor activity modulate the effect of angiotensin II on the ultrafiltration coefficient (Blantz, 1993). The importance of neural modulation of renal function is increased by sodium depletion because intact innervation is needed to achieve maximum sodium reabsorption from urine (di Bona, 1989). There is direct contact between a full range of adrenergic neurons (dopaminergic, adrenergic, α_1, α_2, β_1, β_2 and subtypes) and all major segments of the nephron, including the juxtaglomerular apparatus (Rouse and Suki, 1994). Activation of renal sympathetic nerves causes enhanced secretion of renin and prostaglandins mediated by β_1 and α-adrenergic receptors (Osborn and Johns, 1989). The usual outcome of sympathetic activation is to promote retention of salt and water, but the acute natriuresis caused by renal denervation is probably a surgical artefact (Knox and Granger, 1992). Neural effects on renal function are also likely to prove important in various pathophysiological states, such as hypertension or oedema, but their role remains to be defined (Janssen and Smitz, 1989; Zambraski, 1989).

There is substantial evidence for the ability of changes in sodium within the central nervous system (CNS) to affect sodium excretion (Mouw *et al.*, 1974; Anderson, 1977), whether or not they actually regulate it. In particular, the anteroventral 3rd ventricle (AV3V) area has been suggested as regulator of sodium excretion, sodium intake and arterial pressure (Brody *et al.*, 1978). Two main problems beset the evidence; the plausibility of induced changes in the sodium concentration of CSF and the possibility that responses are simply a result of changes in sympathetic tone and renal perfusion.

Ultrafiltration

Apart from afferent and efferent arteriolar tone, mesangial cells influence ultrafiltration. Contraction of glomerular mesangial cells in response to angiotensin-II or ADH reduces GFR (Kurokawa *et al.*, 1988; Blantz, 1993). The kidney is protected from the excessive effect of these vasoconstrictors on both the mesangium and renal vessels by enhanced local production of prostaglandins and nitric oxide (EDRF, endothelium derived relaxing factor). Extraglomerular mesangial cells may also play an important role in tubuloglomerular feedback (Blantz, 1993; Lorenz *et al.*, 1993).

Just as movement is usually the result of balanced interaction between opposing muscle groups rather than response to contraction, e.g. of flexors alone, it is becoming increasingly clear that 'single-stimulus' effects on renal function turn out to be the result of multiple interaction. For example, the predominant vasoconstrictor effect of angiotensin-II is on the efferent arteriole, maintaining GFR and facilitating reabsorption in the proximal tubule (see below) by reducing the 'downstream' pressure in peritubular capillaries. The direct effect of angiotensin II, however, is to constrict both arterioles, but prostaglandin and nitric oxide overcome the effect on the afferent arteriole (Heller *et al.*, 1988; Blantz, 1993). ANP raises GFR by dilating the afferent arteriole and, at high concentrations, constricting the efferent arteriole (Steinhausen *et al.*, 1988). There is a rich sympathetic innervation of both afferent and efferent arterioles and the juxtaglomerular apparatus, mostly adrenergic but also dopaminergic (Navar *et al.*, 1986; Steinhausen *et al.*, 1988).

The filtration pressure within the glomeruli is high compared with other capillaries, but since the filtration fraction is also high, so is the opposing oncotic gradient. Proteinuria, usually albuminuria as a result of abnormal increases in glomerular permeability, reduces this oncotic gradient, as does any associated hypoalbuminaemia. This means that reduction of GFR caused by glomerular disease may be moderated by its effect on the oncotic gradient.

The balance between glomerular capillary pressure and oncotic pressure provides the drive for ultrafiltration and is related to filtration rate (GFR) by the filtration coefficient (K_f). This depends on the hydraulic conductivity (aqueous permeability) of the filtration membrane and its total surface area. Changes in the surface area available for filtration can result, for example, from mesangial contraction. The filtration membrane consists of the capillary endothelium, the basement membrane and the foot processes of the epithelial cells of Bowman's capsule. Its normal relative impermeability to proteins of a size similar to albumin or larger also depends on their shape and their charge, due to the presence of polyanionic groups. The pressure opposing glomerular filtration within the lumen of Bowman's capsule is normally low, similar to that in the proximal tubule and peritubular capillaries or intrarenal veins (Navar *et al.*, 1986).

There has been considerable controversy, perhaps reflecting species differences, on whether filtration comes to equilibrium before plasma leaves the glomerular capillaries; if so, renal plasma flow becomes a major influence on GFR (Thurau, 1989). In effect, filtration equilibrium prevents utilisation of the full available surface area for fluid transfer, and increased plasma flow, by delaying the attainment of equilibrium along the length of the capillaries, recruits additional surface area (Navar *et al.*, 1986). It seems unlikely that filtration equilibrium is normally important in humans or dogs, unlike rats.

Efferent constriction is most effective in sustaining GFR when it is subnormal. Otherwise, GFR is virtually maximal and the result of efferent constriction is to reduce plasma flow and, by increasing filtration fraction, to increase the oncotic gradient opposing filtration. Loss of efferent tone would, however, reduce GFR. Supranormal GFR is most likely to result from afferent arteriolar relaxation and the associated rise in capillary pressure (Navar *et al.*, 1986). This probably underlies the increased GFR seen early in diabetes mellitus (Michell, 1994a). While angiotensin converting enzyme (ACE)-inhibitors are beneficial for many patients with CRF, there is a danger that they will precipitate acute renal failure (ARF) in patients dependent on efferent constriction induced by angiotensin-II to sustain their GFR (de Jong *et al.*, 1989).

Tubuloglomerular feedback depends on the 'sensing' of solute delivery by the macula densa, which actually responds to reductions in tubular concentration, moreover of chloride rather than sodium (Laragh and Seeley, 1992; Lorenz *et al.*, 1993). The question arises how this can reflect changes in GFR; the electrolyte composition of glomerular filtrate is essentially similar to plasma and insensitive to changes in GFR. The answer lies in the removal of sodium and chloride in the loop of Henlé; this has less impact on the composition of tubular fluid as flow rates increase, i.e. the concentration of sodium and chloride delivered to the macula densa rises with nephron flow rate (Navar *et al.*, 1986). The role of angiotensin-II in tubuloglomerular feedback is not clear. Firstly, the changes in vascular tone are usually opposite to the expected change in renin, e.g. increased distal delivery increases vascular resistance (hence reducing GFR), but decreases renin release. Secondly, apart from circulating angiotensin-II, the kidney can generate angiotensins I and II intrarenally. All the components of the renin-angiotensin system are present in the proximal tubule and possibly in the distal tubule (Moe *et al.*, 1993). Angiotensin may modulate rather than mediate tubuloglomerular feedback (Mitchell and Navar, 1989). The renin-angiotensin system is discussed further in *Chapter 7*. Other important intrarenal mediators include nitric oxide, produced by the glomeruli, and endothelins, notably endothelin (ET)-1 in the renal medulla (Marsden and Brenner, 1991). Nitric oxide also enables the kidney to resist excessive vasoconstriction and may inhibit renin release (Hall and Granger, 1994). Endothelins are not only produced by the kidney but it is also, besides the lungs, the main receptor site (it should be remembered, however, that inactivation of a hormone, as well as its mode of action, may involve specific receptors). Apart from vascular effects, mainly constrictor, endothelins also activate the renin–angiotensin system (Hall and Granger, 1994).

Proximal Reabsorption

While the hepatic portal system is often regarded as the only example in mammals, and a renal portal system is regarded as a feature of other vertebrates, such views are simplistic. Apart from the portal system between the hypothalamus and the anterior pituitary gland, the vascular relationship between the glomeruli and the peritubular capillaries surrounding the proximal tubule essentially comprises a portal system. Increased filtration fraction therefore increases protein concentration in these capillaries and favours reabsorption of sodium; at a given GFR, the higher the filtration fraction, the greater the proportion of filtered sodium likely to be absorbed. Proximal reabsorption is also favoured by efferent arteriolar constriction; this raises upstream (glomerular) pressure but reduces peritubular pressure (Seldin *et al.*, 1991). Part of the reduction in proximal tubular reabsorption of sodium associated with ECF volume expansion is attributable to dilution of albumin in peritubular capillaries (Mendez and Brenner, 1990). This could explain why volume expansion with hyperoncotic albumin decreases urinary sodium excretion, rather than increasing it (Kirchner and Stein, 1994). Maintenance of glomerulotubular balance is facilitated by the fact that proximal reabsorption of sodium and co-transported solutes (such as glucose and amino acids) is promoted by increased delivery, i.e. increased filtered load (Mendez and Brenner, 1990).

The basis of proximal reabsorption is the expulsion of sodium from the basolateral surface of tubular cells. This maintains the low intracellular sodium which allows passive entry of sodium into tubular cells and the associated cotransport or exchange of other ions and molecules, e.g. absorption of glucose, amino acids, bicarbonate, secretion of hydrogen ions (Berry and Rector, 1991). Sodium entry is not by simple diffusion, but depends on the various cotransport or exchange pathways (Giebisch and Aronson, 1986). The pumping of sodium into the interstitial fluid of the proximal tubule creates transient hypertonicity, which is corrected by reabsorption of water from the lumen. While reabsorption is isotonic, contraction of the residual volume of tubular fluid raises the concentration of less readily or non-reabsorbed solutes such as urea and creatinine. Early in the proximal tubule, reabsorption of bicarbonate, glucose and amino acids is more marked than that of chloride, so that, late in the proximal nephron, sodium reabsorption is also facilitated by a chloride gradient (Koeppen, 1990; Berry and Rector, 1991). Much of the reabsorption late in the proximal tubule is via paracellular pathways which are highly permeable to sodium, chloride and water. Proximal tubular reabsorption of sodium and chloride is stimulated by catecholamines or by sympathetic nerves, though the relative importance of α and β receptors is still uncertain (Koeppen, 1990). It is also stimulated by angiotensin, but biphasically, with inhibition at high dose. The effect of angiotensin is direct, as well as reflecting its influence on glomerular haemodynamics (Moe *et al.*, 1993). Dopamine and PTH reduce proximal reabsorption. It is likely that pressure natriuresis involves inhibition of proximal tubular reabsorption of sodium by increased pressure in the interstitial fluid, transmitted from the medulla; this increase, like the natriuresis, is nitric oxide dependent (Cowley *et al.*, 1992). Metabolic acidosis also reduces sodium reabsorption in the proximal tubule (Giebisch and Klein-Robbenhaar, 1993), as does potassium loading. The latter effect may have functional importance; by delivering more sodium to the distal nephron, it facilitates the excretion of potassium under the influence of aldosterone (Laragh and Seeley, 1992).

The importance of pressure natriuresis was originally conceived in the context that it enabled the kidney to function as the long-term regulator of arterial pressure *(Chapter 6)*. It may, however, be equally important in enabling the kidney to achieve more precise regulation of circulating volume (Cowley and Roman, 1989). Even small cumulative errors in the match between sodium intake and sodium excretion will allow large changes in ECF volume. Such errors are probable because mechanoreceptors are likely to adapt to the prevailing conditions and thus as the 'error signal' declines they lose their sensitivity. This would be particularly true of volume receptors on the low pressure side of the circulation. The key to sensitive regulation of circulating volume appears to be the exquisite sensitivity of pressure natriuresis to changes in renal perfusion pressure. Without it, for example, the ability to escape from the salt-retaining effect of excess aldosterone is lost (Cowley and Roman, 1989). Moreover, acute volume expansion with saline fails to provoke natriuresis if renal perfusion pressure and, therefore, medullary interstitial pressure, a key factor in pressure natriuresis, are prevented from rising (Knox and Granger, 1992). The exact site(s) of pressure natriuresis remain a matter of dispute, but the likeliest are the proximal tubule and the loop.

There is greater heterogeneity of proximal tubular function, even within species, than the above account suggests. The early proximal tubule, late convoluted and early proximal straight tubule and the late straight proximal tubule show differences in morphology and permeability characteristics. There are also differences between the proximal tubules associated with superficial or deep glomeruli (Giebisch and Aronson, 1986). Early in the proximal tubule, the electrical potential difference (PD) is lumen negative, i.e. favouring back-leak of sodium, whereas later it becomes positive. The selective net reabsorption of bicarbonate rather than chloride early in the proximal tubule reflects its ability to recycle via CO_2 and carbonic anhydrase. Thus proton secretion, by neutralising filtered bicarbonate, allows it to be reabsorbed from the tubular lumen as CO_2 as well as bicarbonate ions. Intracellular dissociation of carbonic acid makes the equivalent amount of bicarbonate available to tubular interstitial fluid, hence to plasma (see below). The basolateral membrane contains a sodium-coupled bicarbonate transporter which is the main, but not the only mechanism facilitating export of bicarbonate into interstitial fluid (Alpern and Preisig, 1988). Angiotensin affects both the Na^+-H^+ exchanger on the luminal membrane and the $Na^+-HCO_3^-$ cotransporter on the basolateral membrane (Moe *et al.*, 1993).

Back-leakage of sodium, rather than chloride, via the intercellular shunt paths, underlies the difference in the transepithelial permeability of these ions, rather than differences in the permeability of tubular cells (Giebisch and Aronson, 1986). Back leakage also contributes to the reduced reabsorption of fluid during expansion of ECF volume or other changes which lower the oncotic pressure or raise the hydrostatic pressure in peritubular capillaries. This impedes the absorption of fluid from the peritubular interstitial spaces into capillaries and favours its back-leakage into the tubular lumen (Giebisch and Aronson, 1986; Seldin, 1990). Recently, there has been great interest in renal interstitial pressure and its effect on back-leakage, as one of the peritubular physical factors with a decisive influence on the reabsorption of sodium (Knox and Granger, 1992).

Despite the fact that the proximal tubule plays a decisive role in the changes of sodium excretion which result from expansion or contraction of plasma volume, and the evidence for the involvement of natriuretic hormones in renal responses to excess sodium, neither ANP nor endogenous inhibitor of sodium transport appears to regulate sodium reabsorption in the proximal tubule (*Chapters 1 and 7*). Perhaps a central part of the problem is the contradictory demands on proximal tubular reabsorption, according to the context, not just the stimulus. Thus, if filtered load of sodium increases because of volume expansion, the context requires a reduction of reabsorption to maintain external (i.e. body) sodium balance. When filtered load increases because of an increased GFR, the context demands an increased reabsorption to prevent salt depletion. Moreover, one of the direct influences on absorption, efferent arteriolar tone, also influences filtered load. Whatever the response, its outcome has to be viewed in a further context, namely its effect on the delivery of solute, particularly sodium (and chloride) to subsequent segments of the nephron and, therefore, on their function.

Similar conflicts between opposite 'appropriate' responses, according to the context, occur in enteric function (*Chapter 3*). They also occur in thermoregulation, e.g. cool skin may be a warning of low environmental temperature, a response to an adaptive reduction in perfusion or a sign of efficient evaporative cooling. Naturally, every 'context' must be represented, whether to the gut or the kidney, by appropriate neural, humoral or other signals, but as yet we are far from knowing what they are. Meanwhile, appreciation of the importance of the context of a stimulus, as well as its character, arouses two causes for concern:

1. isolated, *in vitro*, or broken cell preparations have only the 'contexts' which we provide;
2. where renal sodium regulation is concerned, one of the main contexts is often unconsidered or uncontrolled; is sodium intake appropriate and, therefore, would we regard ECF volume and GFR as 'normal' *(Chapter 5)*?

Recently the loop has been implicated in a key sodium regulating mechanism. The importance of 'pressure natriuresis' has long been clear, but its basis remained obscure. A major contributory mechanism may be an increase in pressure in peritubular interstitial fluid. This originates from increased pressure in the vasa recta, and the resulting outflow of fluid raises interstitial fluid (ISF) pressure in the medulla and, by hydraulic transmission, also the cortex (Hall and Granger, 1994).

Proximal reabsorption of sodium is also vitally involved in acid–base regulation, in particular, the reabsorption of filtered bicarbonate (Fig. 2.3). This is mainly converted to carbonic acid by protons

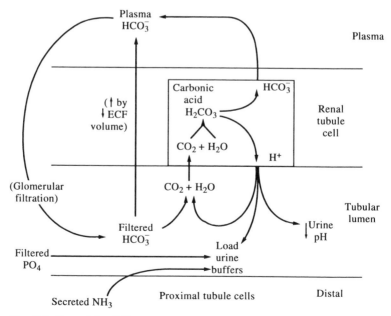

FIG. 2.3. Essentials of H^+ excretion (and regulation of plasma bicarbonate).

secreted into the tubular lumen by the Na–H antiporter. Since CO_2 readily crosses cell membranes, this is a more important route of bicarbonate reabsorption than movement of the anion itself (Maddox *et al.*, 1987). The activity of the antiporter is suppressed by excess sodium intake (Seldin *et al.*, 1991). Within the cell, carbonic anhydrase generates a supply of **hydrogen ions** which can be recycled into urine via the antiporter and **bicarbonate**, which is transported to interstitial fluid (and hence returned to plasma) by cotransport with sodium on the symporter present in the basolateral cell membrane (Aronson *et al.*, 1991). A much smaller contribution to proximal proton secretion comes from an electrogenic H-ATPase (Maddox *et al.*, 1987).

Another aspect of the proximal tubular contribution to acid–base regulation is the production of ammonia buffer which, together with phosphate buffer, may be the limiting factor in acid excretion. Ammonia synthesis depends on delivery of glutamine to tubular cells. This depends on the sodium gradient and potential difference generated by Na–K ATPase. Ammonia synthesis is regulated by glucocorticoids. The adaptive increase in response to an established metabolic acidosis is glucocorticoid-dependent and excess glucocorticoid sufficiently enhances acid excretion to cause metabolic alkalosis (Welbourne, 1990).

Recently, it has been found that acute hyperkalaemia suppresses the reabsorption of sodium, chloride and bicarbonate in the proximal tubule, perhaps by depolarisation of basolateral membranes (Stokke *et al.*, 1993).

Loop of Henlé

The loops, alongside thirst, provide the basis for the regulation of plasma sodium concentration since they allow the excretion of concentrated or dilute urine (water conservation or excretion of excess). In general, species capable of producing the highest urine osmolality are those with the longest loops of Henlé. Thus, while dilute urine may contain as little as 50 mOsm/kg* of solute, that of the Australian hopping mouse can reach a concentration of 9000 mOsm/kg (Jamison *et al.*, 1993). The patient with CRF may not be able to raise urinary concentration much above plasma osmolality

---

*Osmolality is expressed per kq of water (osmolarity is per litre) and is therefore independent of changes of volume with temperature.

(around 300 mOsm/kg). After all, as this is the general concentration of ISF, it does not require any countercurrent multiplication.

Suppose that dietary intake and metabolism require the daily excretion of 900 mOsm of solute. Even the CRF patient will achieve this with the production of 3 l/day of urine, rather than perhaps 1.5 l at 600 mOsm/kg. On the other hand, complete diabetes insipidus would demand the excretion of 18 l of fully dilute urine (at 50 mOsm/kg); that is why the polyuria is so striking. Granted that this could be one bladderful every 40 min, such a patient would envy that hopping mouse with a kidney which could excrete the same amount of solute in 100 ml—less than a wine glass. Since the water needed to excrete the solute would need to be drunk (the contribution from metabolic reactions is minute), the loop is the key to survival in arid environments. For patients with CRF or diabetes insipidus, it is the key to their convenience.

The basis of loop function, the countercurrent multiplier, has been a 'classic' of renal physiology for over 40 years—any student is expected to be able to explain it. Yet, like so much of renal physiology, the fine details, particularly the quantitative aspects, remain contentious.

In its simplest essence (Fig. 2.4), the transfer of solute from B to A, to establish a modest concentration gradient ($B<A$) produces a very high concentration at C because the U-shaped configuration of the loop, with its opposite flows in the descending and ascending limbs, sequentially multiplies the gradient. The reason is obvious; the same process repeated between a and b in Fig. 2.4 provides a solute transfer additional to that between A and B. The concentration at b is therefore higher than that at A. If the ISF surrounding the apex of the loop, at C, has the same high concentration as the urine (which it should, since it is the route for movement of solute between the adjacent limbs), then the loop has created the concentrated medullary environment necessary for final extraction of water from the urine in the collecting ducts when ADH makes them permeable. On the other hand, the urine leaving the loop, beyond B, is fully dilute; the loop has provided the option of fully concentrated or fully dilute urine.

Naturally, this can only work if the solute in the medullary ISF remains concentrated rather than equilibrating with plasma; that is why the medullary blood flow is so low. What is more, the vessels are themselves in loops so that solute tending to wash out of the medulla re-equilibrates with the blood entering the medulla. The blood vessels thus function as countercurrent exchangers, though not as multipliers. Vascular countercurrent exchangers are also used to conserve heat (rather than solute), e.g. in birds' legs or seals' flippers. 'Medullary washout' undermines urinary concentrating capacity. It

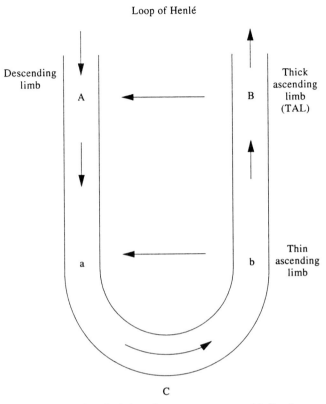

Loop of Henlé

Descending limb

Thick ascending limb (TAL)

Thin ascending limb

A   B

a   b

C

FIG. 2.4. Basic principles of countercurrent multiplication.

is caused by abnormally high medullary blood flow, e.g. as a result of chronic overdrinking (psychogenic polydipsia) or diversion of RBF from the cortex to the medulla in acute renal failure, hence the paradoxical urine of typical ARF; small in volume (as in a dehydrated patient) yet, unlike a dehydrated patient, dilute. ADH reduces medullary blood flow (Jamison *et al.*, 1993), thus it facilitates renal water conservation via effects on sodium transport (TAL), water reabsorption (collecting duct) and maintenance of concentrated medullary interstitial fluid (medullary perfusion). Medullary washout can also potentiate natriuresis, but water diuresis *per se* does not cause natriuresis (Knox and Granger, 1992). The effect of washout is to dilute the sodium reaching the ascending thick limb because the medullary ISF extracts water from the descending limb (see below).

So far, the water conserved from final urine has not returned to plasma. It does so along gradients analogous to those in capillaries, i.e. drawn into circulation along the oncotic gradient provided by plasma albumin and amplified by countercurrent multiplication. The overall efficiency of the system in excreting or conserving water can be expressed as the free water clearance (positive or negative, respectively). These matters, alongside the consequences of abnormal function (hypo- or hypernatraemia) are discussed in *Chapter 8*.

It is necessary now to extend this simple outline to include a little more detail concerning the actual nature of the exchanges of solute and water, the relative importance of sodium and urea as the concentrated solutes and the factors determining the release of ADH and the ability of the collecting duct (CD) to respond to it.

Ascending Thick Limb

Salt extraction in the ascending thick limb (TAL) leaves the urine dilute and the lumen electrically positive. The process involves both Na–K ATPase and secondary active transport of chloride against the electrical gradient. This is achieved by:

1. reduction of intracellular sodium by Na–K ATPase;
2. movement into ICF from urine of two cations (Na + K) accompanied by $2Cl^-$.

The potassium recycles back to the lumen (Winters *et al.*, 1991). This ensures that there is sufficient K^+ in luminal fluid to sustain the process, despite the low initial concentration in glomerular filtrate. There is also recycling of potassium from the collecting duct to the loop (Field and Giebisch, 1989). The Na:K:2Cl cotransporter is inhibited by furosemide. ADH facilitates salt extraction from the TAL by increasing the exit of chloride from the basolateral cell membrane (Winters *et al.*, 1991), though this might be a response to an increased entry of chloride via the cotransporter (Jamison *et al.*, 1993). There is a substantial additional reabsorption of sodium by paracellular pathways. The loop reabsorbs more sodium if delivery from the proximal tubule increases (Koeppen, 1990). The direct effect of ADH on sodium reabsorption in the loop is certainly to increase it yet some reports indicate that ADH increases renal sodium excretion both in sheep and in humans. This is probably the result of working with subjects on high sodium intake, chronic expansion of ECF volume and susceptible, therefore, to natriuresis if ADH further expands ECF volume—this probably explains why this is commonly seen in humans (Sutters *et al.*, 1993), most of whom routinely consume excess salt (*Chapter 6*).

The increased salt concentration of the medullary interstitial fluid extracts water from the descending limb and raises the sodium and chloride concentration of its contents. Thus, since sodium concentration falls in the TAL and rises in the descending limb, it is equivalent to direct transfer of solute, although there is no direct migration of sodium ions.

In the final stages of the collecting duct, in the inner medulla, water conservation also enhances urea movement into the medullary interstitial fluid. The countercurrent **exchange** of the vasa recta tends to keep it trapped there at high concentration (Jamison *et al.*, 1993). As a result:

1. it reduces the osmotic 'resistance' to water removal from urine in the collecting duct because it lowers the resulting urea gradient;
2. it allows extraction of water from the fluid in the descending thick limb, raising its sodium concentration. Therefore, the urine in the thin ascending limb has a higher Na concentration than the surrounding ISF; as a result, sodium is reabsorbed passively along this concentration gradient (Jamison *et al.*, 1993).

The outcome is that sodium is reabsorbed from urine into medullary ISF in both the TAL of the loop of Henlé (by active transport) and the ascending thin limb (by passive diffusion, dependent on urea recycling into medullary ISF). Since the mechanism involves urea, as well as sodium, factors which seriously reduce urea production (e.g. portocaval shunts, severe hepatic disease, or low protein intake) impair urinary concentrating capacity (Levy, 1978; Thier, 1981; Bovee, 1984; Jamison *et al.*, 1993). So does hypercalcaemia, by affecting both the TAL and collecting duct response to ADH (Winters *et al.*, 1991; Jamison *et al.*, 1993; Teitelbaum *et al.*, 1994). Hypokalaemia also causes polyuria, partly by stimulating thirst, partly by impeding the renal response to ADH in the collecting duct and perhaps by reducing Na:K:2Cl cotransport in the TAL (Teitelbaum *et al.*, 1994). Ultimately, potassium depletion damages proximal tubules (Krishna *et al.*, 1994). While urea surrounding the nephrons facilitates urine concentration, high concentrations flowing through the nephrons cause osmotic diuresis (*Chapter 11*) and therefore predispose to water loss and hypernatraemia (Morrison and Singer, 1994).

Regulation of ADH Release

ADH release is increased most sensitively in response to increased plasma sodium concentration, less sensitively in response to hypovolaemia or hypotension and also in response to 'inappropriate' stimuli such as nausea, stress, and various drugs (Robertson, 1993). Some neural stimuli are, in fact, appropriate, allowing anticipatory responses, e.g. reduction of ADH secretion in response to drinking *per se*, without awaiting the results of water absorption. Other neural stimuli long assumed to affect ADH secretion, e.g. stress or pain, may actually do so only via their haemodynamic effects (Zerbe and Robertson, 1994). In humans, ADH secretion can respond to rises in plasma osmolality as small as 0.17% (0.5 mOsm/kg); in contrast, changes of 5–10% in blood volume or blood pressure have little effect. Nevertheless, more severe hypovolaemia becomes a very potent, indeed the overriding stimulus to ADH secretion. In between, ADH secretion doubles with a rise in osmolality of 1% or less, whereas it needs a 10–15% fall in plasma volume.

Above a certain plasma osmolality ADH secretion increases linearly so that the key determinants of the prevailing level in circulation are the threshold of the linear response and its slope. Factors such as hypovolaemia, angiotensin, hypercalcaemia or lithium alter the slope of the line (Robertson, 1993). The normal osmotic stimulus is mainly, but not entirely, a reflection of plasma sodium concentration, i.e. other solutes have a relatively minor effect. Nevertheless, infusions of mannitol, for example, work just as well as sodium. Naturally, the solute must create an osmotic gradient, which is not the case with glucose or urea when their concentration increases slowly enough to allow redistribution into ICF. At receptor level, both with thirst and ADH, what matters is whether they swell or shrink and the resulting change in wall tension. Chronic disturbances of osmolality affect other brain cells by inducing compensatory changes in organic solutes (osmolytes or idiogenic osmoles, *Chapter 8*)—it would be of interest to know whether they also occur in osmoreceptors and their influence on sensitivity. The fact that osmoreceptors respond to changes in plasma osmolality, despite the existence of the blood–brain barrier (BBB) (for example, the BBB excludes urea), substantiates the view that they are located among the 'special permeability areas' of the brain, with a typical ISF composition rather than one governed by the BBB. Interestingly, glucose is a more effective osmotic stimulus when insulin is deficient and, presumably, glucose is consequently less able to enter the osmoreceptors (Robertson, 1993). The effects of glucose, however, are not purely osmotic, thus acute hypoglycaemia, surprisingly, is a potent stimulus to ADH secretion.

While ADH, alongside thirst, provides the main defence against hypernatraemia (*Chapter 7*), additional protection is conferred by 'dehydration natriuresis', although its basis has not yet been defined in detail. Excess ADH causes hyponatraemia, and at very high levels it may cause sufficient ECF volume expansion to increase sodium excretion. It also raises arterial pressure (especially when it is subnormal), but the kidney is protected against excessive vasoconstriction, whether caused by ADH, angiotensin or other agents, or by prostaglandin-dependent mechanisms. Patients on non-steroidal anti-inflammatory drugs (NSAIs) are therefore very vulnerable to the renal effects of hypovolaemia or hypotension (Stokes, 1989). In disease, it is important to remember that the plasma concentration of ADH depends on its rate of metabolic clearance, as well as its rate of secretion; the chief sites of removal are the liver, the kidneys and, in pregnancy, there is additional breakdown by vasopressinases from the placenta. Nevertheless, anephric patients maintain normal plasma ADH concentration.

It is worth re-emphasising that in species such as dogs, where water rather than flavoured drinks is the main source of fluid, thirst rather than ADH provides the primary protection against hypernatraemia both on normal and high salt intakes (O'Connor and Potts, 1988; Cowley *et al.*, 1983, 1986).

Distal Nephron

Structurally, functionally and according to the response to diuretics, the distal nephron can be divided into a number of sites, e.g. early and late convoluted tubule, connecting tubule, cortical, outer and inner medullary collecting duct (Jamison *et al.*, 1993). The latter is the segment in which collecting ducts from adjacent nephrons become confluent. ADH-responsive water reabsorption begins as early as the connecting tubule. While this may seem unexciting, since hypertonic urine cannot be achieved until the high concentrations of medullary interstitial fluid are reached, the actual volume of water conserved in achieving hypertonic urine is only about a quarter of the amount conserved in converting the hypotonic urine leaving the loop back to isotonicity (Jamison *et al.*, 1993). It is important to understand that older references to the 'distal convoluted tubule' referred to the entire segment between the macula densa and the confluence of adjacent nephrons in the collecting duct (Koeppen, 1990). The same region would now include the distal convoluted tubule (DCT) ('early distal tubule'), the connecting tubule (CNT) and the initial collecting tubule (ICT) ('late distal tubule').

Setting aside considerable knowledge of permeabilities and transmembrane potentials and their differences between sub-segments (and species), the essential functions accomplished in the distal nephron are:

1. the ability to produce virtually sodium-free urine;
2. the facilitating effect, via transmembrane PD, of sodium conservation (under the influence of aldosterone) on the secretion of potassium and/or H^+; and

3. in the collecting duct, reabsorption of water and, to a lesser extent, urea.

Hydrogen ion secretion in the proximal tubule may be quantitatively more important in loading the urinary buffers but it is the distal nephron which 'acidifies' the urine, i.e. lowers its pH. It is thus with H^+, as with Na^+, the segment where extremes of concentrations are achieved. Aldosterone stimulates the synthesis of ammonia buffer as well as enhancing proton secretion (Stone *et al.*, 1990).

In the remainder of this account, unless specifically indicated, the only distinction will be between the 'distal nephron' and the 'collecting duct'.

The presence of high concentrations of Na–K ATPase in the distal nephron, along with mineralocorticoid receptors, is consistent with the role of aldosterone as a regulator of sodium reabsorption in this site. Strangely, however, there is more direct evidence for the likely effect of calcitonin and PTH (Koeppen, 1990), hormones which we presently think of as 'affecting' sodium excretion, rather than controlling it. This may be because the physiological tradition is to think of direct feedback loops, i.e. a problem of sodium balance is corrected by a response in sodium excretion. We will see *(Chapter 4)* that similar rigidity has constrained the conceptual approach to the physiological significance of salt appetite. Nevertheless, as we will see in *Chapters 6 and 8*, changes in sodium excretion can cause problems in calcium balance, so it is not inconceivable that the appropriate response should not only involve adjustment of calcium excretion but, perhaps, re-adjustment of sodium excretion by 'calcium regulating' hormones. Sodium reabsorption continues in the CD as well as the distal tubule, under the influence of aldosterone. Indeed, the 'classical' physiology of its mode of action is mainly derived from studies of this segment (Koeppen, 1990). The cortical collecting duct responds to both ANP and the circulating sodium transport inhibitor associated with uraemia (Kirchner and Stein, 1994). The inner medullary collecting duct has ANP receptors. It may be capable not only of adjusting its rate of sodium reabsorption but, with extreme expansion of ECF volume, it may secrete sodium (Kirchner and Stein, 1994).

The role of Na–K ATPase and aldosterone in regulating potassium excretion is at least as important, arguably more important, than its role in sodium conservation. It involves a separate feedback loop, since potassium directly stimulates aldosterone secretion independently of angiotensin II. Indeed, potassium loading suppresses the renin–angiotensin system (Michell, 1978a; Seeley and Laragh, 1990). The key determinants of the ability of aldosterone to enhance potassium secretion are the availability of potassium for secretion from renal cells (hence the competing effects of K^+ and H^+, e.g. when lack of potassium exacerbates metabolic alkalosis by enhancing H^+ secretion; Michell *et al.*, 1989) and also the delivery of sodium. The latter is especially influenced by proximal reabsorption and, therefore, by plasma volume (Laragh and Seeley, 1992). The effect of slight hypovolaemia on proximal reabsorption is offset by the increased aldosterone secretion; if anything, potassium excretion is likely to increase. Severe hypovolaemia, however, undermines the effect of aldosterone by severely restricting the delivery of sodium (Stokes, 1989).

The effects of ADH on the collecting duct have been defined in considerable detail (Brown, 1991). The main route of water movement is via cells rather than paracellular pathways and, since the basolateral surface is always permeable to water, the decisive effect of ADH is to increase the permeability of the apical (luminal) membrane. This may involve components of the cytoskeleton, including microtubules and microfilaments, as well as recycling of water-permeable channels between the cytoplasm and the membrane.

Immaturity of Renal Function

Renal function begins *in utero* with swallowing of amniotic fluid and production of urine which becomes part of the amniotic fluid. Elaboration of urine is detectable in the mesonephros of the three-month old human foetus; proximal secretory pathways and distal absorptive pathways are already demonstrable in cell culture. Within another month the loop is functional, though not fully (Jose *et al.*, 1987). Renal perfusion and GFR are, however, very low because of low blood pressure and high renal vascular resistance (El-Dahr and Chevalier, 1990). This is not surprising, since the placenta has 10 times the blood flow of the foetal kidney and serves many of its regulatory functions (Chevalier, 1994). Renal perfusion is not easy to measure, since even in neonates *para*-amino-hippuric acid (pAH) extraction is incomplete. It can be shown (in sheep) that the volume of amniotic fluid is regulated by changes in both foetal swallowing and urine production (Kullama *et al.*, 1994). In the human foetus, urine output rises from 10 ml/h at 30 weeks to 30 ml/h by 40 weeks and GFR reaches neonatal levels by 36 weeks (Jose *et al.*, 1987).

At birth, total body water is higher than in adults, not simply because of a lower fat content, but as a percentage of lean body mass. This results particularly from an increased volume of plasma and ECF (Fanestil, 1994). Interestingly, levels of digitalis-like natriuretic factor are high in the newborn infant (Ghione *et al.*, 1993). *In utero*, total body water is as high as 90% of bodyweight and, unlike the adult, the foetus has most fluid in its ECF (El-Dahr and Chevalier, 1990). Thus, whether born pre-term or at full term, the early weeks of life are characterised by some degree of diuresis as the 'excess' ECF volume is reduced. ANP levels, initially high, fall during this period (Chevalier, 1993). This is more obvious in pre-term infants. Nevertheless, in humans, immaturity of renal function means that neither conservation of sodium and water nor excretion of excess are efficient in the neonatal period (El-Dahr and Chevalier, 1990). This is especially important in pre-term infants whose renal salt wasting may predispose them to hyponatraemia and to impaired growth (Haycock, 1993). During the first three months, GFR rises rapidly and reaches adult values (per surface area) by two years (Jose *et al.*, 1987).

Evidence from rats and puppies suggests that autoregulation of renal perfusion matures early, though interestingly the range of arterial pressure compatible with normal perfusion is below that of adults (Yared and Yoshioka, 1989). In contrast, GFR is poorly regulated, i.e. if perfusion falls, filtration fraction fails to compensate, if anything it also falls (Yared and Yoshioka, 1989). Nevertheless, glomerulotubular balance is already detectable by mid-gestation (Chevalier, 1994).

Newborn humans and other mammals have high renin levels compared with adults, together with increased angiotensin levels. This does not seem essential for normal renal function as inhibition of angiotensin II formation with ACE-inhibitors does not interfere either with renal perfusion or autoregulation (Yared and Yoshioka, 1989). Though basal levels are high, the renin–angiotensin system is less responsive to hypovolaemia and the glomerular vasculature is less responsive to angiotensin II. Foetal urine is not the main source of the high concentrations in amniotic fluid. This probably originates from the placenta and may serve to raise maternal blood pressure, since the placental circulation is less sensitive to its vasoconstrictor effects (Lumbers, 1993).

The lower concentrating capacity of the immature kidney means that the newborn require more water to excrete their solute load; they also have high insensible losses (dermal and respiratory). Stool loss can also be substantial. The sodium requirement of the newborn infant (2–3 mmol/kg) is well above adult maintenance levels (*Chapter 5*), partly as a reflection of the sodium needed for growth. Excessive sodium excretion in pre-term infants may impede their growth (Herin and Aperia, 1994). Neonates tend to have higher plasma potassium concentrations, perhaps because of poorer sodium delivery to the distal nephron and a blunted response to aldosterone (El-Dahr and Chevalier, 1990).

These findings suggest that, while the newly born have considerable adaptive capacity in their renal function, it is like that of kidneys in compensated chronic renal insufficiency, inflexible in the face of extreme demands. The relatively fixed intake of solute and water prior to weaning minimises the importance of any immaturity of renal function unless there is interruption of intake, dehydration (e.g. due to diarrhoea) or increased water losses associated with fever or high ambient temperature. Additional hazards are posed by incorrectly mixed oral rehydration solutions *(Chapter 3)* or by administration of excessive or inappropriate parenteral fluids.

Conclusion

The kidney plays a pivotal role in long term regulation of arterial pressure, through pressure natriuresis, and is defended against the impact of hypotension by autoregulation of renal perfusion. The stimulus–response relationships of renal physiology are mostly clear, but the intermediate steps, the underlying mechanisms, are still emerging. With the juxtaglomerular apparatus, for example, it is not long since debate raged over whether the stimulus to renin secretion was increased or decreased delivery to (absorption by) the macula densa and whether the stimulus was sodium or chloride. In traditional accounts of renal physiology, the emphasis is on the segmental detail of solute transfer, the gradients, the potentials, the permeabilities, the channels, the transporters, the pathways (cellular or paracellular), the backleaks, etc. Equally important, however, perhaps more so in adaptive responses to disease, is the balance between the function of different nephron segments and this will become very clear in considering the actions of diuretics *(Chapter 11)* or the effects of diarrhoea *(Chapter 3)*.

Perhaps the areas of most rapid growth of interest have been the intrarenal mechanisms affecting renal function, especially the autocrine or paracrine effects of 'traditional' hormones such as angiotensin or prostaglandins and 'newer' regulators such as nitric oxide and endothelin, also the direct effect of nerves upon the nephron. It has been an abiding paradox of renal physiology that evidence for the scope and subtlety of neural control of renal function (Di Bona, 1990) has seemed set to make it a key factor, yet denervated kidneys work well. Hormonal mechanisms are discussed further in *Chapter 7*, but while we now understand a lot about the secretion, receptors and effects of a range of hormones which allow the kidney either to conserve sodium or water or to excrete excess sodium, our knowledge remains incomplete and we may not yet have identified all the relevant hormones.

We have satisfactory, though not necessarily complete, explanations for the ability of the kidney to defend against changes in plasma osmolality and changes in plasma volume caused either by loss of plasma, loss of sodium or sodium loads. These explanations, while physiologically satisfying, are insufficient when we confront clinical situations in which sodium excretion is radically changed without an apparent or appropriate stimulus (*Chapter 8*).

Much of the progress in our understanding of the renal physiology of sodium regulation during the last 20 years was stimulated by two challenges to the then prevailing wisdom, that the mechanisms dealing with sodium conservation were also those which dealt with excess. The refusal to believe this and subsequently the demonstration, by De Wardener, that GFR and aldosterone could not explain the renal response to sodium loading, were decisive, as was his persistent assertion that the 'third' factor involved was not just changes in the peritubular physical forces but the existence of natriuretic hormones (De Wardener, 1978). The 1980s clearly vindicated that view with accelerating progress in establishing the physiological and pathophysiological role of two major groups of natriuretic hormones, the natriuretic peptides and the endogenous cardiac glycosides. Yet neither, as yet, explains De Wardener's critical experiment (*Chapters 1 and 7*; Lichardus *et al.*, 1993).

While sodium is the decisive influence on ECF volume, and ECF volume, notably plasma volume, is a key determinant of both sodium excretion and renal function, our knowledge of the crucial sites where this volume is monitored remains incomplete. That is why, whether we consider the inappropriate retention of either salt (resulting in oedema) or water (resulting in hyponatraemia) in clinical situations, we are thrown back on the euphemism of 'effective arterial blood volume' Michell (1995a). It means the same as 'idiopathic'. It means we do not know, but by naming the problem we can give the impression of being able to control it. As in politics, jargon is a powerful substitute for knowledge and a potent antidote to conflicting facts.

3

Enteric Sodium Uptake

Introduction

The conventional view of mammalian sodium balance is that, regardless of intake, sodium is efficiently absorbed by the gut, any excess is excreted by the kidneys and only trivial amounts remain in normal faeces. Like the kidneys, the colon can reduce sodium output virtually to zero during sodium depletion, thanks to the effect of aldosterone.

Potassium, calcium, magnesium and phosphate are the main organic minerals in faeces with only traces of sodium . . .

(Wrong *et al.*, 1981)

Against this background, it should seem surprising that, whether in animals or humans, the main cause of clinical disturbances of fluids and electrolytes (and, in particular, of sodium depletion) is neither renal nor adrenal dysfunction but diarrhoea. This reflects the fact that the great majority of enteric sodium turnover is not a result of dietary intake, but of the massive outpouring of alimentary secretions and the need for accurate and efficient reabsorption of their contents. Failure rapidly leads to death; 10,000 children die of diarrhoea every day. Imagine the headlines if 100 died in a plane crash on a single day, let alone 10,000 every day. In developing countries, a child under five may have 7–10 episodes of diarrhoea annually and four million will die. Even in 'developed' countries, some 10% of the children in this age group who go to hospital are admitted because of diarrhoea (Rivin and Santosham, 1993).

Matching of enteric secretion and reabsorption of sodium, with both vastly exceeding daily dietary intake, provides an interesting analogy with the problem of glomerulo-tubular balance (Michell, 1986). Indeed, the analogy goes further because, like the proximal tubule, the proximal segments of the intestine are characterised by their ability to transport large amounts of sodium and water, rather than their ability to produce steep gradients of concentration. Moreover, in both cases there is substantial co-transport of sodium with organic solutes such as glucose and amino acids, but no effect of aldosterone. The sodium–glucose cotransport systems of both the nephron and the gut comprise subunits concerned with the translocation process and with its rate (Koepsell and Spangenberger, 1994). These co-transport mechanisms are exploited in the formulation of oral rehydration solutions (ORS) for the treatment of diarrhoea (see below). In contrast, the distal nephron and the colon both respond to aldosterone by creating steep concentration gradients and, in the extreme instance, reduce the sodium concentration of their contents virtually to zero. In the kidney and the gut, as in other tissues, sodium gradients are the basis of water movement and also facilitate the movement of cations (by countertransport) and anions or other solutes (by cotransport). This neglected harmony between renal and enteric function and their closely parallel mechanisms is emphasised by the recent identification of a highly selective amiloride-sensitive sodium channel protein likely to provide the rate limiting step for aldosterone-responsive sodium reabsorption in the kidneys (Ausiello, 1993). It was obtained from the distal colon, where it has the same role.

Unlike the kidneys, the gut shows drastic differences of form and function between species, notably between herbivores and omnivores or carnivores. Substantially these relate to the twin problems of plant diets; the large bulk required to provide adequate nutrients and the strategies needed to deal with cellulose cell walls and other aspects of plant composition. These include expansion of the foregut in ruminants and camels, or the hindgut in horses and rabbits, to provide chambers for bacterial fermentation. In ruminants particularly, the large volume involved provides a potential store, e.g. of water or sodium. Thus sodium-depleted ruminants, by replacing salivary sodium with potassium, convert their ruminal fluids from a sodium-rich to a potassium-rich pool. This is, once again, a response to aldosterone and, potentially, an important protection against hyperkalaemia as well as sodium depletion. Granted that the latter, once severe, predisposes to the former, this is especially important in species such as sheep or goats which confront the need to excrete their large dietary potassium load even when sodium and water are scarce.

It is easy to argue that the adaptations among herbivores are so specialised that they have no implications for other species, notably man. This neglects the fact that for most of their evolutionary history humans and their primate predecessors were probably herbivores rather than carnivores (Denton, 1982; Eaton and Konnor, 1985).

Not only do a number of 'primitive' societies (albeit a vanishing minority) remain so, but a significant and rapidly growing minority of people in 'advanced' societies are reverting to vegetarian diets. In this context, the realisation that, at least in some normal herbivores, the gut rather than the kidney is the major regulator of sodium excretion on moderate sodium intakes (i.e. close to requirement) may have wider implications. Whether in herbivores or carnivores, the kidneys are certainly the main excretory route for **excess** dietary sodium (Michell, 1986; Michell and Moss, 1992). The question arises, therefore, whether in humans on plant diets and low sodium intakes, especially where their parents had similar diets, the regulation of sodium excretion is so predominantly renal as generally assumed. This is particularly important since human sodium intake, notably of primitive communities, is often inferred from urinary excretion alone. This could lead to serious underestimates if, as in herbivores, significant additional quantities of sodium, perhaps the majority, are actually excreted in faeces *(Chapter 5)*.

Once we accept that the gut plays an important role in the normal regulation of sodium balance, a challenging problem arises. There is a conflict between the demands of external sodium balance (i.e. that in normal animals which are not growing, pregnant or lactating, sodium intake should match total loss) and of internal balance (i.e. that secreted sodium should be reabsorbed and accurately matched). The former requires that increased enteric delivery of sodium is a signal for enhanced excretion, whereas the latter requires that increased delivery promotes a parallel increase in reabsorption. There is thus an open field of research to elucidate the mechanisms which underlie enteric sodium regulation and integrate it with renal sodium regulation. Recently a natriuretic factor has been found in intestinal fractions; fascinatingly, this was not in a herbivore but in the cat, a pure carnivore, but similar data have been obtained in rats (Hansson *et al.*, 1993). The liver may provide a key link between enteric and renal function, a view reinforced by recent evidence that it may exert neurally mediated effects on sodium excretion (Morita *et al.*, 1993). The possible role of the liver in sodium regulation is discussed further in *Chapter 4*. Just as there is a growing awareness of partnership between the kidneys and the liver in acid–base regulation, we may increasingly come to appreciate that not only the effects of diarrhoea but the normal regulation of body sodium depend on interplay between renal and enteric mechanisms. Better understanding of these mechanisms is also likely to improve the effectiveness of the treatment of diarrhoea by oral rehydration solutions, by increasing the efficiency of their absorption.

Enteric Sodium Transport

Transepithelial solute transport depends upon pathways involving energy-consuming 'pumps', passive 'carriers' or selectively permeable 'channels'. The primary route may be transcellular, involving both entry and exit, or paracellular through the tight junctions between cells and along gradients of osmolality or charge (Field *et al.*, 1989; Barrett and Dharmsathaphorn, 1994). Regulation of these movements may involve solute concentrations, endogenous intestinal agents [such as vasoactive intestinal peptide (VIP), GIP, glucagon, etc.], intestinal lumen-active agents (nutrients, bile salts and other fatty acids), neural and endocrine influences (both enteric and systemic in origin) and immune-mediated effects on transport (Brown and Miller, 1991). Dysregulatory effects include the diarrhoea caused by excess secretion of peptides such as VIP from certain tumours, also the additional secretion caused by fatty acids in steatorrhoea (Phillips, 1994). Not only the responses of pumps and channels but their numbers contribute to regulation (Schultz and Hudson, 1991).

The main pump transporting sodium is Na–K ATPase, which responds to aldosterone and (in the intestine) to glucocorticoids, while the main carriers affecting sodium movement are those involved in Na–H exchange and Cl–HCO_3 exchange (which may be calmodulin-responsive), Na–K–Cl cotransport (as in the ascending loop of Henlé) and the cotransport of sodium with glucose, amino acids, fatty acid anions and other nutrients (Sullivan and Field, 1991; Barrett and Dharmsathaphorn, 1994). These mechanisms interact, thus the sodium gradient created through active transport by Na–K ATPase [low intracellular fluid (ICF) Na] allows substances to enter cells by passive cotransport with sodium. Equally, increased intracellular leakage of sodium activates sodium extrusion by Na–K ATPase, stabilising ICF sodium concentration but increasing transepithelial transport (increased ICF Na is not, however, the only factor to activate the pump; Schultz and Hudson, 1991). Regulatory effects may potentially be expressed by modulation of receptor binding, intracellular messages, or the transport pathway itself.

Chloride channels are particularly important because the basic cause of enteric sodium loss in secretory diarrhoeas, such as those caused by enterotoxins from *V. cholerae* or *E. coli* (or by prosta-glandins in inflammatory bowel diseases), is a primary hypersecretion of chloride mediated by cAMP or cGMP (Field *et al.*, 1989). Glucose-stimulated sodium uptake remains unaffected (Booth and McNeish, 1993). Interestingly, *E. coli* produces two enterotoxins: one (heat-labile) extremely similar to cholera enterotoxin, with effects mediated by cAMP production, the other (heat-stable) with effects mediated by cGMP. The enteric receptors for the heat-stable toxin are similar to those for ANP, though they do not respond to it (Booth and McNeish, 1993). Nevertheless, the toxin has natriuretic effects in rats (Lima *et al.*, 1992). Recent evidence suggests that there may also be substantial similarity between renal and enteric chloride channels, and that the reason for the widespread abnormalities in secretion associated with cystic fibrosis *(Chapter 10)* is a defect in a gene regulating the production or function of chloride channels (Barrett and Dharmsathaphorn, 1994; Field, 1993). The basic components of chloride secretion are a cotransporter of sodium, potassium and chloride (as in the loop of Henlé; *Chapter 2*) on the basal cell membrane, which supplies the cell with chloride, and a cAMP regulated channel on the apical side, which permits secretion into the gut lumen. The Na and K supplied to the cell by the cotransporter can exit again through the basal membrane, via Na–K ATPase and potassium channels, respectively. They thus recycle in supplying the cell with chloride. Sodium also accompanies secreted chloride but enters the gut through a paracellular route. At cellular level, therefore, secretory diarrhoea and cystic fibrosis involve opposite abnormalities; the former, an overstimulation of chloride channels and the latter a defect in the protein that allows cAMP to activate them (Field, 1993). A crucial difference between renal and enteric chloride channels is that those in the ascending thick limb of the loop of Henlé are basolateral and therefore promote absorbtion, whereas those in the intestine are located on the luminal side and are, therefore, secretory (Field *et al.*, 1989; Madara, 1991).

While particular attention has been focused on chloride secretion, secretory diarrhoeas also involve failure of electroneutral absorption of sodium and chloride, also of active sodium transport. Treatment by oral rehydration engages sodium cotransport with glucose, amino acids and fatty acid anions (bicarbonate has similar effects), i.e. enhancement of absorptive flux. So far, there is less scope for clinical manipulation of excessive secretory flux (chloride): eventually a suitable chloride channel blocker may fulfil this role (Field *et al.*, 1989; Barrett and Dharmsathaphorn, 1994). This might be particularly important in mature ruminants, since ruminal fermentation of carbohydrate to volatile fatty acids results in a virtual absence of hexoses from the small intestine. As the ruminant matures (and is weaned from milk to roughage), intestinal sodium–glucose cotransport activity declines sharply with the declining presentation of glucose; it can, however, be reactivated by renewed exposure to glucose (Shirazi-Beechey *et al.*, 1991).

The specific importance of colonic sodium transport is that, particularly, the distal colon (and rectum) are aldosterone-responsive and can lower faecal sodium to negligible concentrations (at the expense of enhanced loss of potassium); the colon, therefore, potentially has a pivotal role in the regulation of external sodium balance. It will be considered in more detail (including regional and species variations in function) in the context of enteric sodium regulation.

Enteric Sodium Regulation

The gut is not conventionally acknowledged to have an important role in the regulation of mammalian sodium balance, being overshadowed by the kidney. In a substantial proportion of individuals from a variety of species, however, the gut, rather than the kidney, is the main route of sodium excretion unless sodium intake is exorbitant (Michell, 1986; Michell and Moss, 1992). Moreover, where the most direct comparisons can be made, individuals excreting most of their sodium in faeces, compared with those excreting it mainly in urine, are arguably better adapted (Michell *et al.*, 1988; Michell and Moss, 1992, 1994; Michell, 1992a). Thus, they show the lower obligatory sodium loss in pregnancy, a greater salt appetite when sodium solutions are available, a lower water turnover both normally and during dehydration, better maintenance of potassium excretion and less tendency towards hyperkalaemia during dehydration. They respond equally effectively to sodium deprivation.

The idea that faecal sodium excretion can exceed urinary sodium excretion in healthy, rather than diarrhoeic animals seems so heretical that many of those who observed the fact declined to comment on it (Michell, 1986). Yet it is not, seemingly, a new idea.

> *The dung of such animals as feed upon grass or grain doth also contain plenty of common salt.*
>
> (Brownrigg, 1748)

It is not just of academic interest because, as a result of the tendency to regard urine as an adequate means of estimating sodium excretion, animals may be misconstrued as having sodium depletion, simply because their urinary sodium output is low. Thus, when urine alone was collected, Grim and Scoggins (1986) could only account for the excretion of 25% of the dietary sodium; their sheep, however, were on intakes of 25 mmol/day when 75% of sodium excretion would be expected to be faecal (Michell and Moss, 1992). It is very important to realise that the range of intake in which excretion is predominantly faecal is the range consistent with nutritional requirement; urinary excretion predominates when excess dietary sodium is consumed (Michell, 1986; Michell and Moss, 1992). Thus, McSweeney and Cross (1992) calculated requirement for growing goats and cattle as 0.7, 0.4 mmol/kg/day; at 1.4, 0.7 mmol/kg/day (about twice requirement) the predominant route of sodium excretion was faecal. Similarly, Alcantara *et al.* (1980) showed that sodium excretion in weanling pigs was predominantly faecal until requirement was exceeded. It is interesting that estimates of faecal sodium loss in wild moose (Jordan *et al.*, 1973) showed that it exceeded urinary sodium for most of the year, except perhaps in summer when sodium intake was increased (although the loss was attributed to 'undigested' food).

While these observations mainly concern herbivores, it would be premature to assume that they apply solely to them. In particular, herbivores are characterised not only by the variety of their anatomical adaptation to a plant diet, but also by the high potassium loads to which it commits them. Bearing in mind that humans have their evolutionary ancestry in herbivorous primates and remained herbivores throughout the great majority of their history and that an increasing number are becoming vegetarians, I would not assume that a similar enteric role in sodium regulation is completely extinct. The best chance of finding it would be in populations accustomed to high potassium–low salt diets, or in families with parents as well as children who were low-salt vegetarians, because of the possible importance of early exposure to salt in modifying subsequent responses (Michell, 1986; see also *Chapter 4*).

An enteric role in sodium regulation poses the conundrum that maintenance of **external balance** demands that increased sodium load induces increased excretion, whereas most intestinal sodium is endogenous, from saliva and enteric secretions, hence maintenance of **internal balance** demands that enhanced delivery, as in glomerulo-tubular balance, elicits increased reabsorption (Michell, 1986). These and other fascinating questions, such as the factors which integrate enteric and renal sodium regulation, will not be confronted, let alone answered, until there is less universal acceptance of the traditional certitude that mammalian sodium regulation, essentially, is a matter of renal physiology. It is, but not exclusively so.

Enteric Responses to Changes in Sodium Balance

Jejunal sodium uptake is stimulated by angiotensin II in various forms of extracellular fluid (ECF) volume depletion (Levens *et al.*, 1984); conversely, volume expansion reduces intestinal uptake of sodium (Schultz, 1981). As well as angiotensin II and aldosterone, antidiuretic hormone (ADH) and atrial natriuretic peptide (ANP) affect enteric as well as renal sodium absorption, as does the natriuretic peptide from brain (BNP). VIP not only has natriuretic effects, but its secretion is increased by high salt intake (Brown and Miller, 1991; Matsushita *et al.*, 1991; Davis *et al.*, 1992; Argenzio and Armstrong, 1993; Kagawa *et al.*, 1994). A number of other enteric peptides also cause natriuresis, including pentagastrin, cholecystokinin, glucagon (at pharmacological concentrations), whereas neurotensin, at physiological concentrations, causes renal sodium conservation (Unwin, 1989). Interestingly, high doses of angiotensin II promote sodium excretion, rather than sodium conservation, both in the gut (Brown and Miller, 1991; Halm and Frizell, 1991) and the kidney (Koeppen, 1990). Nitric oxide may be an important regulator of ileal electrolyte transport (Rao *et al.*, 1994). Epidermal growth factor increases enteric sodium absorption as well as renal sodium excretion *(Chapter 7)*. It promotes both glutamine absorption and Na–glucose cotransport and may therefore have therapeutic potential in oral rehydration (Salloum *et al.*, 1993; Horvath *et al.*, 1994). Endothelin-3 suppresses jejunal electrolyte absorption, independent of nitric oxide (Chowdhury *et al.*, 1993). Somatostatin promotes enteric sodium uptake and octreotide, a synthetic analogue with a longer duration of activity, has been used successfully to inhibit diarrhoea of a surprising variety of infectious, endocrine and other causes (Maton, 1994). Gut peptides may not only be involved in regulation of enteric sodium uptake, but excessive or ectopic secretion underlies some forms of diarrhoea (Delvalle and Yamada, 1990).

Unlike ileal sodium absorption, distal colonic sodium uptake is unaffected by glucose, but is promoted by aldosterone and also by short chain fatty acids in sheep, goats, humans, but not horses (Michell, 1986; Argenzio, 1991). Unlike the distal nephron, colonic sodium conservation is also stimulated by glucocorticoids via their own receptors, not simply by activating mineralocorticoid effects (Michell, 1986; Binder and Turnamian, 1989); it is, however, electroneutral Na–Cl absorption which is enhanced by glucocorticoids, not the electrogenic reabsorption mediated by Na–K ATPase and responsive to aldosterone. Corticosteroids may thus have favourable effects on diarrhoea which do not depend on their anti-inflammatory properties (Binder and Turnamian, 1989). It is also clear that glucocorticoid effects, especially those in the small intestine, have a number of distinctive features which exclude any possibility of mediation by mineralocorticoid receptors (Powell, 1986). There is also evidence of glucocorticoid-induced renal sodium retention, independent of mineralocorticoid effects, especially at high concentrations, such as those associated with stress (Clore *et al.*, 1992).

On low sodium intakes, the distal colon is a moderately tight epithelium, allowing sodium to be actively reabsorbed from low luminal concentrations against a high concentration gradient; on high sodium intake, the epithelium becomes leaky and sodium absorption is mainly by neutral cotransport with chloride, favoured by high luminal sodium. Obviously, the former is more consistent with the demands of external sodium balance, the latter with those of internal sodium balance, e.g. reabsorption of secreted sodium (Michell, 1986). In pigs, rabbits, horses and sheep, both neutral and electrogenic sodium reabsorptive mechanisms are available in different parts of the colon (Scharrer and Medl, 1982; Argenzio and Clarke, 1989; Argenzio, 1991).

Differences between proximal and distal colon also reflect the demands of microbial fermentation in the former and sodium conservation in the latter (Argenzio and Clarke, 1989). In the proximal colon, sodium is exchanged for hydrogen ions which then facilitate the absorption of volatile fatty acid anions by converting them to the undissociated form (Argenzio, 1993). Aldosterone in horses appears to stimulate neutral cotransport as well as electrogenic absorption of sodium (Argenzio and Clarke, 1989). In rabbits, it is responsible for the circadian variation between 'hard' and 'soft' faeces; in this species faeces are reingested. The importance of Na and Cl absorption in association with fermentation is related to the absorption of undissociated volatile fatty acids (using H^+ exchanged for Na^+) and the availability of HCO_3^- (exchanged for Cl^-) as a buffer anion. The importance of colonic Na^+–H^+ exchange is not restricted to herbivores; the beneficial effect of anions such as acetate on sodium absorption depends, at least in part, on enhanced antiporter activity in the colon (Jenkins *et al.*, 1993).

In ruminants, in contrast, where the main fermentation site precedes the intestine, buffering is provided by salivary bicarbonate (Kay, 1960). Saliva also contains large quantities of phosphate in mature ruminants and this, rather than urinary phosphate, is increased by parathyroid hormone (Michell, 1983a). Although ruminants are herbivores, their large intestines may have more in common, functionally, with carnivores than other herbivores, since sodium conservation, rather than microbial fermentation, is the main issue; even in omnivores, the large intestine is a significant site of fermentation (Argenzio, 1991).

In all species, microbial fermentation of unabsorbed carbohydrate can become an important factor in the metabolic acidosis which usually accompanies diarrhoea. In herbivores generally, the volume of the gut fluids, whether in the rumen or large intestine, is comparatively large. The gut can thus function as a reservoir or, for example with the rapid ingestion of large amounts of dry diets, as a drain on ECF when enteric pooling of fluid occurs. While the substitution of salivary (and hence ruminal) sodium by potassium, in response to aldosterone, is usually envisaged as a means of conserving Na for ECF, it may be equally important as a defence against hyperkalaemia. This is especially important in herbivores, since the diet provides an obligatory potassium load which needs to be excreted or temporarily sequestered when shortage of water or sodium predispose to inadequate urinary excretion (Michell, 1978a).

Diarrhoea

Particularly since the work of Gamble (1954), we are accustomed to considering the impact of diarrhoea on fluid, electrolyte and acid–base balance as probably its most important consequence, and we are conscious of the importance of differential effects on the various compartments of the body fluids. Correction of these disturbances is, therefore, a vital aspect of therapy. As recently as 1965, a pioneering paper on intravenous fluid therapy in the treatment of calf diarrhoea by J.G. Watt led to the caution that it would be better to regard it "as a supplementation of antibiotic and chemotherapeutic treatment". Even now, however, many would still view fluid therapy as symptomatic treatment. This is certainly true of parenteral rehydration, but the success of oral rehydration, which has transformed the management of acute diarrhoea whether in humans or animals, rests on a radically different perception. Diarrhoea is not just a **cause** of fluid and electrolyte disturbances: fundamentally, it **is** a fluid and electrolyte disturbance, varied though the causes may be. Indeed, the causes are usually multiple since the intestine is an internal ecosystem and in those regions which have a normal flora it is protective; pathogens are often secondary rather than primary causes of diarrhoea, capitalising on nutritionally-induced changes in the enteric environment (Michell, 1974a; Cohen, 1991). In children with diarrhoea it is recognised that many will not benefit from antibiotics, especially those with viral enteritis. In some bacterial infections they may prevent systemic spread or reduce faecal output of bacteria, but they can also cause problems such as colitis, bacterial resistance and, with Salmonella infection, increased risk of bacteraemia (Pickering, 1991).

Just as the distal nephron, responding to ADH and aldosterone, can produce urine of small volume, high osmotic concentration and negligible sodium content, despite variations in delivery from the proximal tubule, so the colon can greatly reduce faecal water content and minimise faecal sodium concentration. Moreover, it has a large 'functional reserve capacity', i.e. in the face of increased delivery from the small intestine, colonic sodium and water conservation can increase far above normal. The colon can therefore conceal the consequences of small intestinal pathology by maintaining a relatively normal output of faeces. Perhaps more importantly, faecal consistency or volume can return to normal before full recovery of small intestinal function—an important reason for a cautious and gradual return to a normal diet.

Diarrhoea, therefore, is a failure of net enteric uptake of sodium and water sufficient to exceed the compensatory capacity of the colon (Michell, 1983b).

Clearly, conditions which impair colonic function are more likely to cause severe diarrhoea and it is becoming apparent that this is a supplementary factor in some 'small intestinal' diarrhoeas. While the failure is of net enteric uptake of sodium and water, the net flux is the outcome of far larger contralateral fluxes either within regions or in sequence. The underlying problem could thus comprise hypersecretion, defective absorption, abnormal permeability or abnormal osmotic gradients, alone or in combination. As with bank accounts, small percentage changes in large unidirectional flows have a major impact in changing the resulting balance because it is small compared with the contributing fluxes. A seemingly small percentage improvement in unilateral absorptive flux may therefore minimise or correct an abnormal accumulation of enteric secretion; the factors causing excess secretion, fortunately, do not usually block the absorptive response (Argenzio, 1985). This, together with the fact that, even in severe diarrhoea, much or most of the small intestine remains sufficiently normal to respond to oral rehydration therapy (ORT), underlies its surprising, almost miraculous success against severe diarrhoeas with a variety of causes.

This breadth of potency would not have been anticipated in its origins. For while the early research confronted cholera, still the major cause of human diarrhoea, and *E. coli*, then regarded as the major cause of diarrhoea in young animals, these happen to share remarkably similar mechanisms. In particular, the cause of enteric fluid accumulation is, at least initially, through specific effects on ionic flux (Na^+ or Cl^-) rather than through membrane damage or effects on villus architecture. Subsequently, ORT was successfully applied to an increasing range of bacterial and viral diarrhoeas, thus including conditions where the affected regions of the intestine suffer membrane damage, invasion and gross alteration of villus architecture. The latter includes acceleration of cell turnover, leading to immaturity of secretory or absorptive function (which normally matures as cells move from the crypts to the tips of the villi) as well as stunting and loss of surface area (Field *et al.*, 1989; Sullivan and Field, 1991).

Almost as important as the hypovolaemia caused by diarrhoea is the metabolic acidosis (Fig. 3.1). It results from bicarbonate loss in faeces, impaired renal function, fermentation leading to organic acid production within the gut and perhaps from exchange of excess intestinal chloride for bicarbonate. Ischaemic areas will also generate lactic acid in severe cases with shock. Ischaemia, compromised renal function and acidosis can conspire to produce severe hyperkalaemia, despite the underlying cell deficits of potassium caused by faecal loss and anorexia or withholding of food (Groutides and Michell, 1990). Fluid therapy without adequate attention to replacement of bicarbonate can leave patients rehydrated but still severely ill as a result of their acidosis (Naylor, 1986). The most eloquent testament to the impact of acidosis comes not from a physician but from an American humorist, as long ago as 1919:

> *Life is a struggle, not against sin, nor against the money power, but against hydrogen ions. The healthy man is one in whom these ions, as they are dissociated by cellular activity, are immediately fixed by alkaline bases. The sick man is one in whom the process has begun to lag, with the hydrogen ions getting ahead . . . The dying man doesn't struggle much and he isn't much afraid. As his alkalies give out he succumbs to a blest*

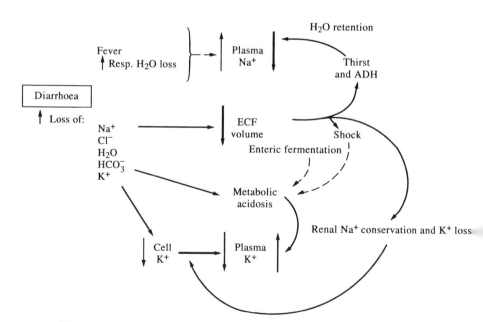

FIG. 3.1. Main effects of diarrhoea on fluid, electrolyte and acid/base balance.

stupidity. His mind fogs. His will power vanishes. He submits decently. He scarcely gives a damn.

(Mencken, 1919)

For its time, let alone any other, the insight is remarkable.

Oral Rehydration

In 1831 Europe was terrorised by a new lethal plague, the 'Asiatic' cholera. Yet before the year was out, William O'Shaughnessy, with the primitive means at the disposal of a physician in those days, had the genius to demonstrate, for all time, that the life-threatening consequences of acute diarrhoea were dehydration, sodium depletion, acidosis and, in severe cases, compromised renal function: within months Dr Thomas Latta had published encouraging results from the use of intravenous sodium bicarbonate in such cases (Michell, 1983b), something which O'Shaughnessy himself had advocated.

> *In severe cases in which absorption is totally superseded . . . I would not hesitate to inject some warm water into the vein . . . I would also, without apprehension, dissolve in that water the mild innocuous salts which nature herself is accustomed to combine with human blood, and which in cholera are deficient.*
>
> (O'Shaughnessy, 1832)

What was his remedy for milder cases?

> *I would expect much benefit from the frequently repeated use of the neutral salts by mouth . . . dissolved in large quantities of tepid water.*
>
> (O'Shaughnessy, 1832)

The rest is history—typical history of scientific discovery. O'Shaughnessy became an FRS—for some long forgotten advance in telegraphy in India. His discovery was ignored and, in the case of parenteral fluids, it was over a century in human medicine and perhaps 130 years in veterinary medicine until its importance was adequately appreciated (Michell, 1983b).

> *By 1832 the rationale and effectiveness of fluid therapy was roughly but adequately demonstrated. But the medical community was not persuaded to accept so bold a therapeutic procedure. It was used only in the advanced stages of the disease and many patients did not survive . . . Sixty years later the efficacy of parenteral fluid therapy in the treatment of cholera was rediscovered . . . but again fluid therapy did not receive general acceptance. German physicians found it ineffective for the now evident reason that the quantity of fluid provided was inadequate.*
>
> (Gamble, 1953)

Indeed, Latta had lost his first patient because the dramatic initial improvement misled him into providing insufficient fluid subsequently.

By the early 1980s, intravenous saline had cut cholera mortality by nearly 30% and the inclusion of bicarbonate (as advocated nearly 100 years earlier) reduced mortality by a further 20% (Rivin and Santosham, 1993). Neither in Third World countries nor on farms, however, could parenteral therapy, with its need for copious volumes of sterile fluid and antiseptic precautions and surveillance of administration, provide a cost-effective remedy for large numbers of patients. Only in the 1960s came the discovery of a way of promoting enteric fluid uptake by oral rehydration, using equimolar isotonic mixtures of glucose and sodium, and thus of implementing O'Shaughnessy's alternative strategy to full effect (although hypertonic glucose–saline solutions had been used orally in the 1940s; Rivin and Santosham, 1993).

The subsequent success of the application of this principle led the WHO, to the surprise even of enthusiasts, to identify ORT as "the most important medical advance this century" (WHO/UNICEF, 1985). Certainly, in terms of lives saved per dollar spent, it could have few rivals and none if animal lives are added to the equation. Moreover, the discovery was timely as it became increasingly clear that antibiotics were ineffective in many forms of diarrhoea, probably most in humans (Griffiths and Gorbach, 1993; Scott and Edelman, 1993), and that their indiscriminate use undermined their future effectiveness as a result of transmissible bacterial resistance. If a posthumous Nobel prize were feasible, one wonders who could rival O'Shaughnessy, granted the correctness of his conclusions, the magnitude of their impact and the century-and-a-half in which their full implications, despite the primitive means at his disposal, remained ahead of their time.

O'Shaughnessy could not have foreseen the pivotal importance of sodium–glucose cotransport in successful oral rehydration, yet his emphasis on sodium remains valid, as will emerge below. It is sad to realise, through anecdote and publication, that there are those even today who, despite their qualifications, fail to grasp the importance of the principles underlying successful ORT, simple though these are. As a result, the belief persists that almost any old fluid by mouth will do, including proprietary health drinks, which may create additional disturbances (Bucens and Catto-Smith, 1991). Both in developing and developed countries, ORT is lamentably underused (Rivin and Santosham, 1993). More important, diarrhoea is probably treated better in less developed countries, because it is taken more seriously, and the issues are more widely understood (Fig.3.2). In

Saving lives with salt and sugar

THIS CHRISTMAS, Save the Children is seeking to reduce the number of young lives lost from preventable diseases like diarrhoea.

More than 8,000 children suffering from diarrhoea die every day.

Yet according to Nicholas Hinton, the charity's Director General, many of these lives could be saved for as little as 10 pence, the cost of a sachet of Oral Rehydration Salts.

The salts, a simple mixture of sodium and sugar, prevent children from dehydrating, the most common cause of death from acute diarrhoea. Over the past few years, the effectiveness of these salts has been proven as they have helped to save thousands of lives.

Hinton continues, "The sachets cost as little as 10 pence and are simple to use, which makes them easy to include in our health programmes."

Urgent Appeal

Save the Children's Christmas Appeal aims to raise funds to buy more Oral Rehydration Salts and other vital healthcare resources and send them to children most in need.

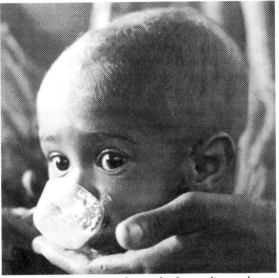

A simple solution can save thousands of young lives each year.

Photo by MIKE WELLS

Hinton adds, "Just £10 can help save the lives of 100 children. What better gift could you give a child than the gift of life?"

If you would like to make a donation to help Save the Children with their work, please return the coupon below to Christmas Appeal, Dept. 4050517, Save the Children, FREEPOST, London SE5 8BR.

Yes, I want to give the gift of life this Christmas

Please accept my gift of: ☐ £30   ☐ £15   ☐ £10   ☐ Other £_____

Name Mr/Mrs/Ms: _____ Address: _____

_____ Postcode: _____

I enclose my: Cash ☐   Postal Order ☐   Cheque ☐   Giro No. 5173000 ☐   CAV ☐

Or charge my:
Access ☐   American Express ☐   Card No. ☐☐☐☐☐☐☐☐☐☐☐☐☐☐☐☐

Visa ☐   Diners Club ☐   CAF Card ☐   Signature_____ Card Expiry Date_____

Save the Children ♼

Please return to: Christmas Appeal, Dept. 4050517,
Save the Children, FREEPOST, London SE5 8BR
Or dial 071-701 0894 with your credit card details. Thank you.

Registered Charity no. 213890

FIG. 3.2. Cost effective mass therapy: 10 p per life. Reproduced with kind permission from Save The Children (Christmas 1994).

developed countries, the ORS developed for calves may well be superior to some of those prescribed for humans. Certainly, the British National Formulary ORS is less effective than the WHO solution in both normal humans and diarrhoeic rat models (Hunt *et al.*, 1992). It is thus important to examine the principles underlying oral rehydration and its efficacy and to explore, with due caution, some of the perspectives derived from comparative medicine, in particular, those which will illuminate the pivotal role of sodium and the need for something more than restoration of renal function to adequately correct metabolic acidosis.

It is also pertinent to re-examine one of the traditional assumptions of therapy for diarrhoea—the importance of 'resting' the intestine, minimising the substrates available to cause malabsorption or bacterial proliferation. The cost is malnutrition, an enhanced threat of hypoglycaemia and, perhaps, an adverse effect on villus form and function, heightening the risk of renewed diarrhoea when the patient is transferred from ORT to a more normal diet. The current focus of attention in oral rehydration is the means of sustaining nutrition **alongside** rehydration (Brown, 1991; Rivin and Santosham, 1993), either by maintaining milk intake in parallel or by using nutritional oral rehydration solutions based on rice or polysaccharides which provide calories without the immediate 'osmotic penalty' of glucose (Rivin and Santosham, 1993). One of the benefits of continued breast feeding, apart from provision of antibodies, may be an optimal approach to small, frequent feeds, rather than intermittent intake of larger volumes (Rivin and Santosham, 1993). Again, these issues are not as new as they seem.

> *Requirement for parenteral provision of water and the extracellular electrolytes, sodium and chloride, in diarrhoeal disease depends largely on the extent to which food can be taken.*
>
> (Gamble, 1954).

There is sometimes resistance to oral rehydration solutions on the grounds that they may worsen the diarrhoea. If so, it scarcely matters, provided that they improve ECF volume and composition. It means that they are not fully efficient, but few therapies are. It may undermine the confidence of those using them, but should not do so provided that it is explained that what matters is the effect on the patient, not the effect on the faeces (Ludan, 1988; Avery and Snyder, 1990; Brooks *et al.*, 1995)

Principles and Criteria of Efficacy

The cornerstone of the original WHO solution for oral rehydration was a composition close to 100 mmol/l Na^+ and 2% glucose. This achieved an equimolar Na:glucose ratio and, because the sodium was accompanied by 100 mmol of anions, was approximately isotonic (300 mmol/l).

The commitment to isotonicity raises several problems:

1. Diarrhoea causes potassium depletion through faecal loss, urinary loss (in response to aldosterone) and food deprivation. In order to accommodate any potassium, the sodium content needs to be reduced (thus WHO-ORS originally had 90 mmol/l). Despite their cell deficits of potassium, severely diarrhoeic patients can be hyperkalaemic as a result of metabolic acidosis (which displaces K^+ from cells) and compromised renal function (Michell, 1988a; Groutides and Michell, 1990). Nevertheless, with re-expansion of ECF volume, correction of acidosis and improvement of renal function, hypokalaemia can develop rapidly as a reflection of the underlying cell deficits (Michell *et al.*, 1989). Most ORS, whether human or veterinary, probably do not adequately confront the issue of potassium replacement.

2. Glucose concentration cannot be increased unless sodium is reduced and this alters the Na:glucose ratio. This assumes, however, that the optimal Na:glucose ratio is known for the species in question, rather than experimental animals. In particular, it assumes that the appropriate ratio for diarrhoeic as opposed to normal subjects is known. One key difference is that, whatever the ratio in the solution administered, diarrhoeic gut is likely to contain additional sodium whose absorption might be facilitated by additional glucose, whereas the intestine is most unlikely to contain 'excess' glucose.

The glucose content of a conventional ORS is so low that it cannot be significantly enhanced unless the solution is made markedly hypertonic. For example, 1 l of a WHO-type ORS has an energy content equivalent to only 100 ml of calf milk-replacer (Michell, 1983c; Argenzio, 1985). In humans, the hazard of a hypertonic ORS is that it will draw sufficient water into the gut, prior to absorption, to cause hypernatraemia (Michell, 1988a). In calves, this seems to be much less of a hazard, perhaps because they are less likely to be nursed in warm, dry surroundings which further increase insensible water losses (Michell, 1989a). Calves, therefore, can tolerate ORS with a high concentration of glucose (Michell, 1994b). The fear of hypernatraemia, together with the misplaced belief that it results from excess sodium rather than water loss, provided much of the pressure for lower Na concentrations in ORS for humans; particularly those outside Third World countries and exposed to less severe forms of diarrhoea (Michell, 1988a; El Dahr and Chevalier, 1990;

Santosham and Greenough, 1991). This question is examined further. First, however, two other attributes of an ORS are considered; the ability to correct acidosis and the influence of amino acids.

Amino Acids

The original WHO solution contained glycine, which also promotes sodium uptake by cotransport, but it was subsequently dropped because the additional benefit did not justify the additional cost (Michell, 1994b). More recent evidence suggests that it confers little convincing benefit and may be deleterious, though not drastically so: alanine may be preferable (Rivin and Santosham, 1993; Michell, 1994b). On the other hand, glutamine potentially confers unique benefits, since it not only promotes sodium uptake (Soares *et al.*, 1991), but may also have favourable effects on villus form and function (Inoue *et al.*, 1993; Powell-Tuck, 1993; Van der Hulst *et al.*, 1993; Zhang *et al.*, 1993; Ko *et al.*, 1993). Glutamine is the primary metabolic fuel of the small intestine and stimulates oxidative metabolism as well as both electrogenic sodium transport and cotransport with chloride (Rhoads *et al.*, 1990, 1992). It is also the main nucleotide precursor for intestinal cells (Dechelotte *et al.*, 1991; Souba *et al.*, 1992). Given enterically (but not parenterally), it promotes both intestinal glucose absorption and hepatic glucose uptake (Gardemann *et al.*, 1992), a property which might allow better utilisation of higher glucose concentrations in an ORS. The beneficial effects of glutamine on sodium uptake are maintained in piglets with rotavirus enteritis (Rhoads *et al.*, 1991) and include stimulation of ileal sodium uptake in exchange for H^+ (Argenzio *et al.*, 1994). The problem is that it is expensive in the concentrations which seem necessary (30 mmol/l) and that it may require more sustained use than the 48 h during which a conventional ORS provides sole therapy. The limiting factor, as already indicated, is the energy deficiency of a conventional ORS; with a nutritional ORS, more extended use is feasible and could thus allow more time for glutamine to show its potential. We are currently investigating these possibilities in calves.

Bicarbonate Precursors

There is evidence in humans that ORS without bicarbonate precursors can correct acidosis simply by replacing circulating volume and thus restoring adequate renal function (Elliott *et al.*, 1988); it is most improbable, however, that this would apply to severe diarrhoea (e.g. cholera) and it is certainly not the case in calves (Michell *et al.*, 1992), nor is it invariably true in humans (Rautanen *et al.*, 1994). The same study in calves also emphasised the pivotal role of sodium as a constituent of an effective ORS. Metabolic acidosis impedes renal sodium conservation and should, therefore, be corrected *(Chapter 2)*. Excess precursor is unlikely to be harmful unless the excess is gross, because the bicarbonate yield is gradual and, in any case, the kidney resists metabolic alkalosis unless this defence is undermined by hypovolaemia or potassium depletion (Michell *et al.*, 1989). Indeed, since the tendency for the ileum and colon to absorb more sodium during hypovolaemia, alongside the presentation of excess chloride during diarrhoea, may enhance bicarbonate losses from these sites (Phillips, 1994), it might be argued that excess chloride and inadequate bicarbonate precursor in an ORS would potentially exacerbate this trend. An additional reason for the importance of bicarbonate precursors is that by correcting acidosis they also correct hyperkalaemia. Acute hyperkalaemia suppresses the reabsorption of sodium, chloride and bicarbonate in the proximal tubule (Stokke *et al.*, 1993) and the detrimental effect of this on a volume-depleted, acidotic patient is self-evident.

The losses of bicarbonate in faeces are best represented as Na + K − Cl because the bicarbonate is titrated by organic acids generated by bacterial fermentation. Normally, these anions would be absorbed in the colon and would thus promote sodium uptake and provide a metabolic source of bicarbonate. Indeed, organic anions such as citrate, acetate, propionate are used as bicarbonate precursors in ORT. The reason for favouring precursors rather than bicarbonate is to avoid neutralising the protective effects of gastric (in ruminants, abomasal) acidity and, where milk is also consumed, the possible effect of pH on clot formation in the stomach or abomasum.

Sodium

Sodium content is the focal issue in the formulation of an effective ORS (Michell, 1983c; Von Hattingberg, 1992). There are several reasons:

1. cotransport with glucose provides the gradient which sustains absorption of water;
2. because it is the osmotic skeleton of ECF, it is the sodium which ensures that water, once absorbed, remains in ECF where it is most urgently needed, rather than being drawn into cells; hypovolaemia, not lack of ICF, underlies collapse of circulatory and renal function;
3. hyponatraemia, if present (as it usually is in calves; Michell *et al.*, 1992) needs to be viewed for its implications as well as its

effects. It is likely that to cause clinical symptoms the severity must exceed a fall of 15 mmol/l. Lesser falls, however, imply not only that ECF is more dilute than normal but that, as a result, water has moved into cells. In other words, there is a degree of hypovolaemia which is superimposed on that caused by the external losses and exacerbates their effect; this is associated with the inevitable and unfavourable redistribution of water which accompanies hyponatraemia.

Two conclusions seem clear:

1. unless the water entering ECF is accompanied by 145 mmol/l Na, some at least will be lost to ICF;
2. since there are always insensible losses of water, this concentration would probably be too high for a conventional ORS. Nevertheless, solutions with Na concentrations substantially below 90 mmol/l should only be entertained where there is evidence in that species of either very high insensible or faecal water loss (rather than sodium) or of enhanced absorption of sodium, at lower concentrations—which seem improbable.

It is worth emphasising that:

1. a high sodium ORS, whether or not it is more effective, is unlikely *per se* to cause hypernatraemia;
2. the issue is not, for example, whether cholera causes more sodium loss than temperate diarrhoea. The determinants of the outcome of diarrhoea, and particularly its effect on plasma sodium concentration, do not simply depend on the relative proportion of salt and water losses, as they would in an inert system. They also depend on the adaptive responses of thirst and renal function, both of which push hypovolaemic animals towards hyponatraemia (*Chapters 2 and 8*). The inescapable obligation is that every litre of ECF volume restored must contain 145 mmol of Na^+ unless some, at least, is to be drawn into cells, thus sustaining hypovolaemia.

Some of the evidence on optimal sodium concentrations for ORS relates to the effect of the absorption on **water** absorption, often in normal rather than diarrhoeic intestine. Sometimes there is a genuine conflict between the optimum for uptake of sodium and of water; thus a recent comparison of a hypotonic ORS with WHO-ORS emphasised its beneficial effects on water absorption, especially in rat models, while underplaying the comparable or superior effect of WHO-ORS on sodium absorption, especially in a rotavirus model (Hunt *et al.*, 1992). We would argue that optimal sodium uptake takes precedence, not least because that is what will keep water where it is most needed, after absorption; our data in calves reinforce this view (Michell *et al.*, 1992).

While primarily formulated for the treatment of diarrhoea, the range of applications for ORT continues to grow, including correction of deficits associated with exertion and reduction of the total requirement for intravenous fluids, even in conditions where parenteral fluids remain the primary approach (Michell, 1994b). Although the absorption of ORS is minimally affected by exertion, there may be a conflict between the optimal formulation for absorption and palatability (Burke and Read, 1993; Koepsell and Spangenber, 1994).

Sequestration: 'Third Space'

It is not difficult to understand the impact of external losses associated with enteric disease on body fluids. Intuitively, however, it may seem surprising that obstruction, without external losses (e.g. due to vomiting) can have very similar effects (Michell, 1974; Phillips, 1994). The reason is that obstruction intervenes between sites of secretion and absorption and thus produces pooling within the intestine. While the fluid is not yet lost from the body, it is certainly subtracted from ECF; surgical removal of distended and swollen intestine may then produce an irreversible loss of the sequestered fluid.

The specific causes of enhanced fluid loss associated with enteric obstruction (and distension) include reduced absorption, increased secretion, increased permeability of the intestine and its capillaries (hence protein leakage and osmotic gradients undermining fluid uptake), together with direct effects of toxins (Phillips, 1994). Mechanical effects, due to torsion or foreign bodies, may cause increased venous or lymphatic pressure. If hypovolaemia or enterotoxaemia cause shock, this also compromises intestinal structure and function (Michell, 1989b). Obstruction can also be functional (ileus) due to abnormal smooth muscle activity associated with toxins or electrolyte disturbances (e.g. hypocalcaemia or hypokalaemia). Peritonitis readily causes ileus, adding enteric sequestration of water and electrolytes to that within the peritoneal cavity. This is particularly dangerous since the large surface areas of both the intestine and the peritoneum, which normally favour fluid uptake, become sources of fluid loss.

Enteric disease is not the only cause of ECF depletion by sequestration of fluid in an abnormal site ('third space') but it is perhaps the most potent in its effects. Oedema *(Chapter 8)* actually expands ECF, but by abnormally extending the 'overflow' function of interstitial fluid (ISF), at the expense of plasma volume. This, of course, is only possible, since it is the expansion of the major compartment from the minor compartment of ECF, so long as the kidney successfully defends plasma volume which, therefore, may not be detectably diminished. Whether this or a primary expansion of plasma volume, resulting in an overflow into interstitial fluid, underlies the major causes of generalised oedema is discussed further in *Chapter 8*.

The effects of sequestration, rather than external losses, may characterise acute diarrhoea. Thus, while the clinical impact of dehydration is often related to clinically-derived impressions that it represents a loss of bodyweight of 5% (scarcely symptomatic), 10% (obvious signs) or 15% (severe), the relationship between clinical severity and measured loss of bodyweight is far more variable. Groutides and Michell (1990) observed that calves could die of diarrhoea with losses little more than 5% of bodyweight: while this might reflect effects of toxaemia, it more probably results from severe pooling of fluid within the intestine—the fluid that would have appeared in diarrhoeic faeces had the calf lived long enough. This, together with the compensatory powers of the colon, underlies the pitfalls of interpreting the clinical impact of diarrhoea, or the progress of the patient, primarily on changes in faecal output or consistency (Carpenter, 1987; Naylor, 1990; Ribeiro and Lifshitz, 1991; Brooks *et al.*, 1995). Indeed, the object of oral rehydration is to restore the patient's ECF and circulatory function so that the natural defence mechanisms within the intestine, including both immune responses and the natural flora, can overcome the cause of the diarrhoea (Ludan, 1988). There are even those who would see diarrhoea as part of that defence, diluting toxic materials, opposing adhesion of bacteria or absorption of toxins and promoting the expulsion of harmful gut contents (Ludan, 1988; Cohen, 1991). That is perhaps an academic, if not an extreme view, but the important point is that an ORS may successfully rehydrate a patient even if it also increases the diarrhoea. This may undermine patient compliance or the market acceptability of the solution, but it need not detract from its effectiveness: no ORS yet formulated is 100% absorbed and the unabsorbed fraction will be lost in the diarrhoeic faeces (Greenough and Khin-Maung, 1991; Ribeiro and Lifshitz, 1991; Brooks *et al.*, 1994). One of the benefits of specific use of appropriate antibiotics in certain forms of diarrhoea may be to accelerate the restoration of normal faecal output (Arduino and Dupont, 1993) and similar benefits may be obtained from the use of polysaccharides rather than glucose (Alam *et al.*, 1992; Thillainayagam *et al.*, 1994).

Iatrogenic Disturbances

These occur more frequently than we know or admit, because the relevant measurements are often not made and the adverse consequences are attributed to the disease rather than the treatment (Michell, 1994b). Thus, acidotic calves treated with ORS lacking adequate bicarbonate precursor may well be satisfactorily rehydrated, but still profoundly ill because of their acidosis (Naylor, 1986; Michell *et al.*, 1992). Indeed, this is probably worsened by bicarbonate-poor expansion of ECF volume, i.e. dilutional acidosis (Michell *et al.*, 1989). Acid–base and electrolyte disturbances are also readily caused by continued surgical aspiration of gastric or intestinal contents, as they are by enterostomies (Phillips, 1993).

Parenteral fluids more readily cause iatrogenic disturbances if their composition, dose or speed of administration is inappropriate. This is particularly true when renal function is compromised (hence the ability to correct for excessive volumes or inappropriately formulated fluids), but also with respiratory disease (since respiratory responses provide the immediate source of compensation for primary metabolic acid–base disturbances; Michell *et al.*, 1989; Narins *et al.*, 1994). The most important risks, perhaps, are over-rapid provision of potassium or bicarbonate, or over-expansion of ECF volume, plasma volume in particular. Volume for volume, the latter is most likely with artificial colloids (since they do not overflow into interstitial fluid) and is most hazardous where renal function, cardiac function or pulmonary integrity have been compromised, since the greatest danger is pulmonary oedema (Michell, 1989b; Phillips, 1993).

Less widely appreciated are the risks of low sodium fluids in hypovolaemic patients. Firstly hypovolaemia, as already discussed, predisposes to hyponatraemia (Meyer-Lehnert and Schrier, 1990). Secondly, water deficits are best replaced orally unless vomiting or surgical considerations undermine or contra-indicate oral rehydration. The indication for parenteral water (5% glucose and its variants) should be severe hypernatraemia. Commonly, the low sodium solutions are maintenance solutions, not intended for correction of hypovolaemia. Since hypovolaemia (and ECF depletion) can only be effectively corrected by solutions with adequate 'osmotic skeleton', i.e. sodium concentrations reasonably close to 145 mmol/l, low sodium solutions provide inadequate therapy unless used in unnecessarily large volumes, which heighten their main detrimental effect of exacerbation of hyponatraemia (Michell *et al.*, 1989; Michell, 1994e).

Conclusion

The gut is not only the main cause of fluid and electrolyte disturbances, and recently the most important route for their correction, but a vital contributor to the regulation of normal sodium balance. The importance of this contribution is, as yet, insufficiently appreciated and the underlying mechanisms, particularly those which harmonise enteric and renal function, scarcely understood. They may, however, offer a far more potent application of the principles initially encountered in oral rehydration.

In the gut, as in the kidney, there is increasing awareness of the importance of neural factors in modulating ion transport—the ability of denervated kidneys to function satisfactorily, e.g. following transplantation, does not detract from the importance of nerves in fine-tuning normal function. A neural component has been identified in the enteric response to cholera toxin (Booth and McNeish, 1993). The influence of stress on enteric responses has long been appreciated, but it is not easy to disentangle effects on motility, perfusion and transport. Nevertheless, the influence of nerves in linking transport both to enteric stimuli and to influences from the central nervous system is becoming clearer (Brown and Miller, 1991); the importance of such influences in overall regulation of extracellular volume remains to be addressed.

Just as the physiological factors regulating pain and sodium transport came to be understood on the basis of mechanisms originally exploited by drugs such as opiates (which also influence enteric electrolyte transport) or cardiac glycosides, factors which could lead to far more effective ORT could be the same as those which eventually enlighten our understanding of enteric sodium regulation.

4

Behavioural Regulation of Sodium Intake

All vegetable and animal food contains considerable quantities of chlorine and sodium. Why do these quantities not suffice, and why do we add common salt?

(Von Bunge, 1902)

As a general thing the amount of salt desired is exactly related to the amount required.

(Eskew, 1948)

General Issues

The appetite for salt is widespread and powerful among mammals, least so in carnivores, though it is not entirely absent (Carpenter, 1956; Michell, 1989c; Schulkin, 1991). Ingestion of salt begins well before birth, from amniotic fluid, and the peripheral mechanisms underlying salt taste are already functional, moreover the composition of amniotic fluid is affected by maternal sodium intake (De Snook, 1937; Phillips and Sundaram, 1966; Bradley and Mistretta, 1975). Maternal salt intake during gestation affects the fluid intake and the salt appetite of the offspring as adults (Contreras and Bird, 1986; Nicolaidis *et al.*, 1990). Spontaneous preference in newborn rats is for high concentrations of sodium, falling to adult levels within three weeks; even two-week old rats increase their intake of salt solutions (but not milk) when sodium-depleted (Moe, 1986).

Salt is one of the very few substances, perhaps the only one, detected by taste with high specificity and reliability in humans and, probably, animals. Indeed, among the many attempts to classify tastes over the centuries, and the quest for qualities analogous to primary colours, one of the most consistent features has been the inclusion of a 'salty' category. For example, Bravo (1592), Linnaeus (1751), Haller (1793) and Hermann (1880) had taste classifications invoking four to 11 categories, but the common ground was recognition of saline, acid, sweet and bitter (Boring, 1942; Michell, 1969). This could, of course, reflect dietary experience of salt, of sour or sweet fruit and of alkaloids, respectively, rather than a fundamental attribute of taste. Certainly salt is one of the few essential dietary constituents to occur widely in isolation as well as being a component of food. Thus herbivores, as well as humans, take the opportunity to seek salt and carnivores take it unavoidably by ingesting blood and tissue fluid. Milk also contains substantial amounts of sodium, though this varies with species and stage of lactation; it is increased by mastitis.

An Assumed Need, or a Real One?

The appetite for salt is widely perceived as natural, beneficial, adaptive, protective against sodium deficiency. While it can be impressively enhanced by gross sodium depletion, it is pervasive in animals without any likelihood or evidence of sodium deficits (Michell, 1976a, 1977). The American settlers used tracks left by buffalo spontaneously seeking salt. Although the supposedly beneficial effects of salt have been known at least since classical times *(Chapter 5)*, there is little evidence, prior to the eighteenth century, of widespread use of salt for agricultural animals, except in Germany (Multhauf, 1978). Among patients striving to reduce sodium intake for clinical reasons, its implacable persistence is akin to an addiction (Michell, 1978a, 1984a,b); it is undeterred by scarcity, poverty or price (Multhauf, 1978), let alone health advice.

The dominant drive underlying research on salt appetite has been the hypothesis, often the assumption, that avidity for sodium is governed by the physiological need for it.

"It is only under **special** conditions and particularly in animals on vegetarian diets, that the **normal** diet may not contain **enough** sodium. These animals seek out and ingest salt in amounts and concentrations which they would not **normally** accept" (Nachman and Cole, 1971; bold type added). Special, by what criteria? Enough for whom? Normal in whose eyes? The evidence assumes the hypothesis. Similarly, there has been a tendency to equate low sodium environments with sodium-deficient inhabitants (Denton, 1973) without considering their requirements or their obligatory losses, both of which may be low enough to allow them to cope (Michell, 1978a). Indeed, these assumptions go back to the roots of the concept of homeostasis (bold type added):

> **If there is need for salt in the body,** *as there may be, for example in herbivorous animals whose diets contain more potassium than the body requires, the phenomenon of* **salt appetite appears.**
>
> (Cannon, 1932)

As a result, the relationship is often a circular argument, with salt appetite the only evidence of 'sodium deficiency' and the assumption that mineral licks are taken for sodium, whatever else they contain and however adequate the dietary sodium (French, 1945; Michell, 1977).

If, for example, a 50 kg sheep needs 0.1 mmol/kg/day of sodium and has a daily food intake of 1 kg, it will obtain the necessary 5 mmol (115 mg) of sodium from a diet containing as 'little' as 0.01% sodium. In Britain, even very low sodium grasses contain twice as much sodium (Griffiths and Walters, 1966; Michell *et al.*, 1988). It has also been assumed too readily that pica is a sign of sodium deficiency and that compulsive licking, urine drinking, etc. are forms of salt appetite. Such behaviour is seen in cattle with clear evidence of adequate sodium intake (Van de Kerk, 1968; Orfeur, 1985).

Salt appetite is perhaps the best example of nutritional wisdom— the ability to selectively increase intake of a dietary constituent when there is greater need. The earliest example of such 'specific appetite' was vitamin B, but this turned out to be learning of the alleviating effects of diets containing more vitamin B when rats were experiencing the symptoms of deficiency (Rodgers, 1967). Rats are particularly adept at nutritional learning; for example, learned aversion ('bait shyness') protects them from ever re-sampling a substance whose taste they associate with sensations of illness. Such protection is particularly important in a species which cannot vomit. Learning is almost inevitably the explanation with vitamin B, since it does not occur in isolation and, so far as we know, it has no distinctive taste. Salt appetite appears to be innate (Wolf, 1969), i.e. present in individuals unlikely to have experienced sodium depletion even *in utero* and displayed immediately in response to deficits, without the need for a delay to experience alleviating effects of ingestion. Nevertheless, single instances of deficit can induce a lasting enhancement of appetite (Frankmann *et al.*, 1986). Moreover, appetite may be modified by learning, even by rats observing one another's food preferences (Galef, 1986).

Learned aversion and learned preference are by no means unknown in humans and everyday experience reminds us that many adult favourites in food and drink are rejected when first sampled. The same is often true when salt is first offered to individuals from low-salt cultures which do not normally encounter it (Dahl, 1972; Fregly, 1980). There are also various communities with a substantial sodium intake from hunting and fishing who use no added salt, may not even have a word for it and generally dislike it, should it be offered. A 'low salt' culture in which salt is disliked is the Saharan

Bedouin; they do not salt their food, but their drinking water is a hypotonic sodium solution. By combining the intake of sodium with water, they avoid an important hazard of salt consumption in an arid environment—failure to find adequate water to avoid hypernatraemia (Paque, 1980). The fact that salt is perceptible not only in deposits or in seawater, but in blood, sweat and tears, allowed plenty of opportunity for conditioned responses, for example, in primitive communities, during the hunt or the kill.

Cultural factors reinforce a superficial belief that the physiological importance of sodium appetite is self-evident; both history and geography are rich in allusions to salt, in the New World and the Old (Salt Lake City, Salzburg). Whether from salt deposits, mines or brine, salt production has long been both a key industry and a source of state wealth (Multhauf, 1978). As a source of trade and friction, it occupied the place for 5000 years which oil seized in this century; indeed, early oil technology capitalised on techniques for brine pumping and oil, initially, was often an inconvenient contaminant of this vital source of salt (Eskew, 1948). The payment of the original 'salary' was in salt, the Roman *Via Salaria* brought it to the city from the salt works at Ostia; salt has continued in use as currency almost to the present day and in Marco Polo's day it was, literally, worth its weight in gold (Fregly, 1980). The importance of salt as a prized commodity, the basis of solemn covenants and religious ceremonies or the subject of religious allusions, a substance sufficiently precious to be taxed (hence Gandhi's march to the sea and, in pre-revolutionary France, the *gabelle*, the most hated of all taxes), produced only under licence from the Crown (hence place names in England such as Harwich, Nantwich, etc.*) and to be adulterated (like drugs in our own time) to the extent of 'losing its savour', all adds up to a testament of a profound need for it. If we craved only what we need, life would be simpler and healthier; the widespread delight in peppermint does not reflect depressed concentrations of serum peppermint. Carlson (1916) suggested that, despite its widespread prevalence, salt appetite might occur irrespective of 'need':

> We may assume that a certain amount of salt flavour is pleasant to these animals . . . hence they will eat salt wherever found, irrespective of their actual need.

The 'null hypothesis' which haunts the shadows behind research on salt appetite is that it is simply hedonic (Young, 1966), truly

---

*The 'Wich-house' was where the brine was boiled.

'a matter of taste'. The only proof of such a position would be a failure ever to relate sodium intake to aspects of the physiological need for it. The problem in the literature is quite different—a plethora of data purporting to establish such links. They range from anecdotal observations—animals showing amazingly powerful drives and extreme behaviour in consuming mineral licks—to sophisticated experiments. Often, mineral licks are assumed to be rich in sodium, when the possible role of other constituents is ignored, so the behaviour itself is the main 'evidence' of sodium deficiency. Indeed, although the assumption is more subtly assimilated, **the circular argument that salt consumption itself gives evidence of sodium deficiency flaws the interpretation of many experiments.** For example, increased sodium intake during pregnancy is not necessarily attributable to a sodium-based need, if the increment vastly exceeds the combined needs of the dam and conceptus (Michell, 1994d). The preoccupation with 'need', to the extent of regarding 'need-free' intake as mere 'preference', rather than true salt 'appetite' (Grossman, 1990) seems all the more remarkable when, clinically, it is likely that far more people are harmed by excessive, rather than inadequate, salt intake (Michell, 1978a).

Experimental Approaches

In examining experiments on salt appetite, a number of issues should influence our interpretation. **How was sodium available?** Within the diet, added to it or as solutions for drinking? **How was appetite assessed?** Spontaneous addition from the salt shaker over extended periods, or urinary sodium excretion (in humans)? Spontaneous intake of sodium solutions in animals either by short-term exposure tests with little actual ingestion or preference tests comparing intake with water over 48-h periods (with reversal of containers at 24 h to allow for position habit)? Operant conditioning with sodium as the cue or the reward? Changes in receptor sensitivity or electrical activity in specific areas of the brain? Ability to identify sodium in various solutions present in appropriately designed tongue-drop tests (in humans)?

The list is not exhaustive, but it clearly encompasses very different indices of 'appetite'. In particular, experiments showing positive drive, i.e. real work and overcoming of obstacles or penalties to obtain salt are not necessarily comparable with those simply measuring 48-h intake—the latter are obviously influenced by post-ingestional effects, some of which may be osmotic. Equally, heightened sensitivity cannot be equated with heightened avidity; the example is obvious with pain. Reduced sensitivity could be the basis for greater consumption. In either case, the ability to taste salt may not be the everyday determinant of how much is taken.

How was need assessed: how were deficits induced? Two issues are paramount; the speed and severity of the deficits where they are artificially induced and the soundness of the evidence for deficiency when it is assumed. Thus, experiments involving the induction in a single day of deficits equivalent to the demands of an entire pregnancy do not offer a reliable guide to the everyday regulation of sodium intake, any more than asphyxia, while relevant, offers refined insights into the control of respiratory minute volume.

Elevation of Na–K ratio in urine, faeces, saliva, sweat (human), is often cited in support of sodium deficiency, but it merely reflects a higher concentration of aldosterone in plasma. Adrenal secretion of aldosterone is influenced by potassium (and to a much lesser extent by ACTH) as well as sodium, moreover the latter acts via the renin–angiotensin system *(Chapter 7)*, i.e. sodium deficiency is the usual but not the only stimulus; what is detected is reduced renal perfusion (and delivery of sodium to the distal nephron; *see Chapter 2*). The attitude to aldosterone also depends on assumptions about normal levels and normal sodium intake *(see Chapter 5)*. Simply providing a 'low-salt' diet is a most unreliable way of inducing sodium 'deficiency', since renal and faecal losses are so readily reduced virtually to zero in healthy animals.

Different models of sodium depletion are not necessarily comparable. Thus extravascular colloid depletes ECF volume by redistribution of fluid, without causing external deficits. Diuretics affect electrolytes other than sodium *(Chapter 11)* and have vascular as well as renal effects: it is interesting, therefore, that their effects on taste neurons are quite different from those of dietary sodium deprivation (Tamura and Norgren, 1993). One of the most widely-used models of sodium 'deficiency', especially in the early literature, is adrenalectomy. Despite some anecdotes about extraordinary salt appetite in patients with adrenal insufficiency, it is not a feature of human Addison's disease (Henkin *et al.*, 1963). The appeal, to psychologists, was the fact that it usually depresses the plasma sodium concentration. This is, however, a poor and indirect index of sodium status, since it is primarily a reflection of water balance *(see Chapter 1)*. Thus, hypovolaemia is one reason for hyponatraemia in adrenal insufficiency, but probably via enhanced water intake, enhanced antidiuretic hormone (ADH) secretion and reinforcement of the latter by removal of two important effects of adrenal glucocorticoids; normally they suppress ADH secretion and, by maintaining the impermeability of the collecting duct, they also suppress the response to it *(Chapter 8)*. In fact, adrenal insufficiency has a range of effects other than hyponatraemia and, even if the deficiency is purely of mineralocorticoid (aldosterone), hyperkalaemia is as characteristic as hyponatraemia.

While children with Addison's normally do not crave salt, we are yet to explain, after 60 years, why children with diabetes mellitus may crave inordinate amounts—1500 mmol/day (McQuarrie *et al.*, 1936), over three times the level associated with a 40% prevalence of strokes in Japan *(Chapter 5)*. The children were, indeed, hypertensive and their pressures, interestingly, were controlled by increased potassium intake.

Early Studies of Salt Appetite

The intensive study of salt appetite dates essentially from the early 1930s with the pioneering studies of McCance (1936, 1938) on the consequences of experimentally induced sodium deficiency and, in particular, those of Richter, based on his discovery (1936) that adrenalectomy intensified salt appetite in rats. He and his collaborators went on to show that:

1. rats offered the choice between water and salt solutions now drank the latter at concentrations which they previously avoided;
2. the preference related to the cation rather than the anion;
3. mineralocorticoid replacement restored salt appetite to normal; on the other hand, in non-depleted rats, or at high dose, it intensified the craving;
4. pregnancy enhanced salt appetite (Richter, 1943).

A recurrent theme in research on salt appetite, from the earliest days, was an interest in its sensory basis. Experimental approaches included:

1. Two-bottle preference tests vs water, as already described. The development of a preference, however, involves perhaps two separate taste thresholds ('detection' of a difference from water and 'recognition' of what the difference is), as well as motivation to take the detected substance and responses to the after-effects of ingestion (e.g. osmotic flow of water into the intestine). In addition, since water provides the comparator, changes in sodium preference can result from independent responses of thirst rather than salt appetite (Michell, 1980a).
2. Short-exposure tests to minimise the effects of satiety; these included measurements of lick-rates, T-maze choices and operant techniques, i.e. where the animal voluntarily makes a discrimination in order to obtain the reward associated with a correct choice (or to avoid 'negative reinforcement').

3. Various psychophysical tests in humans, e.g. tongue drop discrimination tests, with the benefit of language.
4. Electrophysiological studies, originally with whole nerves, latterly with single units in taste receptors, taste nerves or central areas along the taste pathway. Most of these suffered from two drawbacks; the effects of anaesthesia and the fact that perception involves central interpretation as well as peripheral transduction and conduction.
5. Classical neural ablation studies to define areas impeding salt responses after lesioning.

The focus of most these studies was whether intensification of salt appetite involved enhanced sensory acuity (and, if so, whether in taste pathways or other areas) or enhanced motivation. Considerable scope for conflicting data developed from the growing range of methods for:

1. induction of deficiency (e.g. diuretics, peritoneal dialysis, subcutaneous colloid solutions);
2. assessment of changes in either taste or salt appetite.

In particular, even in conscious humans, taste tests varied according to whether the tongue was static (e.g. in a chamber) or mobile, whether the stimulus was static or allowed to flow, and whether rinses were used which might range from distilled water to solutions resembling saliva (Dethier, 1980). Indeed, it was soon realised that **changes in salivary sodium were a potential cue to the existence of sodium depletion, however, they could not provide an essential basis for the enhancement of salt appetite because they were aldosterone-dependent**. Thus adrenalectomy, unlike other forms of sodium depletion, would not reduce salivary sodium, but it reliably enhanced salt appetite, both in rats and sheep (Denton, 1982). An important observation was that of Henkin and Solomon (1962) showing that adrenal insufficiency alters human taste sensitivity, but that this depends on glucocorticoid, rather than mineralocorticoid effects. In short, it became increasingly clear that **there were great dangers in equating adrenalectomy with sodium depletion or preference with taste**. Moreover, while Haycraft had suggested as early as 1886 that salty taste was a feature of metals in Group 1 of the Periodic Table, and Nachman (1963) and Denton (1965) had shown that rats and sheep could be deceived by lithium, Falk (1965) suggested that the problem with such comparisons was knowing which concentrations should be compared, i.e. based on similar acceptability rather than similar composition. Experiments with selective dietary depletion show that salt appetite is a response to sodium rather than chloride depletion (Muntzel *et al.*, 1991).

Despite the consistent focus on sodium depletion, Tosteson *et al.* (1951) examined salt appetite in hypertensive rats and found that it was depressed; while this is not consistently true either in rats or humans, it is interesting that this pre-dated the main stream of rat studies by Dahl (1958) and his collaborators (Dahl *et al.*, 1962), which were a decisive influence on the growing suspicion of a link between salt intake and hypertension (*Chapter 6*).

Studies in Sodium-Depleted Sheep

A division between 'classical' studies of salt appetite, mainly in humans and rats, and those in sheep, is somewhat arbitrary. Nevertheless, nothing since the original studies of Richter so intensified interest in salt appetite, or added so much to our detailed knowledge of its properties, as the burgeoning series of studies of sodium-depleted sheep by Derek Denton and his collaborators. This began with their discovery, in the mid-1950s, that the huge volume of normal parotid saliva in ruminants, and its high sodium concentration, offered a highly controllable, quantifiable method of causing sodium depletion by unilateral diversion of one parotid duct and collection of its outflow (Denton, 1956, 1957). This had minimal direct impact on other systems; merely one salivary gland was affected, leaving the others unimpaired and there was no interference with endocrine glands.

It rapidly became clear that **sodium-depleted ruminants could preferentially ingest sodium solutions in amounts related to their deficits, correcting for changes in concentration, and could do so rapidly**, i.e. with little time to experience the effects of absorption (Denton *et al.*, 1959; Bell and Williams, 1960; Denton and Sabine, 1961). Neither sodium concentration in plasma nor the rumen linked sodium depletion to salt appetite (Beilharz and Kay, 1963). Subsequent developments are extensively reviewed by Denton (1982).

In one of the earliest papers in the series (Denton and Sabine, 1961), three important facts emerge, apart from the conclusions drawn:

1. The ability to adjust volume of intake to compensate for change in concentration is not necessarily as 'targeted' as it appears. In these experiments it was exhibited over a concentration range of 120–940 mmol/l, over which normal sheep, without any deficits, also reduce their intake as concentration increases (Michell, 1969).

2. The deficits incurred by the sheep were about 490 mmol/day; while the depleted sheep matched their intakes to their deficits

with satisfying accuracy, other normal sheep had very similar intakes (7/10 sheep averaging 460 mmol/day).

3. Deficits of this magnitude and more, incurred in 24–48 h, represent hyperacute and severe sodium depletion: granted that the maintenance requirement is unlikely to exceed 0.1 mmol/kg/day, the deficit is equivalent to the loss by a 50 kg sheep of three months' maintenance intake in a single day. Even the demands of pregnancy may amount to no more than 12 mmol/day (Michell *et al.*, 1988), in which case the loss of 490 mmol in 24 h imposes a demand equivalent to six weeks' gestation.

Already, three features of the development of research on salt appetite are clear:

1. **preoccupation with mechanisms rather than physiological role**;
2. **focus on salt appetite as a defence against depletion**; little attention to 'need-free' intake or, indeed, salt appetite as a potential cause of excess salt ingestion;
3. **indifference to the physiological plausibility of the deficits induced**.

The state of knowledge is almost oppositely balanced to the position in renal physiology: broadly, we have a clear picture of what the kidney contributes towards homeostasis, but still a bewildering lack of consensus concerning the detailed mechanisms which enable it to fulfil this role. With salt appetite, we have a profusion of descriptive detail concerning its *properties*, but an incomplete and unconvincing picture of the significance of its contribution to the regulation of body sodium. Since this book is concerned with the latter, rather than with sensory physiology or physiological psychology, this problem will regain our attention towards the end of the chapter. First, however, some additional consideration of the properties and mechanisms of salt appetite, as opposed to its role, is necessary. This must include the possibility that 'need-free' salt appetite may not only be non-homeostatic, but the basis of a salt load which is not harmlessly and efficiently excreted: i.e. a factor potentially contributing towards hypertension. It is convenient to consider mechanisms broadly under 'neural' and 'humoral' categories, the latter reflecting changes in extracellular fluid (ECF) volume or composition, not restricted to endocrine responses alone. The remainder of the chapter thus comprises:

1. neural aspects;
2. humoral aspects:
3. salt appetite and hypertension;
4. biological role of salt appetite.

Both with neural and humoral aspects, particularly the latter, it is important to realise that **everything that affects sodium appetite or excretion does not necessarily regulate it**. Partly, the distinction rests on whether the factor is *likely* to operate in physiological circumstances (are concentrations plausible; would the substance reach receptors or regulatory areas?), also on whether the factor *responds* to changes in sodium balance as well as mediating them (is the feedback system complete?), but is heavily influenced, often subconsciously, by our current prejudice as to what does or does not contribute to the regulation of body sodium. For example, the natriuretic effects of parathyroid hormone (PTH) (Koeppen, 1990; Bertorello and Katz, 1993) may not simply be side effects but, if bone sodium is mobilised as well as bone calcium, a way of excreting this endogenous excess of sodium, just as the accompanying excess phosphate is excreted under the influence of PTH (Michell, 1976b, 1983a). The example is chosen deliberately because, as we shall see, relationships between calcium and salt appetite are far closer than any current hypothesis can satisfactorily explain.

Another confounding factor found more frequently (but not exclusively) in neural or behavioural rather than endocrine research on salt appetite is the tendency to confuse hyponatraemia with sodium deficiency (or hypernatraemia with salt excess). As already discussed, and subsequently reiterated *(Chapter 8)*, plasma sodium concentration is primarily an index of disturbances in water balance; sodium depletion (through hypovolaemia) may predispose to hyponatraemia, but the latter is an insensitive and unreliable index of the former. Similarly, the consequence of excess salt intake is expansion of extracellular fluid (ECF) volume and stimulation of thirst and antidiuretic hormone (ADH), with hypernatraemia occurring only when these latter responses are impaired.

1. Neural Aspects

There are many ways in which neural responses could be involved in salt appetite, for example in coordinating information from peripheral receptors (including chemoreceptors or ECF volume receptors), by influencing motivation, coordinating salt-seeking behaviour (both its initiation and its termination), by identifying that salt is present in food or fluids, or by enabling it to act as a positive or negative reinforcement (reward or deterrent), by monitoring body sodium or coordinating the information from sites which do; there could also be effects secondary to neurally mediated changes in the excretion or internal distribution of sodium.

Granted that the most physiologically important activity of sodium, and the largest proportion of body sodium, is concerned with the volume and composition of ECF, one would expect neural coordination to be activated, directly or indirectly by changes in ECF. These do not explain all attributes of salt appetite (hence the reservoir hypothesis of Wolf—see below). Of all places, however, one would least expect changes in the central nervous system (CNS) *(Chapter 1)* to be sensitive indices of changes in sodium balance, because their impact on neural function is potentially devastating and because cerebrospinal fluid (CSF) Na is peculiarly well defended against changes in extracellular electrolyte concentration (Michell, 1978a). The exception, however, is the 'special permeability' areas of the brain, which are exposed to changes in ECF composition by virtual absence of a blood–brain barrier (BBB) and which, therefore, seem likely to be prime candidates as CNS sites able to monitor aspects of sodium balance (Michell, 1975a). Recent evidence in cattle, both normal and sodium-depleted, suggests that, indeed, responses of salt appetite to angiotensin depend on such sites, whereas thirst responses involve centres with an intact BBB (Blair-West *et al.*, 1989; Weisinger *et al.*, 1993).

Hitherto, we have more awareness of the role of endocrine, rather than neural responses, in the regulation of sodium excretion *(Chapter 2)*. It is therefore easier to see hormones as indices of sodium balance, or current responses to its disturbance, as well as potential mediators of changes in salt appetite. Thus, for example, while adrenergic responses are clearly involved in the fine tuning of sodium excretion, no one could interpret sodium balance from even a detailed analysis of sympathetic activity or α- or β- adrenergic responses, in the way that they could from a profile of angiotensin II, aldosterone, atrial natriuretic peptide (ANP), active sodium transport inhibitors (ASTI)/ endogenous digitalis-like inhibitors (EDLI) *(see Chapters 1 and 7)* and ADH. It may yet prove, however, that, until now, we have insufficient understanding of the role of CNS centres and peripheral nerves in directly mediating renal function, apart from effects mediated by arterial pressure or renal perfusion. Direct effects on electrolyte transport do exist, both in the kidney and in the gut *(Chapters 2 and 3)*.

Neural effects on salt appetite may be conveniently divided into those involving chemoreception, mainly taste (either peripherally or centrally) or other mechanisms.

Taste

Most of the work on taste in animals has involved acute anaesthetised preparations with their obvious and multiple drawbacks. In particular, changes in taste may arise from central perception as well as peripheral transduction and transmission of information.

Pfaffmann and Bare (1950) were the first to indicate that peripheral taste thresholds were unaltered in adrenalectomised rats. While it is true that the means did not differ, five out of six adrenalectomised rats had thresholds **above** all but two of the seven controls. This aroused no comment because the expectation was that enhanced salt appetite would relate to **increased** sensitivity. It is particularly interesting in view of subsequent evidence that dietary sodium deprivation reduces sensitivity in those taste units most sensitive to salt (Contreras and Kosten, 1986). On the other hand, it has long been realised that normal rats can be trained to discriminate salt solutions below the reduced preference threshold of adrenalectomised rats, without direct effects on their perceptual acuity, i.e. motivation may be the crucial factor (Harriman and MacLeod, 1953; Koh and Teitelbaum, 1961). This may reflect changes in the sensitivity of solitary tract neurons which are associated with the 'reward value' of tastes and respond to changes in sodium balance (Schulkin, 1991).

Taste, as opposed to post-ingestion effects, is important, however, as rats feeding via a chronic intra-gastric cannula do not show a sodium preference (Borer and Epstein, 1965). It is also clear that gastric loads reduce oral salt intake but are less satiating. This is not simply a matter of by-passing oral receptors, because infusions into the hepatic portal circulation have the same effect as oral preloads, moreover, oral sensation alone is not sufficient to sustain normal intake; sham drinking of sodium increases intake (Schulkin, 1991). Salt intake is particularly dependent on amiloride-sensitive sodium channels (Schiffman *et al.*, 1987) and this may explain the absence of spontaneous salt preference in the 'Fischer' (F344) strain of rats (Midkiff *et al.*, 1987; Bernstein, 1993).

Hepatic Chemoreceptors

There is a recurrent suspicion in sodium regulation that the liver plays an important and undervalued role. Much of the evidence is based on observations that portal infusions have specific effects on sodium regulation, whether in increasing sodium excretion (Carey *et al.*, 1976; Unwin *et al.*, 1985) or decreasing sodium ingestion. Intuitively it is obvious that the liver is an ideal site to monitor sodium delivery from the intestine, just as it is for immune surveillance and detoxification of the products of digestion, before they reach the systemic circulation. Arguably, however, the liver is also an ideal site for an ECF volume receptor, since the hepatic veins are readily affected by venous overfill—a matter discussed further in the context of oedema *(Chapter 8)*. Thus, the liver is a potential site of chemoreceptors or baroreceptors involved in sodium regulation.

There has also been the suspicion that the liver may have a degree of control of renal function, perhaps via a specific hormone (glomerulopressin; Davidson and Dunn, 1987). The importance of such an interaction is more obvious once it is appreciated that regulation of body sodium is not just a matter of renal physiology but the outcome of synergistic mechanisms involving the kidney, the gut *(Chapter 3)* and behaviour. Thus, recent evidence from dogs indicates that the renal excretion of dietary salt loads involves a reflex which is abolished by hepatic denervation (Morita *et al.*, 1993).

In contrast, hepatic vagotomy does not impair the satiation of salt appetite caused by sodium depletion (Frankmann and Smith, 1993). Yet portal infusions of sodium activate taste areas within the brain (Hermann *et al.*, 1983) and hepatic denervation alters salt perception (Deems and Friedman, 1988). Moreover, earlier evidence indicated that salt appetite was reduced by hepatic vagotomy (Contreras and Kosten, 1981; Tordoff *et al.*, 1987). Unfortunately, such data conflicts also characterise the hepatic role in renal sodium excretion (Kirchner and Stein, 1994), hence it remains a concept which, for all its attractiveness, requires further substantiation before it can be accepted. Hepatic receptors have also been implicated in CNS regulation of potassium excretion (Rabinowitz and Aizman, 1993).

Other Brain Mechanisms

Classical neurophysiological techniques, involving records of neural activity, effects of specific lesions and delineation of neural circuits, could impinge on salt appetite for a variety of reasons, related to pathways concerned, for example, with taste, motivation, reward, sensory discrimination, arterial pressure, blood volume, renal function and hormones involved in the regulation of sodium and water balance. As already seen, a number of hormones may have localised effects within the brain, functioning as transmitters (e.g. angiotensin II, ADH, oxytocin) as well as their more familiar systemic effects. The use of molecular biology, together with hormone binding studies, to define the distribution of specific receptors has greatly refined knowledge in this area (Schulkin, 1991).

Salt appetite appears to be controlled by mechanisms in the upper brainstem and limbic system. The parabrachial and rostral nuclei of the solitary tract, the lateral and ventromedial hypothalamus, the zona incerta, septal area, median nucleus and amygdaloid complex all influence sodium intake, whereas the hippocampus does not (Grossman, 1990). Particular interest has centred on the anteroventral 3rd ventricle (AV3V) area, the hypothalamic region lying anteroventral to the third ventricle, since it has attracted attention as a potential coordinator of a variety of responses related to sodium balance (Johnson, 1985; McKinley *et al.*, 1986). An important role for forebrain structures located along the lamina terminalis has also been suggested (Zardetto-Smith *et al.*, 1993).

Care is needed in interpreting neural effects since, for example, dorsolateral hypothalamic lesions may have opposite effects on spontaneous sodium intake and salt appetite caused by adrenalectomy or mineralocorticoids (Grossman, 1990). Indeed, it is likely that separate neural circuits are involved in controlling salt appetite responses to different stimuli, e.g. angiotensin and aldosterone (Schulkin, 1991). It is interesting that, despite the fact that most measures of sodium appetite are based on solutions rather than dietary salt, and the potential confounding between neural effects on salt appetite and fluid intake, there is little overlap between the brain areas implicated in thirst and those regarded as likely mediators of sodium intake. The main exceptions are the AV3V area and the area postrema (Grossman, 1990; Schulkin, 1991). The latter is not only a 'special permeability' area, but one which receives projections from the hepatic innervation.

2. Humoral Aspects

The pioneering studies of Richter already established the two main endocrine contexts in which salt appetite has been studied; changes in adrenal cortical secretion (and related hormones) and changes during the reproductive cycle, particularly pregnancy. Although angiotensin II exerts a major influence on aldosterone secretion, it also became an additional focus of interest as an influence on thirst and, potentially, on salt appetite, initially through the work of Fitzsimons (1979).

Adrenal Hormones and Angiotensin

Among the many pertinent observations by George Wolf was the fact that mineralocorticoids [e.g. deoxycorticosterone acetate (DOCA) in rats, aldosterone] not only reduced the enhanced salt appetite caused by adrenalectomy but, at higher doses, induced salt appetite, even in adrenalectomised animals (Wolf, 1965). Richter had previously seen this in intact animals, but it was Fregly and Walters (1966) who showed the effects were truly biphasic, even in intact animals, i.e. suppression of appetite by low doses. Biphasic effects are not unusual with salt retaining hormones; angiotensin II is salt retaining at low doses, natriuretic at high doses (independent of aldosterone) and aldosterone promotes increased sodium excretion if excessive salt retention expands ECF volume *(Chapter 2)*. Such renal effects are readily explicable (though not necessarily definitively explained) by the multiple sites at which renal sodium excretion is controlled and the variety of hormones involved, including those responsible for the excretion of excess salt. DOCA also causes salt appetite in sheep (Hamlin *et al.*, 1988) and various other species, though not all (Schulkin, 1991).

The sodium transport hypothesis of salt appetite (see below) attributes the enhancement by adrenalectomy to secondary effects, notably hyperkalaemia, with enhancement by mineralo corticoids as their direct effect, mediated by their stimulatory effect on Na–K ATPase. Certainly, neither aldosterone nor its effects on salivary Na:K ratio are necessary for salt appetite, since it occurs, classically, in adrenalectomised animals of various species. Thus, the possibility that the contrasting effects of low and high dose mineralocorticoid could be explained by high levels of angiotensin in adrenalectomised animals and suppression of angiotensin by excess aldosterone (Fitzsimons, 1986) does not alter the fact that, whatever factors explain the salt appetite of adrenalectomy, salt appetite also occurs with intact adrenal function. The same problem confronts the concept that 'hormonal synergy', between angiotensin and aldosterone, underlies salt appetite (Epstein, 1986; Sakai, 1986). This concept also collides with the fact that in sheep, unlike rats and rabbits (Denton, 1982), pregnancy does not increase salt appetite, even when aldosterone and angiotensin secretion are clearly increased (Michell and Moss, 1988; Michell *et al.*, 1988; Quillen and Nuwayhid, 1992) and despite the fact that high concentrations of mineralocorticoid stimulate salt appetite in sheep (Hamlin *et al.*, 1988).

Low dose captopril enhances angiotensin II (A-II) in the brain and stimulates appetite (the enhancement of A-II is by increases in circulating A-I which result from the inhibition of its further conversion; brain A-I also rises but can still be converted). High dose captopril (or direct intracerebral administration) reduce salt appetite (Grossman, 1990). The evidence that angiotensin, *per se*, can stimulate salt appetite in the absence of natriuresis is strongest when the hormone is administered directly into specific brain areas in various species; including rodents, sheep and pigeons (Weiss, 1986; Schulkin, 1991). Thus, while there are those who see angiotensin as a virtually comprehensive response to ECF volume depletion (*Chapter 7*), enhancing sodium conservation directly and indirectly (via aldosterone), and water conservation through thirst (and, less convincingly, by promoting ADH secretion), the question hanging over certain aspects of this concept, including salt appetite, is the importance of angiotensin in normal or physiologically plausible circumstances. These must, however, extend to stimuli such as acute haemorrhage, in which the vasoconstrictor effects of angiotensin undoubtedly contribute to the defence of arterial pressure *(Chapter 7)*.

Many of the early experiments on adrenalectomy were confounded by the fact that glucocorticoids, as well as mineralocorticoids, were affected. The potential for interaction is substantial and complex. Firstly, in preference tests where sodium intake is measured against water intake, effects on water balance via interactions with ADH are important. Secondly, there is overlap of receptors between glucocorticoids and mineralocorticoids *(Chapter 7)* and reciprocal effects on receptor binding (Schulkin, 1991; Clore *et al.*, 1992; Hulter, 1994). Glucocorticoids promote salt appetite, with or without mineralocorticoids, according to species (Denton, 1982; Schulkin, 1991), thus explaining the effects of stress or ACTH (except that the latter, in rabbits, appears to have independent effects). Adrenal insufficiency in humans causes increased taste sensitivity which is not restricted to sodium, and which is restored to normal by glucocorticoids rather than mineralocorticoids (Henkin and Solomon, 1962; Henkin *et al.*, 1963). Since mineralocorticoids correct the increased salt appetite of adrenalectomy, it would appear that alterations of taste sensitivity may facilitate it but are not essential. In humans with normal adrenal function, glucocorticoids do not affect salt taste sensitivity (Wong *et al.*, 1993).

Reproductive Hormones

Females generally have a greater salt appetite than males (Schulkin, 1991; Flynn *et al.*, 1993; Stellar, 1993). Salt appetite falls at oestrus, both in sheep and in rats (Antunes-Rodrigues and Covian, 1963; Michell, 1975b; Danielsen and Buggy, 1980). The oestrous fall and luteal rise reinforce the sodium retention seen during the early luteal phase and the natriuresis associated with oestrus in sheep (Michell, 1978b). Sodium intake at oestrus is further depressed by reduction of food intake. Oestrogen, in sheep, is natriuretic and depresses food intake but, like progesterone, has no effect on salt appetite (Michell, 1980a; Michell and Noakes, 1985). Progesterone is not natriuretic in sheep, probably because plasma aldosterone is normally low (Michell and Noakes, 1985; Michell *et al.*, 1988). Reduction of food intake *per se* increases sodium preference in sheep, mainly by reducing water intake (Michell, 1980a). Thus, not only does the change in salt appetite at oestrus reinforce other responses, it occurs independently of changes in food intake, oestrogen or progesterone. The reason for these concerted changes in sodium balance, i.e. the biological advantage, remains unknown (Michell and Noakes, 1985).

The biological advantage of enhanced salt intake during pregnancy and lactation is obvious and such behaviour was originally observed in rats by Barelare and Richter (1938). Similar behaviour occurs in rabbits (Denton, 1982) and mice (McBurnie *et al.*, 1988), though the salt taken vastly exceeds that required (Denton, 1982; Michell, 1994d). Pseudopregnant rabbits also increase their salt appetite; progesterone does not affect salt appetite in rabbits, but it increases the otherwise small stimulatory effect of oestrogens in this species (Denton, 1982). Oestrogens also stimulate salt appetite in guinea-pigs (Middleton and Williams, 1974), but in rats they are inhibitory (Thornborough and Passo, 1975). Both oxytocin and prolactin increase salt appetite in rabbits, independent of other effects on sodium balance (Denton, 1982); prolactin also increases salt craving and thirst in humans (Horrobin *et al.*, 1971). In rats, the increase of salt appetite caused by pregnancy and lactation persists beyond weaning of the litter, though at a somewhat diminished level (Frankmann *et al.*, 1991).

Striker and Verbalis emphasised the role of oxytocin as an **inhibitor** of salt appetite, relating this initially to plasma concentrations (Grossman, 1990) but subsequently emphasising the role of oxytocinergic neurons within the brain. There are, however, doubts concerning the specificity of the effects of oxytocin and the fact that an oxytocin receptor antagonist decreased rather than increased salt intake (Grossman, 1990; Schulkin, 1991). In contrast, these antagonists do enhance angiotensin-induced salt intake (Blackburn *et al.*, 1992).

The most remarkable aspect of salt appetite during pregnancy is the fact that it does not increase in sheep even during low dietary sodium intake (Michell and Moss, 1988): this is discussed below.

Other Hormones

Endorphins may influence salt appetite (Cooper, 1986), as may dopamine (Cooper and Gilbert, 1986), but in neither case is the significance of such responses in the control of sodium intake clear. Both ADH and growth hormone have no effect on salt appetite, at least in rabbits (Denton, 1982).

3. Salt Appetite and Hypertension

Abrams *et al.* (1949) and subsequently Fregly (1959) found that hypertensive rats showed an aversion to sodium. This depends, however, on the cause of the hypertension. Thus, in rats with a salt-dependent tendency to develop hypertension, voluntary salt intake is below that of normal rats (Wolf *et al.*, 1965), but spontaneous hypertensive rats (not salt-dependent) had a greater sodium preference than normotensive controls (McConnell and Henkin, 1973). 'Goldblatt' hypertension (caused by increased renin secretion in response to unilateral or bilateral renal arterial occlusion) reduces sodium preference, but this may be preceded by an increase (Vijande *et al.*, 1986).

In humans, data on salt-taste thresholds in hypertensive patients are inconsistent (Michell, 1978a), but two-choice preference tests, analogous to those in rats, show a greater preference in essential hypertensives (Schechter *et al.*, 1974). The difference in preference did not relate to differences in taste acuity, since neither detection nor recognition thresholds were affected; preference was reduced by antihypertensive therapy, but this result was obtained from a small subset of patients and was not statistically significant (Henkin, 1980). These observations are supported by those of Bernard *et al.* (1980), who found that various salt solutions had similar taste intensities for controls and essential hypertensives but hypertensives differed in finding higher concentrations more acceptable.

It is thus conceivable but not proven that hypertension is not only exacerbated by higher salt intake but, at least in some patients, associated with a greater tendency to take salt. Granted that sodium requirement is far below many estimates and that the decisive influence of salt on hypertension may concern the range 10–100 mmol/day rather than much higher intakes *(Chapters 5 and 6)*, relatively small differences in salt preference could have disproportionately important effects on the age-related rise in blood pressure. Over a lifetime, these could amount to the difference between remaining normotensive or eventually requiring treatment.

Having considered the properties of salt appetite and some of the factors which may control it, there remains the question of its biological role, and its place in the hierarchy of mechanisms available for the regulation of body sodium.

4. Biological Role of Salt Appetite

Normal metabolic processes are possible without the adding of salt to natural foodstuffs. Why then do we eat salt?

(Kaunitz, 1956)

If we want to consider physiological challenges to sodium regulation, as opposed to experimental stimuli, the two obvious sources of increased demand, i.e. potential causes of deficit, are enteric disease and pregnancy. The former has not been studied in the context of salt appetite, though two obvious questions are:

1. can diarrhoeic animals, given a choice of sodium and glucose

    solutions, select a combination of concentrations which seems appropriate according to our knowledge of the principles of oral rehydration *(Chapter 3)*?

2. do animals with a consistently higher spontaneous ('need-free') salt appetite confront diarrhoea at any significant advantage compared with animals whose customary sodium intake is lower?

It is logical, before considering pregnancy, to look at the oestrus cycle which, at least in sheep, has an important lesson to offer. For whatever reason, oestrus is a time of sodium loss, whereas the luteal phase is marked by sodium retention—the opposite of the case during the human luteal and follicular phases (Michell, 1979b). But the systems for sodium regulation act consistently, in concert, to raise or lower body sodium (Michell, 1978b). We do not see salt appetite 'compensating' for enhanced renal or faecal sodium loss, but all three reinforcing one another. It seems obvious in the context of overall regulation of total body sodium. But 25 years ago, when we observed the changes in salt appetite, one journal referee was all too ready to attribute them to the need to 'compensate' for luteal natriuresis caused by progesterone acting as a competitive antagonist of aldosterone (as in humans) and the need to offset the salt retention caused by oestrogens (in women). The essential lesson, however, was not only the scope for species diversity (oestrogens are natriuretic in sheep: Michell and Noakes, 1985), but the fact that in a multiple servo-control system, mechanisms must reinforce one another, not cancel each other out, if adaptive responses are to occur. Thus, one would expect ANP to depress salt appetite rather than cause a 'compensatory' increase in sodium appetite and this is what happens, although it may be central rather than peripheral levels of ANP which matter (Michell, 1988b; Schulkin, 1991). The possible role of the closely related peptide originating from the brain (BNP) remains to be defined (Imura *et al.*, 1992). The other lesson, from ruminant pregnancy, is that, in animals on marginal intakes, balance may be on a seasonal rather than a daily basis, with demands met from reserves during pregnancy and lactation, and reserves repleted between weaning and conception (Michell, 1989d).

    The mystery of salt appetite in pregnancy, at least in sheep, is the mystery of Sherlock Holmes's dog in the night; not what it does but the fact that it does nothing. Pregnancy on a low sodium diet activates renal–adrenal–colonic conservation of sodium, yet salt appetite remains unaffected by this most physiological and vital challenge to sodium balance. Perhaps it forces us to confront the danger of regarding the biological role of salt appetite—indeed of sodium itself—in such a restricted fashion. It also reminds us of a shrewd but now much ridiculed insight of Von Bunge (1902), namely that the species in which salt appetite is most obvious, the herbivores, are those with high potassium diets. Subsequently, this link was viewed simplistically as potassium-induced sodium deficiency and the dismissal of this led to the loss of the insight (Michell, 1978a). But is there something else about potassium, perhaps especially in combination with low sodium intake, which poses a problem; is that, rather than sodium deficiency alone, the evolutionary *raison d'être* for salt appetite? Otherwise could animals not manage on low-sodium diets by simply reducing their urinary and faecal sodium losses virtually to zero, under the influence of aldosterone? One of the beneficial effects of restoring sodium intake may be to promote excretion of potassium, not only directly but through the associated rise in ANP, which also promotes potassium excretion (De Nicola *et al.*, 1993). The other main defence against hyperkalaemia, the ability of Na–K ATPase to move potassium into cells, particularly in muscle, is weakened during periods of semi-starvation by a reduction of pump sites (Kjeldsen *et al.*, 1986), hence resumption of intake of an inevitably high-K diet is a particular hazard if renal excretion is prejudiced.

Na, K and the Sodium Transport Hypothesis of Salt Appetite

Among sheep with differing degrees of spontaneous salt appetite, i.e. without evidence of sodium deficiency, those with the greatest sodium preference were those with the higher plasma potassium concentration (Michell, 1976a). If we consider the main renal problem posed by the combination of inadequate sodium intake combined with a high potassium diet, it is that enhanced proximal tubular conservation of sodium delivers less sodium to the distal nephron. As a result, potassium excretion is constrained, despite raised concentrations of aldosterone. Indeed, in herbivores, aldosterone may be viewed as a particularly important defence against hyperkalaemia, not just conserving sodium from urine, faeces and saliva, but enabling potassium to be excreted or sequestered in the gut (Michell, 1978a). Aldosterone secretion is directly stimulated by hyperkalaemia, without the involvement of the renin–angiotensin complex (*Chapter 7*). Perhaps the distinctive feature of pregnancy is to provide an additional site where large amounts of potassium can be sequestered without causing problems—the conceptus. Plasma potassium is reduced in pregnant women (Webb *et al.*, 1993). The other problem posed by a chronic activation of aldosterone secretion, particularly in animals on marginal intakes, is the increased expenditure of ATP by a major consumer of energy, the sodium pump, Na–K ATPase (Michell, 1989d, 1994e).

The **sodium transport hypothesis of salt appetite** (Michell, 1970, 1971, 1973, 1974, 1978a) suggested that sodium intake was related to both sodium and potassium balance and that this link was provided by a system which monitored the activity of Na–K ATPase in a specific receptor site. Salt appetite would thus be enhanced by those factors which would stimulate Na–K ATPase activity *(Chapter 1)*, e.g. a rise in intracellular Na, a fall in intracellular K, a rise in extracellular K, increased secretion of aldosterone, insulin, etc.; it also offered explanations of some hitherto anomalous features of sodium appetite, including a number related to potassium. Appetite would be expected to fall when Na–K ATPase activity is suppressed, e.g. by circulating transport inhibitors associated with uraemia (Michell, 1991) and does so in response to both nephrectomy and ureteric ligation (Fitzsimons and Stricker, 1971). A key feature of the hypothesis is that Na–K ATPase inhibition would be expected to be natriuretic, yet to suppress appetite despite the enhanced salt loss. Moreover, intracellular sodium is increased by potassium depletion; the 'reservoir' concept of salt appetite, which sought to relate it to intracellular changes (Wolf *et al.*, 1974; Denton, 1982) would expect suppression of appetite by increased ICF sodium, whereas the transport hypothesis expects an enhanced appetite due to the stimulatory effect on the pump; potassium depletion increases salt appetite (Michell, 1978a). Indeed, the reservoir hypothesis always was implausible, since intracellular sodium is normally low and bone sodium only minimally responsive to sodium balance. Subsequently, further data have appeared which are consistent with the sodium transport hypothesis (Michell, 1979c; Vivas and Chiaraviglio, 1987; Michell, 1994d).

Paradoxically, De Nicola *et al.* (1992) relate mineralocorticoid arousal of salt appetite to suppression of Na–K ATPase activity, whereas the usual effect of these hormones is to stimulate it *(Chapter 1)*. Definitive data remain to be found, particularly direct evidence of a link between hyperkalaemia or potassium-loading and salt appetite, and evidence that such a link results from Na–K ATPase activity in a critical site. Part of the problem with potassium infusion experiments is that potassium suppresses angiotensin II (Seeley and Laragh, 1990), whereas in natural sodium deficiency hyperkalaemia coexists with elevated concentrations of angiotensin II. For the moment, the evidence of a relationship between salt appetite and potassium remains indirect but suggestive (Michell, 1994d).

Recent work on natriuretic hormones adds a new dimension to the sodium transport hypothesis of salt appetite. If the adrenal cortex is not simply the source of the main retaining hormones (aldosterone), but also of endogenous inhibitors of sodium transport *(Chapter 1)*, there is a new interpretation for the oldest experimental observation on salt appetite—the effect of adrenalectomy. Quite simply, it removes a source of a hormone which, according to the transport hypothesis, should depress salt appetite. This assumes that there is tonic secretion of the hormone but, since rats are usually kept on exorbitantly high levels of dietary salt *(Chapter 5)*, the assumption is likely to be correct; indeed, there is evidence that adrenalectomy reduces circulating levels of endogenous ouabain in rats (Blaustein, 1993).

Conclusion

We may need to explore more widely yet. Even in species such as rats, where salt appetite does increase during pregnancy or lactation, the magnitude of the increase far exceeds any plausible increase in the actual demand for sodium (Michell, 1994d). We should balance our preoccupation with effects of sodium deficits on salt appetite with equal curiosity and investigation directed towards the factors which underlie the considerable differences in spontaneous salt appetite among apparently normal animals. The most challenging data for a sodium-restricted rationale for salt appetite are those showing that calcium deficiency is a potent stimulus to sodium intake and that it is not mediated by the obvious sodium or calcium-regulating hormones (Tordoff *et al.*, 1993). It may be relevant that calcium is an important factor influencing the renin–angiotensin system. The potential effects of calcium deficits on Na–K ATPase are complex. PTH is an inhibitor, as is a rise of calcium in broken cell preparations; in intact cells, however, increased intracellular calcium concentrations within the physiological range stimulate the pump, though high concentrations remain inhibitory (Bertorello and Katz, 1993). If anything, PTH decreases salt appetite (Tordoff, personal communication). Nevertheless, an enhanced salt appetite in a calcium-depleted animal is an exceedingly strange response, since the likely outcome is an increased urinary loss of calcium *(Chapter 9)*.

What seems clear is that to truly understand the importance of salt appetite we need to consider much more rigorously the biological role of sodium itself, particularly its interactions with other univalent and divalent ions and with associated enzyme systems. Why should this be so; why should salt appetite respond, for example, to physiological demands imposed by potassium metabolism as well as those arising from sodium itself?

Perhaps, fundamentally, because potassium "is of the cell but not of the sap" and because salt appetite is both founded on and part of the mechanism which maintains this ancient and necessary status quo.

(Michell, 1978a)

5

Physiological Basis of Nutritional Requirement for Sodium

Introduction

Subjectivity, Science and Sodium Intake

> The a priori *necessity for a constant supply of water is thus evident. But it is otherwise with salt . . . The supply of salt in considerable quantities is not a necessity for the adult.*
>
> (Von Bunge, 1902)

Discussion of sodium intake lies at an intersection between science and tradition, habit and addiction, need and pleasure, observation and anecdote, reason and myth. The prevailing assumption has been that sodium is physiologically beneficial, that excess is harmlessly excreted and that a liberal intake was therefore desirable. This is perhaps reflected in the sustained and vigorous resistance to the concept of natriuretic hormones, from the early intimations of De Wardener in the 1960s until the discovery of atrial natriuretic peptide (ANP); there was no such problem as excess sodium and elimination of surplus intake was simply a matter of reducing sodium conservation. The central problem of sodium regulation was viewed as the defence against deficiency. Similarly, the focus of research on salt appetite has been its role in protecting against depletion (*Chapter 4*), whereas clinically, nutritional sodium deficiency is difficult to induce in healthy adults outside pregnancy or lactation. Yet perhaps 15–20% of adults will suffer a clinical condition exacerbated in many but not all cases by a lifetime of excess salt intake: hypertension (*Chapter 6*).

Unfortunately, attitudes are as important as evidence in the debates on salt intake. For example, granted that glomerular filtration rate (GFR) is the basis of renal function, that proximal reabsorption of sodium profoundly influences various aspects of renal function

105

(including metabolic work) and that both are sensitive to sodium intake, how can it be that the great majority of publications on renal physiology have no discussion of whether the level of sodium intake was appropriate? Indeed, in many it is unstated. The most recent example, in a journal venerated for its excellence by many British physiologists, is a paper which discusses the natriuretic effects of epidermal growth factor (EGF) and their impact on plasma sodium concentration (Gow and Phillips, 1994); the metabolisable energy content of the diet is given, but sodium content and sodium intake are unspecified. The answer lies in the assumption that provided there is **sufficient** sodium, it scarcely matters how much.

With many questions relating to the physiology of sodium, however, the answer you get depends on the sodium intake at which you ask the question. For example, in sheep on sodium intakes close to published requirements for pregnancy, there is a paradoxical increase in urinary sodium loss during the final weeks, i.e. at a time when accumulation in the conceptus is increasing and the accelerated demands of lactation are imminent (*Chapter 7*). One conclusion would be that sheep need a greatly increased sodium intake in late pregnancy. But sheep on much **lower** intakes do not have this defect in renal sodium conservation, moreover, their aldosterone secretion, like most other species, increases as pregnancy advances (Michell *et al.*, 1988). Previously, sheep were thought to be unusual in having no such rise; this was actually because the published requirement provided such an exorbitant excess of sodium that aldosterone secretion was suppressed, even during pregnancy.

Since sodium intake affects extracellular fluid (ECF) volume and renal function, experiments on unspecified intakes are, essentially, unrepeatable experiments. The physiology of sodium is, perhaps, the only field in which unrepeatable experiments are so widely published in respected journals.

Sodium Requirement and Normality, an Elusive Relationship

For at least 2000 years there has been widespread belief in the value of feeding salt to animals and the beneficial effects on fertility and milk yields are noted by Plutarch, Pliny and Virgil (Orr, 1929). The reason, obvious to Orr, was that sodium and chlorine "are needed in definite amounts". Definite? "There is almost universal agreement that the amount of sodium required by human beings at any stage of development is unknown" (Michelson, 1982). In the literature on almost any species one can find statements, decade after decade, complaining that there is great uncertainty concerning the nutritional requirement for sodium (Michell, 1984a, 1985a, 1989c). The problem was recurrently raised but seldom addressed; the attitude was that, provided the intake was liberal, the actual requirement scarcely mattered. The paradox has, therefore, persisted that for this vitally important dietary constituent, and despite the ease with which relevant measurements can be made, the most consistent feature of statements of sodium requirement has been their uncertainty. Nevertheless, the general trend among successive estimates in various species has been downward revision (Michell, 1985b, 1989c).

A central difficulty is the definition of normality. If you believe that sodium depletion carries risks but excess sodium is harmless, you would want to see, among the criteria of normality, signs of adaptation to high salt intake (e.g. those associated with low levels of aldosterone secretion; high Na:K ratios in urine, faeces, sweat, saliva, etc.). On the other hand, if you regarded excess sodium as a hazard, you might wish to see signs of adaptation to low sodium intake (low Na:K ratios in urine, etc.). ECF or plasma volume are not helpful because, while they fall with reduced sodium intake, they increase with high intake, so the question, once more, is what is regarded as normal? Some argue that the higher levels of ECF volume are desirable because they confer protection against conditions associated with hypovolaemia (Swales, 1994). This argument might be acceptable if there were no adverse effects associated with high sodium intake, but even then it would be necessary to show that the degree of volume expansion conferred significant protection against deficits of the magnitude encountered clinically; as Dahl (1958) suggested, this seems unlikely.

The fundamental problem is that we neither know whether there is a 'set-point' for sodium regulation nor what it is. Hollenberg argued (1980, 1982) that it theoretically represented the amount of sodium in the body when in neutral balance on zero intake (and hence zero excretion). Bonventre and Leaf (1982) rejected the idea of a single set point and emphasised, like Walser (1985a,b) the fact that body content and excretion of sodium depend on intake. Thus as sodium intake increases, a new steady state is reached with increased excretion and increased ECF volume; an analogy is a bucket with a hole low down: with increased filling, the level rises and so does the speed of the leak, but at low levels, the leak stops altogether. Brier and Luft (1994) examined the excretory 'half-life' of a given sodium intake as intake rose or fell; in the latter case, the expected first-order exponential was obtained, i.e. $t^{1/2}$ remained constant. With salt loads, however, $t^{1/2}$ depended on the intake. As with so many aspects of sodium regulation, the answers obtained depend on the level of sodium intake at which the question is asked. There is thus no such thing as a 'normal' body sodium, since it depends on intake. It follows that a rigorous definition of requirement must incorporate the lowest steady-state values and intakes which are safely compatible with good health (Michell, 1989d). One steady-state is particularly distinctive, however, namely the intake at which sodium vanishes from urine (Simpson, 1988). This need not mean that the subject is sodium depleted, as usually assumed, any more than the absence of any overspill means that a reservoir is inadequately filled (Michell, 1989d).

The concept of a reservoir reminds us that input and output broadly need to balance over a season, not necessarily on a daily basis, provided the reservoir is never dangerously low, i.e. unable to meet an unexpected demand. Although interstitial fluid can act as a reservoir (and overspill) for plasma, it is not clear to what extent functional reservoirs for ECF sodium exist, perhaps in bone or gut fluids, in various species. Intuitively, if we designed the system, we would be at least as concerned to monitor the level of the reservoir as the rate of outflow. Astonishingly, our understanding of sodium regulation does not include whether, where, or how this is done.

The terms 'high' and 'low' are treacherous in the context of sodium intake. Their use is often entirely arbitrary according to the feelings of the particular scientist. What is needed is a comparison with a stated requirement (and its source). Thus, there are experiments in which so-called 'low' sodium intakes are consistent with requirement while 'high' intakes are so astronomical as to demand what relevance they have to plausible experience (Michell, 1994a). As a single example, '1% salt' diets are often regarded as 'normal' and anything less as 'low' in rats. If a rat receives a diet containing 1 g NaCl per 100 g of food and consumes 20 g/day, then assuming a bodyweight of 0.3 kg, its daily dose of sodium is 11.4 mmol/kg, equivalent to 800 mmol/day in a human. In Akita in Japan, where the prevalence of hypertension reaches 40% at the age of 45 and the death rate from its cerebral effects reaches three per 1000 population aged between 30 and 60, the salt consumption is 460 mmol/day (Denton, 1982) almost double the 'average' intake on 'Western' diets, but merely half that of rats on 1% salt. Some normal diet! But on such grotesquely high sodium intakes rests much of the science based on rats (Michell, 1984a).

Units can also be treacherous in the interpretation of dietary content in relation to requirement. Requirement should be specified in mmol of sodium, not of salt, per kg bodyweight per day. Content should similarly be specified as mmol of sodium per weight of diet consumed; not as a percentage of sodium or salt and not per dry matter. Further information may be helpful, for example, since dogs are fed according to their energy needs, requirement (or content) can **additionally** be expressed in relation to energy content (e.g. mmol/kcal). Without such information, clinicians who wish to advocate a 'low salt' diet, i.e. one which satisfies requirement but avoids gross excess (e.g. in chronic cardiac disease) would not know how to identify one. In research, valuable efforts to assess the risks or benefits of a wide range of sodium intakes in experimental animals (Ely *et al.*, 1990) are undermined when the data relate to dietary content, rather than the actual daily 'dose' of sodium ingested (Michell, 1994f).

Attitudes, Beliefs and Prejudice

Sodium intake elicits attitudes more appropriate to beliefs and heresies than scientific hypotheses. The extraordinary subjectivity which is tolerated in refereed publications concerning dietary salt is well illustrated in the following views concerning possible benefits of a reduction in routine intake:

> *Moderate reduction in sodium intake may have a minor effect upon blood pressure, perhaps roughly equal to the effect of diuretics.*
>
> (Anon, 1978)

But diuretics have had a long run as front-line therapy for hypertension and, over a lifetime, they are expensive and well-known to have side effects which concern patients, if not their physicians (Michell, 1984a). Therapeutic strategies for the control of hypertension must consider side effects and their impact on quality of life, as well as pharmacological efficacy, especially as the condition is often asymptomatic when treatment begins (Testa *et al.*, 1993). Quite tolerable degrees of salt restriction can greatly reduce the need for diuretics (Beard *et al.*, 1982).

> *It seems at least possible that salt intakes as low as these may have an important influence on blood pressure, although they are clearly not feasible and possibly hazardous in most cultures.*
>
> (Swales, 1988)

The problem is that intakes which look low, compared with customary intakes, may still be well above requirement, indeed, customary intake is probably 10 times that obtained at earlier stages of human evolution (Eaton and Konnor, 1985; Michell, 1989d). Thus, unless such discussions are anchored around requirement, they become both arbitrary and dangerously misleading. In the analysis of the recent 'Intersalt' study of diet and blood pressure, four out of five centres with median intakes below 100 mmol/day were excluded because they were atypical. Yet if we assume a representative bodyweight of 70 kg, this amounts to 1.4 mmol/kg/day. If, on the other hand, we accept that mammalian sodium requirement in healthy adults is unlikely to exceed 0.6 mmol/kg/day, except during pregnancy or lactation (Michell, 1989d), the arbitrary nature and pernicious effects of this exclusion become obvious; in centres with sodium intakes above 100 mmol/day the prevalence of hypertension exceeded 18%, whereas on intakes below 0.8 mmol/kg/day the prevalence was 5% or less (Michell, 1989d).

Counterarguments in defence of high salt intake are often based on the dangers of sodium intakes well **below** requirement (Michell, 1989c) or the assertion that humans somehow differ from other mammals. If so, the implication is that they are somehow uniquely inept at conserving sodium from urine and there is no evidence of this in normal people (Michell, 1994f).

There is, therefore, great benefit from an appreciation of the range of sodium requirement in a variety of mammals in conferring a sense of perspective on judgements of human sodium intake. The question, however, is whether requirement is susceptible to rigorous definition. In the case of maintenance requirement, the answer, subject to suitable safeguards, is 'yes'. The additional requirements for growth, pregnancy, lactation (and, in some species, extreme exertion) are more difficult to establish and more likely to vary between species.

Definition of Requirement; Factorial, Empirical—and Pitfalls

A **factorial** approach to nutritional requirement (Aitken, 1976) takes account of obligatory loss and any additions needed to allow for individual variability, or suppositions that dietary sodium is less than 100% available for absorption (Michell, 1989d). Maintenance requirement provides the best basis for comparisons between species because of obvious differences in growth rates, pregnancy and lactation. A recommended feeding level or dietary allowance may exceed requirement for various reasons, including palatability, but the distinction needs to be clear.

An alternative approach to definition of requirement is based on the effect of different levels of intake on selected physiological parameters (**empirical** approach; Aitken, 1976). This is obviously more arbitrary since it involves judgement of the normal value for these criteria. As it turns out, both approaches tend to furnish exaggeratedly high estimates of maintenance requirements for sodium (Michell, 1985a, 1989c,d). Two key papers in understanding this problem are quite old (Strauss *et al.*, 1958; Morris and Peterson, 1975); one concerns humans, the other concerns sheep. They provide valuable insights into the ease with which estimates of maintenance requirement are inflated by factorial or empirical approaches, respectively; neither is sufficiently widely appreciated, although Strauss *et al.* essentially confirm and refine an observation made almost 100 years earlier by Carl Ludwig (cited by Briggs *et al.*, 1990):

The amount of urinary chloride is not only dependent on the chloride content of the diet on a given day but also on previously ingested chloride. This is evident most clearly if one changes from a low salt to a high salt diet. Less is excreted in the first few days after the change than later when the new diet has been maintained for several days. The converse is true with the reverse change.

Maintenance requirement is the minimal daily intake required to maintain health in a normal adult outside pregnancy or lactation. It is, therefore, a reflection of **obligatory loss**, i.e. the lowest total daily loss of sodium which the animal can sustain (not only in urine and faeces but also dermal secretions, saliva, as appropriate). The obvious way to determine this is to have animals on intakes which are clearly adequate, i.e. liberal, and then withdraw all supplementary sodium, leaving the animal on a basal diet which provides the minimum feasible intake of sodium.

Strauss *et al.* (1958) observed the way in which renal sodium excretion responds to abrupt reductions in sodium intake. This differs from the swift reduction of urinary sodium concentration (virtually to zero) when there is actual depletion of body sodium. Instead, there is an exponential fall, over a number of days, which leads to a negative sodium balance; for a range of intakes this equates with a single day's excess intake on the preceding baseline diet. In an experiment on obligatory loss, however, it would form part of the period in which an 'unavoidable' daily loss of sodium was calculated. For example, Devlin and Roberts (1963) observed substantial urinary sodium losses in lambs suddenly deprived of 4 mmol/kg/day of sodium and these losses account for the very high maintenance requirement attributed to sheep by the Agricultural Research Council (U.K.) (ARC) (1980). What actually happened was that the sodium loss in a month was 417 mmol, but 190 mmol were lost in the first six days, probably as a result of the exponential fall already noted (Michell, 1985a). As a result, obligatory loss was calculated at 0.9 mmol/kg, whereas beyond the first six days it was 0.3 mmol/kg. This estimate was probably still inflated by two other factors:

1. neglect of the sodium in drinking water—which will contribute to the excreted sodium;
2. the assumption that obligatory loss relates directly to bodyweight. This is essentially an unexplored assumption; it seems plausible, in a species where adult body size spans a vast range, e.g. dogs, that the obligatory sodium losses of a St Bernard probably exceed that of a Chihuahua. We would expect, intuitively, that those of an elephant exceed those of a mouse. Once an animal is weaned (and its gut, as well as its kidneys, are functionally mature) it

is not necessarily true that a 60 kg individual has three times the obligatory loss of a 20 kg individual. It might be, if, for example, each achieved the same low sodium concentration in excreta but produced a greater bulk. If, however, each achieved a very similar obligatory loss, it would appear to be three times greater (in relation to bodyweight) if it were based on the smaller individuals.

Fundamentally, the problem is a technical one: if the concentration of sodium in urine and faeces is so low as to be hard to measure, the main determinant of the computed loss becomes the large multiplication factor arising from urinary and faecal bulk. Larger animals clearly have a greater water requirement (greater evaporative loss) and a greater bulk of faecal residues but the minimum attainable loss of sodium in urine and faeces need not increase in parallel.

The problem with the empirical approach to requirement is the criteria of normality. As already discussed, these depend on whether they are taken to include signs of physiological adaptation to low or high sodium intake respectively. Morris and Petersen (1975) observed that even during the high demands imposed by lactation, intakes of 20 mmol/day had no adverse effects on ewes or their lambs, whereas to maintain a high salivary Na:K ratio required over four times as much sodium. Similarly, Sinclair and Jones (1968) showed that sodium levels required to maintain a high salivary Na:K exceeded those needed to sustain normal growth. The 20 mmol/day estimate from Morris and Petersen is not necessarily a minimum, simply the lowest they used. If we accept the fact that 30 kg ewes on 37 mmol/day sodium intake are in positive balance (L'Estrange and Axford, 1966) and that peak milk sodium concentration is likely to be 27 mmol/kg (Ashton and Yousef, 1966; Perrin, 1958), the maintenance requirement is below 0.5 mmol/kg.

The magnitude of maintenance requirement is particularly important in factorial estimates because it colours the estimates for growth, pregnancy and lactation, whereas in the empirical approach these estimates are independently determined. Thus, our own data (Michell *et al.*, 1988) suggest that sheep on as little as 0.1 mmol/kg/day can sustain two consecutive pregnancies and lactations without adverse effects on their lambs or themselves. In fact, they probably draw on their reserves but only to an extent that is readily replenished between the time the lambs are weaned and the next pregnancy (Michell, 1989d). Clearly, maintenance requirement must be less and our estimates of obligatory loss were below 0.05 mmol/kg/day. This contrasts with published estimates of obligatory loss in sheep as great as 1.1 mmol/kg/day, on which their maintenance requirement is based (ARC, 1980). On the former, and the assumptions above, a 30 kg ewe producing 0.8 kg milk/day needs a maximum of 23 mmol/day: on the latter estimate she would need 55 mmol/day.

Unmeasured Losses

One other pitfall of the factorial approach to maintenance requirement deserves emphasis; it assumes that all losses are measured. Firstly, unmeasured spillage of food or excreta does not cancel out; both reduce the apparent obligatory loss. More important, there is a tendency, especially with humans and rats, to assume that the urinary excretion of sodium is so predominant that faecal loss need not be measured. Whether or not this is acceptable as a simplified estimate of dietary intake in humans, it is unacceptable in estimating obligatory loss in any species. This is particularly true because the evidence from sheep (Michell and Moss, 1992) suggests that faecal sodium excretion is more likely to be significant when sodium intake is low, as it has to be for determination of obligatory loss. Neglect of faecal sodium excretion will underestimate loss and, therefore, maintenance requirement in a factorial experiment. In an empirical experiment it may well lead to over-estimates because it will appear that the animal has a very low sodium excretion despite a substantial sodium intake; this spurious renal retention of sodium could readily be interpreted as a sign of sodium depletion. Thus Grim and Scoggins (1986) were surprised that their sheep excreted only 25% of their ingested sodium on intakes of 25 mmol/day, whereas our data suggest that on this level of intake, the majority of the ingested sodium would have appeared in the faeces, had these been collected (Michell and Moss, 1992).

Comparative Aspects of Sodium Requirement

The object of this section is not to provide estimates of the sodium requirements for maintenance, growth, pregnancy and lactation for a range of species, together with critiques of their validity. Instead, it is to critically explore estimates of maintenance requirement, since it offers the best basis for a comparative approach to mammalian sodium requirement. Perhaps the earliest serious attempt to quantitate sodium intake was that of Milne-Eduards in 1850 (Multhauf, 1978); his observations in Parisian institutions for orphans and the elderly, also English public schools, arrived at a customary intake (not a requirement) of 4.5 kg p.a.: 3 mmol/kg/day. He was aware that added salt was only a part of this total.

Much of the debate about human salt intake in relation to hypertension concerns whether it can safely be brought below

100 mmol/day (*Chapter 6*). The extent of our ignorance on human sodium requirement is emphasised by the realisation that there is little to suggest that humans need as much as 10 mmol/day and evidence to suggest that less than 1 mmol/day can suffice; a span of two log units. For some vitamins and minerals the ratio of requirement:toxicity is about 1:10 (Michell, 1984a) and the burden of proof should really move to those who continue to defend sodium intakes in the range 100–300 mmol/day as normal and harmless. It is essential that both intake and requirement are clearly expressed—in mmol/kg/day—avoiding the ambiguities of grams of sodium vs grams of salt and the uncertainties of percentage of diet, whether wet or dry weight. We have already seen that 'customary' intake in humans can span the range 1–800 mmol/day and our attitudes to salt would probably be more rational if we viewed dosage, as for drugs, in log increments, e.g. 1, 10, 100, 1000 mmol/day (Hollenberg, 1983), instead of attempting to discriminate between the effects of 150 rather than 300 mmol/day.

Palaeolithic salt intake was probably below 30 mmol/day (Eaton and Konner, 1985); assuming 70 kg bodyweight, this is about 0.4 mmol/kg/day. The extreme example of low sodium intake comes from the Yanomamo Indians, who appear to be able to manage on less than 1 mmol/day, even during pregnancy or lactation, i.e. 0.02 mmol/kg/day (Oliver *et al.*, 1981; Carvalho *et al.*, 1989). It is possible that this is an underestimate, firstly because these readily occur when intakes are based on urinary excretion or descriptive dietary records, rather than measured intakes (Stamler, 1993). Secondly, it seems possible that humans on vegetarian diets, with low sodium intakes, might excrete sodium predominantly in their faeces, like many herbivores (*Chapter 3*). It is also possible that people on marginal intakes, like herbivores, need to balance intake and requirement on a seasonal rather than a daily basis, drawing on reserves during times of demand such as lactation or pregnancy and replenishing in-between weaning and conception; calculations based on data from wild moose and experimental sheep support the feasibility of such a strategy (Denton, 1982; Michell *et al.*, 1988; Michell, 1989a). Unlike humans, however, herbivores, particularly ruminants, have a substantial store of sodium (other than bone) on which they can draw, namely the sodium of gut fluids which can be replaced by potassium under the influence of aldosterone (*Chapter 3*). The Yanomamo supplement their plant diet with irregular supplies of fish, game or insects but never add salt. Their breast milk is similar in sodium content (5–10 mmol/l) to other women and this is consistent with evidence in animals which suggests that the concentration of sodium in milk is unaffected by low intake; sodium depletion may, however, reduce milk yield (Michell, 1985a). European women restricted to 20 mmol/day (0.3 mmol/kg/day) during pregnancy had less increase in cardiac output, plasma volume and body fat and they showed less reduction in peripheral resistance; nevertheless, their blood pressure, also the birthweight of their offspring and their placental weights were no different from those of women on routine salt intake (Steegers *et al.*, 1991).

Even rapidly growing infants probably require less than 1 mmol/kg/day (Fomon, 1967, cited by Dahl, 1972); breast milk readily provides this (Dahl, 1968; Mayer, 1969). Allowing for losses in sweat and stool, Dahl (1958, 1968) suggested that adult sodium requirement was unlikely to exceed 8 mmol/day, i.e. 0.11 mmol/kg/day, and NAS (1980) put it as low as 2.5 mmol/day, even in young children. Tobian (1979) put an upper limit of requirement as 43 mmol/day (0.6 mmol/kg/day)—similar to the mammalian upper limit suggested by Michell (1989d). Nevertheless, granted that 8 mmol/day is adequate for growing children (Hunt, 1983), 0.11 mmol/kg/day seems more likely. Great caution is required with pre-term infants, however; their requirements are much greater and failure to provide them may impede growth (Haycock, 1993). There is little reason to believe that normal, healthy adults require more than 10 mmol/day (0.14 mmol/kg/day) for maintenance and this may well be sufficient to allow for demands associated with pregnancy, growth or increased sweating (Michell, 1989d). Since arterial pressure only rises with age in high-salt cultures, the upper limit of requirement might correspond to the highest level of intake which avoids such a rise (Michell, 1989d).

Extreme sweating may allow the increased output to cancel out the reduced sodium content induced by aldosterone; unnecessary use of salt tablets, in the absence of extreme exertion, may simply postpone this adaptive reduction. The demands of pregnancy add little more than 3 mmol/day, even allowing for expansion of maternal ECF volume (NAS, 1980). There are indications that some humans may respond to sodium intakes at or below 20 mmol/day with increases in arterial pressure, blood lipids or cholesterol (Muntzel and Drueke, 1992; Ruppert *et al.*, 1994), thus a cautious estimate of sodium requirement could be pitched above 0.3 mmol/kg/day. Nevertheless, those whose arterial pressure was salt-sensitive did not show the adverse effect of sodium restriction on blood lipids (Ruppert *et al.*, 1994). This reflects the problem that there is probably individual variation and that a more accurate expression of nutritional requirement might be to state the probability that a particular intake is adequate for a stated percentage of the population (Beaton, 1988). Otherwise, the majority will tend to be salt loaded for the benefit of an outlying minority (Michell, 1989d).

Rats

As already indicated, attitudes to sodium requirement are most cavalier among many of those working with laboratory rats and, as a result, there is a long tradition of 'routine' salt intakes which, in mmol/kg/day, would equate with some of the highest documented in humans.

Although most rats are unintentionally chronically salt loaded, some have been the subject of experiments supposedly showing adverse effects of low sodium diets. Typically, however, these have used intakes **below** the nutritional requirement, thus a 0.01% NaCl diet (Ott *et al.*, 1989) is likely to provide little more than 0.1 mmol/kg/day). Similarly, data showing that salt depletion lowered the resistance of rats to haemorrhage (Gothber *et al.*, 1983) utilised intakes of less than half the maintenance requirement (Michell, 1989d). Rats with added complications due to uninephrectomy or renal artery stenosis show elevated arterial pressure in response to salt restriction (Mohring *et al.*, 1976; Seymour *et al.*, 1980); but these experiments involve even lower intakes, well below requirement and circumstances in which a hypertensive response to angiotensin would not cause surprise (Michell, 1989d). Moreover, attempts at repetition have been unsuccessful (Rojo-Ortega *et al.*, 1979).

What, then, is the likely magnitude of sodium requirement in rats? In considering this, it is important to remember that unlike most species, rats tend to be used well before they are fully grown (Mohring and Mohring, 1972; Michell, 1989d). Their 'maintenance' requirement therefore needs to include an allowance for growth and we would expect it to be higher than the true maintenance requirements (i.e. for adults) in other species.

Brensilver *et al.* (1985) found that if sodium intake fell below 250 μmol/day, weight gain slowed although growth continued and sodium balance remained positive. Similar estimates, corresponding to about 0.05% Na^+ in the diet, come from the work of Grunert *et al.* (1950), Louis *et al.* (1971) and Toal and Leenan (1983). Indeed, the findings of Ganguli *et al.* (1970) and the National Research Council (NRC, 1978) suggest that even pregnancy and lactation can be accommodated at these levels, though more recent results (Bursey and Watson, 1983) suggest that slightly more is needed (about 0.17% Na^+). Prepubescent rats need somewhat more (300 μmol/day) to sustain growth (Fine *et al.*, 1987). Nevertheless, if growing rats (0.4 kg) can manage on 250 μmol/day, the maintenance requirement is 0.6 mmol/day or less. Interestingly, this is only slightly above the level at which sodium becomes undetectable in rat urine (Brensilver *et al.*, 1985).

Cats

Little is known about the sodium requirement of cats; the 1978 edition of *Nutrient Requirements* could not state one, giving a 'recommended level' instead; the 1986 edition shows little change (Michell, 1989d). One publication (Brewer, 1983) cites a requirement of 0.88% salt in the diet, which turns out to be based on a citation chain originating in some 1941 data—from poultry (Michell, 1989d).

Dogs

Since dogs seldom get hypertension, even with renal disease, the main reason for a low salt diet would be in cardiac cases. On grounds of palatability and acceptance, it is advisable to moderate sodium intake sooner rather than later, before the patient feels too ill. Learned aversion makes it likely that novel diets introduced at a time when the patient feels ill may be rejected as the patient associates them with feelings of malaise.

The most recent National Research Council estimate suggests a requirement of 0.4–0.5 mmol/kg/day (NRC, 1985). Despite some early estimates of obligatory loss which seem erroneously high, it is likely that they are of the order of 0.05 mmol/kg/day, comparable with sheep (Michell, 1989c,d). Empirical findings suggest a probable maintenance requirement within the range 0.2–0.7 mmol/kg/day (Michell, 1989c,d). Since dogs are usually fed according to energy content of the diet, requirement corresponds to 0.67 mmol/100 kcal/day. Whenever sodium requirement is stated according to content rather than consumption, it is important to remember that it is the actual intake which matters. Thus, a patient on a low salt diet is assumed to eat it and care must be taken if appetite is a problem.

Ruminants, Horses and Pigs (Michell, 1985a,b)

The sodium requirements for pigs (ARC, 1981), whether for maintenance, pregnancy or growth, rest on extremely flimsy foundations, some dating back over 60 years. In 1981, as in 1967, the ARC considered that it was obvious that studies of mineral requirements of pigs were urgently needed. Like Godot, they are still awaited! The data of Alcantara *et al.* (1980) indicate that 1.3 mmol/kg/day is sufficient for growing pigs of 27 kg gaining around 0.5 kg/day. Assuming, from their data, that 21 mmol of sodium is needed for 0.5 kg of weight gained, this would suggest a maintenance requirement of 0.52 mmol/kg/day, in line with other mammals.

There is little evidence on which to base a maintenance requirement for horses beyond the data of Meyer (1980), which suggest that obligatory loss is probably less than 0.65 mmol/kg/day. NRC (1989) puts the likely maintenance requirement in the range 0.65–1.1 mmol/kg/day, while admitting that precise recommendations cannot be made. In horses, unlike most herbivores, sweat losses add considerably to sodium requirement.

The suggested maintenance requirement for cattle (ARC, 1980) of 0.3 mmol/kg/day is much more reasonable, as are those for lactation (Michell, 1985a,b). The maintenance requirement is less than half that suggested in the previous edition of the same publication, 15 years earlier. The data for growth are highly questionable, resting on pre-1930 analysis of Hereford and beef Shorthorn bullocks and 1974 observations in German Black Pied bulls. Worst of all, the estimates for pregnancy are extremely insecure, resting on pre-1930 analyses of calf composition and some very dubious assumptions (Michell, 1985a).

Lactation is not discussed here, except for two important points. Firstly, modern dairy cows have a milk output which has been drastically increased by selective breeding even in recent decades; unlike beef cattle, therefore, or other ruminants, dietary sodium can be low enough to cause clinical sodium depletion (Michell, 1985a; Harris et al., 1986). In cattle, as in sheep, low sodium intake does not affect the offspring unless it is severe enough to depress milk output. The offspring are favoured by the fact that, unlike urine, faeces, sweat or saliva, milk sodium concentration is unaffected by aldosterone (Morris and Petersen, 1975; Whitlock et al., 1975; Safwate et al., 1981). Milk sodium concentration rises substantially in mastitis (Linzell and Peaker, 1972). Glucocorticoids reduce the sodium concentration of human milk but this is probably an indirect effect mediated by prolactin (Keenan et al., 1983).

With sheep, the suggested maintenance requirement is greatly exaggerated (Michell, 1985a) and this inflates the recommendations for growth, pregnancy and lactation. There are five main problems and they are instructive:

1. estimates of faecal loss are based on cattle;
2. estimates of urinary loss are based on animals recently transferred from extremely high sodium intakes and probably still excreting some of this excess (as already discussed);
3. scaling-up of obligatory loss with body weight; larger adults probably do not have larger obligatory losses unless the size range within a species is enormous (as already discussed);
4. 'obligatory' sodium excretion includes unmeasured amounts of sodium present in drinking water;

5. sodium excretion at zero intake is extrapolated from excretion at much higher intakes. The relationship is not linear; aldosterone secretion is suppressed at very high intake, increasing greatly at very low intake. It is not the only factor determining sodium excretion, but probably the main determinant of obligatory loss since it can reduce sodium excretion virtually to zero in urine, faeces, saliva (and, in some species, sweat). As sodium intake falls below a particular level, sodium may vanish abruptly from urine (as already discussed).

A further problem emerges from the basis for the published requirement for growth in sheep (ARC, 1980; Michell, 1985a). Firstly, although there are seven sources, there are only three author groups, of which one is quoted for unpublished data. Of the two main sources, one fails to state a sodium intake and the other uses the extraordinarily high level of 10 mmol/kg/day—conceivably 100 times the requirement even for pregnancy in sheep. In both cases, what is measured is the sodium retention per weight gain. In neither case is there any reason to believe that this represents the figure for sheep with normal weight gain on lower but adequate sodium intake. In short, we do not know how much is perhaps going into bone or expansion of ECF (or gut fluids) simply because it is surplus. The conclusion that the requirement is 52 mmol/kg of weight gain is unsubstantiated and data from Field *et al.* (1968) and Joyce and Rattray (1970) suggest a more likely range of 10–36 mmol/kg of weight gain.

Perhaps the most difficult problem in determining sodium requirement in herbivores is that there is a balance between urinary and faecal sodium excretion, which varies with individuals, but which favours faecal sodium excretion at low intake *(Chapter 3)*. It is therefore essential that both forms of excretion are measured in the same individual and genuinely at low intake, not extrapolated from higher levels. Otherwise, two major artefacts will affect the estimate of obligatory loss:

1. since excess sodium is excreted in urine rather than faeces, manipulation of excretion vs intake at levels well above requirement will produce a line which, if extrapolated to low intake, will greatly overestimate the urinary loss;
2. if estimates of obligatory faecal loss are done separately they will be *added* to the urinary loss, whether the urinary estimates are reasonable or inflated. Thus if at low intake some sheep have an obligatory loss of 2 mmol/day mainly in faeces, but others have an obligatory loss of 2 mmol/day mainly in urine, there are clearly no sheep with both. If both measurements are made in the same

animals, this is obvious. If not, however, and the urinary estimate is inflated and from separate experiments (say 4 mmol/day) it will be added to the faecal estimate, without realising that the animals with such high 'obligatory' losses in urine are those least likely to lose a further 2 mmol/day in faeces. It could, therefore, happen that a true obligatory loss of 2 mmol/day appeared to be close to 6 mmol/day.

Such considerations may help to explain why an obligatory sodium loss of 1.1 mmol/kg/day is attributed to sheep (ARC, 1980), whereas we found obligatory losses in urine and faeces of 0.06 mmol/kg/day in pregnant sheep on low sodium intakes, despite the increased output of urine and faeces associated with pregnancy (Michell *et al.*, 1988). Dermal sodium losses in sheep are minimal since the skin secretion ('suint') is potassium-rich. The daily dermal loss is unlikely to exceed 0.04 mmol/kg/day (Aitken, 1976). The argument for a much lower maintenance requirement is supported by the fact that we sustained sheep successfully through two consecutive pregnancies and lactations, with weaned lambs indistinguishable from control sheep, on intakes of 0.1 mmol/kg/day. If this is adequate for pregnancy, it must be more than adequate for maintenance (Michell *et al.*, 1988).

The published requirement for pregnancy is further inflated by estimates of sodium incorporation into maternal tissue on generous salt intakes; the fact that ECF expands during pregnancy is beyond dispute, but the question is, how much sodium is needed to maintain normal development of the conceptus? Related questions are: how much extra-uterine sodium retention is necessary as opposed to simply occurring when surplus sodium is available *(Chapter 7)*? If the composition of amniotic fluid is altered by lower sodium intakes, is that necessarily an adverse effect? I doubt that satisfactory answers are available for any species, let alone sheep. Yet the question may become increasingly important. If the view that sodium is an important risk factor for hypertension prevails *(Chapter 6)* and if habitual salt intake begins to fall, questions may well surface about the sodium requirement for human pregnancy and the possible differences in women with pre-eclampsia or pre-existing hypertension. At the very least, sheep would seem to be able to manage pregnancy in a very similar way to Yanomamo women; the implications for women in general remain to be investigated. The answers will not come from experiments on pregnancy in rats (Barron, 1987) using 'standard' diets which contain sufficient salt to provide for the entire sodium retention of pregnancy every 2–3 days (Michell *et al.*, 1988). The need for further work on sodium requirement is reiterated with monotonous regularity, decade after decade, and separately for all species, yet no one seems to listen, least of all those who would have to encourage or fund such research (Michell, 1985a).

Conclusion

We know far less than we profess about the nutritional requirement for sodium, whether in farm animals, laboratory animals or people.

The dog definitely needs calcium, phosphorus, sodium and chloride.
(Michaud and Elvehjem, 1944)

Very little has been done on the requirement for Na and K in the dog.
(Shaw and Phillips, 1953)

It is somewhat surprising, then, to realise how rudimentary is our knowledge of the animal's requirement for salt which is so important in its economy.
(Russell and Duncan, 1956)

Inadequate data are available to set a minimal requirement for sodium.
(NRC, 1974)

Despite the large quantities of salt annually consumed by ruminants in the United States, there is little quantitative data on the requirement of sheep either for sodium or for chloride.
(Morris and Petersen, 1975)

Specific recommendations on appropriate levels of dietary sodium are difficult to make at present.
(Gaskell, 1985)

Because of limited data on specific requirements for sodium . . . precise recommendations cannot be made.
(NRC, 1989)

As a result:

1. the designs of experiments, or their interpretation, are frequently flawed;
2. clinicians or patients concerned to achieve a 'low' salt intake, i.e. safely consistent with requirement but free from a substantial surplus, face great difficulty in defining quantitatively what it means;
3. there continues to be an unquestioning assumption that, since sodium is so vital physiologically, and so efficiently excreted, requirement is unimportant: what is needed is a generous excess to 'play safe'. 'Customary' intakes are accepted as 'normal' without realising how far above normal they lie or their possible

effect even on 'normal' renal function, let alone the hormonal 'climate'.

Basically, the assumption is that, even if excess sodium is ingested, thirst and natriuretic responses promptly and lastingly solve the problem, without any long-term effect. The validity of this assumption is examined in the next chapter, primarily (but not exclusively) in the context of hypertension.

6

Comparative Aspects of Salt and Hypertension

Introduction

The triangular link between renal disease, salt and hypertension is one of the oldest valid clinical insights, originating over 2000 years ago in China:

> *When the pulse is full and hard . . . the illness dominates the kidneys and has its seat therein.*
>
> Shunyu, I (150 BC)

> *If large amounts of salt are taken, the pulse will stiffen or harden.*
> Huang Ti Nei Ching Su Wen★ (cited by Swales, 1975)

In contrast, the first measurements of mammalian blood pressure (in horses and dogs) by Hales (1733) were some 18 centuries later. It took nearly another 100 years until Poiseuille with the mercury manometer made direct measurement of arterial pressure in animals a practical proposition rather than a heroic venture. Herrison, in 1833, was the first to devise a non-invasive (indirect) method of measuring blood pressure. The sphygmogram, invented in the 1860s, enabled Mahomed to investigate hypertension, using both occluding pressure and waveform as criteria, and to emphasise the relationship with renal disease. The combination of an occluding cuff and a mercury manometer, which still underlies routine blood pressure measurement in general practice, is barely 100 years old, originating with Riva-Rocci in 1896, and satisfactory indirect measurements of diastolic,

---

★The exact dating of this quotation is impossible since it is ascribed to the Yellow Emperor who may have been mythical. It is no more recent than 500 BC (Veith, 1966).

as opposed to systolic pressure, awaited the twentieth century and the work of Korotkov (O'Brien and Fitzgerald, 1991; Gallagher and O'Rourke, 1993).

We thus see a parallel with fever which, like the character of the pulse, was the subject of clinical interpretation long before it could be quantitated (Michell, 1982a). Once measurements became possible, there was a temptation to seize upon 'normal' values without sufficient consideration of how variable they may be or whether elevations of arterial pressure or of temperature necessarily require treatment.

Hypertension is like being seven feet tall; it means that you are unusual, not that you are ill, and in both instances the anomaly may well be less harmful than the means of correcting it! Nevertheless, abundant clinical evidence suggests that beyond some limit, elevated arterial pressure is a risk factor for a variety of cardiovascular diseases. It is the precise definition of these limits which remains elusive and the subject of continuous refinement.

Granted that in various countries, whether in America, Australasia, Asia, Africa or Europe, some 15% or more of the adult population is hypertensive (Bohr and Dominiczak, 1991; Woo *et al.*, 1992; Salako, 1993), that salt intake is excessive and that the age-related rise in pressure is a reflection of excess intake *(Chapter 5)*, it seems remarkable that there is such resistance to the idea of encouraging, not coercing, a moderation of dietary sodium intake (Michell, 1984a).

Dogs, like humans, experience customary sodium intakes which are well above requirement (Michell, 1989c). Since the pioneering studies of Goldblatt *et al.*(1934) they have been classic experimental animals for the study of hypertension. Yet the really interesting feature of canine hypertension may be the fact that, despite their susceptibility, dogs are naturally rather resistant to it, and perhaps to its effects, even when they have chronic renal failure (Michell, 1994a). In part, this may reflect their resistance to the effects of salt loading, unless they are also stressed (Anderson, 1989; Michell, 1989c, 1993). In dogs, unlike humans, the majority of hypertension is secondary, rather than primary (essential), although there is an experimental breed line of dogs which provides a useful model of essential hypertension (Michell, 1993; Bovee, 1993; Papanek *et al.*, 1993). In the majority of these dogs, the hypertension is not salt-sensitive. Despite the low prevalence of hypertension, canine blood pressure rises with age, probably a result (though, in this species, not necessarily a harmful result) of chronic ingestion of excess salt (Michell and Bodey, 1994).

The link between arterial pressure, sodium and renal function is not just a matter of disease: fundamentally, the kidney is the main long-term regulator of arterial pressure and it does so by sharply accelerating sodium excretion when pressure increases. Thus salt loads are harmlessly excreted through a transient and mild elevation of arterial pressure which leads to 'pressure natriuresis' (Guyton, 1992b), or so the theory goes. It is noticeable, however, that the supporting data relate to extremely high salt intakes (requirement $\times 10$–$\times 60$), whereas other experiments (also in dogs) show that such increases are associated with chronic expansion of extracellular fluid (ECF) volume and elevation of arterial pressure (Michell, 1992b). Moreover, the effects of sodium retention are not simply those of an expanded ECF volume, but probably secondary effects on intracellular sodium, especially in arterioles. Sodium may also influence arterial pressure through changes in catecholamine secretion (Gill *et al.*, 1988). In hypertension, the blunted pressure natriuresis causes an increased arterial pressure which is adaptive in the sense that it overcomes the impaired sodium excretion and maintains sodium output (Hall and Granger, 1994).

While the effects of 'sodium' and 'salt' on hypertension are widely regarded as synonymous, there is evidence to support the assumption that it is indeed sodium chloride, rather than other sodium salts, which particularly affects blood pressure (Kurtz *et al.*, 1987; Weinberger, 1987; Luft *et al.*, 1990; Boegehold and Kotcher, 1991). Nevertheless, no one suggests that sodium is even the only nutritional factor in hypertension, let alone that salt exacerbates every case of hypertension; there are salt-sensitive patients who are more susceptible to its effects (Luft *et al.*, 1991). Originally it was thought that 'low renin' hypertensives had elevated body sodium and expanded ECF volume, whereas the high renin group have a reduced plasma volume and their blood pressure is not influenced by salt intake. More important may be the renin response to changes in intake, with 'non-modulators' failing to suppress renin as sodium intake is increased. Non-modulators respond appropriately to colloid expansion of plasma volume, i.e. the insensitivity of their juxtaglomerular apparatus is sodium-specific (Walker and Padfield, 1994). Electrolyte disturbances are also a consequence, as well as a possible cause, of hypertension (Williams and Dluhy, 1994). These matters will be discussed further.

It was Sir George Pickering who was the decisive influence in establishing the view that hypertension was not, fundamentally, a disease which was present or absent in a particular patient but rather a clinical judgement that their arterial pressure, in their circumstances, was likely to predispose them to adverse effects (a quantitative, rather than a qualitative concept of hypertension). He also emphasised that normality was a relative, not an absolute concept. Unfortunately, salt restriction was "to his mind quite unjustified" (Pickering, 1961). Yet, almost in the same breath, he held out the hope that "the environmental factors determining the rate of rise of pressure with age will ultimately be recognised and assessed". Among these, one is now implicated beyond reasonable doubt; excess dietary salt (Michell, 1989c, 1994f). These issues are now addressed in greater detail.

What Matters about Hypertension? (Michell, 1993, 1994a)

It is far from obvious what matters about arterial pressure. Systolic peaks? An inadequate fall at diastole—in which case high heart rates matter? An inadequate nocturnal fall? Diurnal peaks, perhaps associated with stress, or the prevailing level? (Zachariah and Sumner, 1993). Is 'white coat' hypertension (the tendency for pressure to be elevated until the measurement routine becomes familiar) an artefactual increase or an index of a predisposition towards hypertension even during mild anxiety and thus an indicator that true hypertension may follow (Burnier et al., 1994)? Fatigue in structural materials often reflects the number of cycles as well as the amount of distortion (Nichols and O'Rourke, 1990), so that the continuing debate concerning the prognostic significance of systolic as opposed to diastolic hypertension is probably simplistic (Bulpitt, 1990; Bruce et al., 1993; Gallagher, 1993; Sagie et al., 1993).

Then there is the assumption that sounds or oscillations transmitted through the arterial wall and surrounding tissues, and the changes associated with a measured occlusive pressure, are an adequate representation of the stresses on the arterial wall resulting from the arterial pressure. Quite apart from the importance of correct cuff size and first class instruments used with impeccable technique, there is the possibility that extremes of limb conformation (obesity, muscularity) could affect the data (Prineas, 1991), as could changes in the arterial wall, independent of arterial pressure.

On the other hand, direct arterial measurements, while regarded as 'gold standard', are the subject of excessive trust. In conscious, untrained subjects they will certainly be elevated by apprehension, if not pain, unless chronic implantation techniques are used. The type of transducer, its placement (preferably in the catheter tip) and the hydraulic properties of any linking tubes are crucial; so is the influence of bubbles, partial occlusions by micro-thrombi or malposition of the catheter tip against the vessel wall (Nichols and O'Rourke, 1990; Rafterie, 1991; Sykes et al., 1991). Last, but not least, the pressure depends on the artery in which it is measured—contrary to the impression of most clinicians and physiology students—hence direct vs indirect comparisons inevitably compare different pressures. Even the position along the artery matters, because the measured pressure does not simply result from a wave transmitted from the heart but its enhancement or damping by reflections, particularly from major branch-points. In humans, pressure even differs between left and right arms (Hunyer, 1991).

The pitfalls of data gathering are reinforced by those of interpretation. Normal limits can be set with reference to standard deviations, but these include technical imprecision as well as biological variables which require consideration: should there be allowance for age, weight, race (or breed)? The common assumption that data are normally distributed about the mean is neither reasonable nor true (Feld *et al.*, 1990). Intuitively, the selection pressures against marked hypotension will have a far more immediate influence than those against a comparable degree of hypertension and indeed, in both dogs and people, the distribution is not 'normal' (Gaussian) but 'log normal' (Zhang and Popp, 1994; Bodey and Michell, 1995). What matters is not the threshold of abnormality, in this statistical sense, but the level at which the risk of disease becomes increased or the benefits of therapy become clear. The reliability, even of such clinical definitions of hypertension, however, still depends on the imprecision arising from the factors already discussed. Another factor remains, perhaps the most perplexing.

The more we learn from chronic ambulatory pressure measurements, whether recorded or transmitted by telemetry, the more obvious it becomes that, far from being the closely regulated 'set-point' of feedback theory, arterial pressure is fascinatingly variable. Indeed, within the same day, a normal person may have a diastolic pressure above their systolic pressure at a different time (Cornelissen *et al.*, 1992). Yet individuals yield measurements which represent them reasonably consistently on repeated occasions and 'tracking', i.e. a fairly constant rank order, emerges from studies where groups of individuals have been followed over extended intervals, even decades, especially if this starts in adolescence rather than early childhood (Harlan *et al.*, 1973; de Santo *et al.*, 1988; Mahoney and Lauer, 1989; Dillon, 1991; Muntzel and Drueke, 1992). How can this be, when even a series of readings, over several minutes, is little more representative than conclusions based on a few minutes' observations of the exposed height of a rock above the wave-troubled surface of the sea, at a particular stage of the tide, on a particular day in a particular month? Is it possible that in a clinical setting the routine of measurement provides a reasonably standard distraction and thus tends to reduce variability?

There is a lot more to hypertension than a systolic pressure above 140 mmHg or a diastolic above 90 mmHg on three consecutive occasions, or whatever similar criterion is in vogue (Hunyer, 1991; Sever, 1993). For example, young, healthy adults, with 'normal' blood pressures (WHO criteria) but abnormal circadian rhythms showed left ventricular thickening (Cornelissen *et al.*, 1992); the rhythm of arterial pressure may be more important than its magnitude in influencing both sodium excretion and the effects of hypertension (Frattola *et al.*, 1993; Ruddy *et al.*, 1993). In diabetes mellitus, changes in the circadian rhythm of pressure may precede the development of hypertension (Hassan *et al.*, 1993). In which case, what is it that matters about sodium intake; what aspect of hypertension does it supposedly affect? The question of mechanisms will be discussed below but the immediate answer is very simple. On a high salt intake, blood pressure, both systolic and diastolic, rises with age to an extent which, at worst, is associated with a greatly increased prevalence of strokes and, at best, still brings individuals who would otherwise be untroubled by their arterial pressure into the range at which therapeutic intervention needs to be considered (Michell, 1994a). There are few exceptions to this trend and even less which bear scrutiny. On a low salt diet, blood pressure scarcely rises with age; there are no exceptions to this trend. In gauging exposure to high or low salt, what matters is:

1. lifetime exposure;
2. possibly, exposure at critical periods early in life or even prenatally (Dahl, 1972; Di Nicolantonio *et al.*, 1987; Contreras, 1989, 1993); partly these effects may reflect behavioural modifications concerning the taste or acceptability of salt *(Chapter 4)*.

The definition of 'high' or 'low' intake, unless it is compared with nutritional requirement, is as arbitrary as debating whether the Eiger is a high mountain: this subjectivity afflicts the interpretation of both clinical and experimental data, as already discussed *(Chapter 5)*.

As already noted, not all hypertensives are salt-sensitive; indeed, even those that are include interesting subsets such as the 'non-modulators' (Williams and Hollenberg, 1991). These have a heritable defect in either local angiotensin II production or the receptor response, which blunts the effects of exogenous angiotensin II or salt loading by restricting renal sodium excretion; they fail to show the rise in renal blood flow which normally occurs with increased sodium intake. In effect, they have an intrarenal excess of angiotensin II and thus, paradoxically, they benefit from angiotensin-converting enzyme (ACE)-inhibitors, whereas one would expect these to have least effect in patients sensitive to the effects of high salt intake. Laragh (1989) has suggested that nephron heterogeneity may undermine excretion of excess sodium in hypertensives with underperfused nephrons not only excreting less but, by production of angiotensin, restricting sodium excretion by other nephrons. Others have implicated suppression of dopaminergic activity (also reduced secretion of kallikrein and prostaglandin E) as a cause of restricted sodium excretion in essential hypertension (Iimura *et al.*, 1993).

Whatever the role of sodium, other factors contribute to the initiation or maintenance of hypertension. There is intense interest in the role of nitric oxide as a regulator of vascular tone originating within vascular endothelial cells. Essential hypertensives have subnormal ability to generate nitric oxide when limb blood flow is reduced and inhibitors of nitric oxide synthesis raise arterial pressure in experimental animals. Some of the beneficial effect of ACE-inhibitors may result from indirect facilitation of nitric oxide production as well as reduction of angiotensin formation (Moncada and Higgs, 1993). Diminution of the effectiveness of nitric oxide as a vasodilator in essential hypertension is not universal and may well be a secondary effect (Luscher, 1994; Cockcroft *et al.*, 1994). Like vascular remodelling (Gibbons and Dzau, 1994), its importance probably lies in sustaining hypertension and predisposing to its adverse effects, rather than in initiating it.

Salt is thus by no means the only factor which matters in hypertension, but it does matter, and the reasons form the focus of the remainder of this chapter. First, however, some of the other nutritional influences on hypertension are outlined in order to provide a context and to emphasise the interactions between cations which may contribute to elevation of arterial pressure. A recent review of guidelines on the management of hypertension from the U.S.A., Canada, U.K., New Zealand and the WHO, generally advised sodium restriction, whereas none advocated changes in the intake of other cations (Alderman *et al.*, 1993). This may simply reflect the need for a wider appreciation of the growing evidence for the interdependence of their effects.

Nutrition and Blood Pressure; Influences other than Sodium

Apart from specific aspects of dietary composition, obesity is associated with increased arterial pressure both in humans (Stein and Black, 1993) and in dogs (Michell and Bodey, 1994). The distribution of body fat, not just the degree of obesity, affects the impact on blood pressure. Obesity increases the sensitivity of arterial pressure to dietary sodium intake (Rocchini, 1994). Weight reduction reduces blood pressure independent of decreased salt intake and the effect is not restricted to hypertensives. Nevertheless, salt restriction potentiates the effect of reduced calorie intake. Weight loss caused by exercise is more effective than that caused solely by diet (Anderson, 1993). Although the effects of obesity have attracted most attention, arterial pressure correlates with body mass through a wide range and independent of potential confounding factors such as age, alcohol or electrolyte intake (Mikawa *et al.*, 1994).

Obese humans are less able to excrete salt loads, partly because they secrete less atrial natriuretic peptide (ANP) in response to them (Licata *et al.*, 1994). The link between obesity and hypertension may also lie in changes in stroke volume and cardiac output, sympathetic activity, sensitivity of the renin–aldosterone axis and a combination of hyperinsulinaemia and insulin resistance. Both obesity and essential hypertension are typically accompanied by insulin resistance (Barbagallo and Resnick, 1994). Insulin causes renal sodium retention and also affects sympathetic activity and promotes potassium uptake and sodium extrusion by cells (de Fronzo, 1981; Landsberg, 1992; Tuck, 1992; Stein and Black, 1993; Shimamoto *et al.*, 1994; Weder, 1994). Resistance to the hypoglycaemic effects of insulin can leave patients sensitive to its sodium retaining effect (Rocchini, 1994). Diabetics have an expanded exchangeable sodium pool, mainly interstitial fluid rather than plasma. This could result from increased sodium–glucose cotransport in the proximal tubule as a result of glycosuria (Weder, 1994). As well as affecting $Na^+–K^+$ exchange, insulin increases $Na^+–H^+$ countertransport, a marker of essential hypertension (see below). It also causes resistance to the natriuretic effect of ANP (Semplicini *et al.*, 1994). Obesity, like stress, undermines the natural resistance of dogs to salt loading (Granger *et al.*, 1994; Michell, 1994a), however, they seem to resist the hypertensive effects of insulinaemia in various circumstances (Brands and Hall, 1992; Hall *et al.*, 1992). Rats become hypertensive with hyperinsulinaemia, but this is independent of effects on sodium (Brands *et al.*, 1992). While the links between insulin and hypertension continue to provide a focus of active debate, it is important to realise that patients chronically exposed to very high levels of insulin, as a result of tumours, are not usually hypertensive and in normal humans there is little relationship between insulin and arterial pressure (Hall *et al.*, 1992).

Lipids and Fibre

Neither overall lipid intake nor the proportion of saturated fat convincingly affect human blood pressure (Anderson, 1993), but fish oils rich in omega-3 fatty acids, despite their high calorie content, reduce blood pressure (Appel *et al.*, 1993). Like their beneficial effects in protecting compromised renal function from further decline

(Moorhead, 1991; Donadio *et al.*, 1994), the basis of the beneficial effects of lipids on blood pressure may result from associated changes in prostaglandin production (Iacono and Dougherty, 1990). The basis for reductions in hypertensive blood pressure associated with increased fibre intake remains obscure, although evidence of benefit appears in prospective studies of both men and women (Witteman *et al.*, 1989; Ascherio *et al.*, 1992).

Magnesium and Potassium

There are epidemiological and experimental data to suggest a relationship between dietary magnesium depletion and an increased prevalence of hypertension (Ryan and Brady, 1984); evidence of antihypertensive effects from magnesium supplements is less persuasive (Cappuccio *et al.*, 1985; Altura and Altura, 1987; Whelton and Kalg, 1989; Widman *et al.*, 1993). Similar evidence is obtained from rats (Anderson, 1994). It is possible that the effects of magnesium include indirect effects on intracellular calcium, sodium and potassium, since both Ca ATPase and Na–K ATPase are magnesium-dependent enzymes (Solomon, 1987). If we accept that inhibition of Na–K ATPase predisposes to hypertension (see below), this provides a possible basis for the effect of magnesium depletion *(Chapter 9)*. Maternal intake of magnesium, potassium and calcium affects the blood pressure of the infant (McGarvey *et al.*, 1991; Grobbee, 1994). Magnesium, among the plasma electrolytes which relate to blood pressure, is the most strongly heritable (Williams *et al.*, 1991). Recently, it has been suggested that, regardless of the effects of insulin, diabetics are predisposed to hypertension because hyperglycaemia reduces cell magnesium and increases intracellular calcium (Barbagallo and Resnick, 1994). The combination of a diet low in sodium with supplementary magnesium and potassium reduced blood pressure in patients with mild to moderate hypertension (Geleijnse *et al.*, 1994)

Antihypertensive effects of potassium were originally suggested by Addison. Subsequently, McQuarrie *et al.* (1936) found that high potassium intakes reduced blood pressure in hypertensive children with intense salt appetite *(Chapter 4)*. Dahl demonstrated beneficial effects in his salt-dependent rat model of hypertension (Dahl *et al.*, 1972). He speculated that the effect might be to correct excessive arteriolar sodium, perhaps resulting from suppressed Na–K ATPase activity. Apart from natriuresis, potassium supplements also reduce peripheral resistance and renin release (Stein and Black, 1993; Maxwell *et al.*, 1994). Potassium depletion may underlie the hypertension of aldosterone excess (Krishna, 1994). There is substantial evidence for an association between higher potassium intake and a reduced prevalence of hypertension in humans. Evidence of benefit from deliberate supplementation is less consistent (Langford, 1991; Stein and Black, 1993), but supplements do reduce the need for antihypertensive medication, mainly through natriuresis; apart from its effects on blood pressure, potassium may also confer additional protection against strokes (Anderson, 1993). Potassium depletion caused by diuretics may blunt their antihypertensive effects (Maxwell *et al.*, 1994). This may reflect the expansion of ECF volume associated with potassium depletion. Indeed, since the hypotensive benefits of potassium supplementation are not seen if sodium intake is restricted, it is likely that the effects of potassium on blood pressure are mediated, at least in part, by changes in sodium balance (Krishna, 1994).

Calcium

Inadequate calcium intake (associated, for example, with soft water) predisposes to hypertension in humans (Michell, 1976b). Hypertensives are prone to hypocalcaemia and hyperparathyroidism; nevertheless, their intracellular calcium concentrations tend to be increased, perhaps by reduced Ca ATPase activity (McCarron and Morris, 1986; Resnick, 1992). This is particularly the case in platelets rather than other blood cells, which is important because they share various similarities with vascular smooth muscle cells (Kaplan, 1988). The role of parathyroid hormone (PTH) itself is enigmatic; in pharmacological doses in experimental animals it has long been known to be a vasodilator (Pang, 1994). Yet it has pressor effects in plausible concentrations, both in humans and animals, and patients with primary hyperparathyroidism are frequently hypertensive until their PTH levels are corrected (Kaplan, 1988). Cats, as well as humans, with hyperparathyroidism tend to be hypertensive (Kobayashi *et al.*, 1990) and PTH levels are raised in some rat models of hypertension, perhaps as a result of hypocalcaemia secondary to salt retention (Rusch and Kotchen, 1994). PTH also inhibits Na–K ATPase, at least in the kidney (Bertorello and Katz, 1993). If this effect is more widespread, it would add PTH to the sodium transport inhibitors potentially involved in hypertension.

Recently, it has been suggested that hypocalcaemia causes increased production of a newly recognised parathyroid hormone of unknown structure which raises blood pressure (PHF: parathyroid hypertensive factor), perhaps by raising intracellular calcium, whereas PTH, *per se*, is hypotensive, as is the PTH-related peptide (Pang *et al.*, 1994; Shan *et al.*, 1994). This may resolve the paradox that patients with primary hyperparathyroidism are hypertensive despite reduced intracellular calcium in some cells (Lind *et al.*, 1994). The key question is what happens in arteriolar cells. PHF levels are increased in hypertensives, especially those with salt-sensitive hypertension (Pang, 1994).

Aviv's hypothesis (1994) is that increased intracellular calcium not only underlies hypertension but also renal sodium retention and that, in muscle, it enhances performance and therefore had survival value in evolution. Calcium reduces the hypertensive effect of sodium (Hamet *et al.*, 1991). Excess sodium intake predisposes to calciuresis, in parallel with the excretion of the unwanted sodium (Shortt and Flynn, 1990; McGregor and Cappucio, 1993; Cappucio *et al.*, 1993). Thus, hypertensives often have increased calcium excretion, despite an elevated plasma PTH concentration; perhaps the latter is a response to sodium-induced loss of calcium in urine (Kaplan, 1988). Calcium supplements have antihypertensive effects in most trials, though not all (McCarron, 1989; Cutler and Brittain, 1990; Grobbee, 1994) and there are many possible reasons, including effects on sympathetic activity (Hatton and McCarron, 1994). Evidence from genetically hypertensive rats also suggests that calcium deprivation increases arterial pressure, whereas calcium supplements reduce it, though responses vary between strains (Dumas *et al.*, 1994). Moreover, despite the evidence that the direct effects of PTH are vasodilator, it opposes the antihypertensive effects of calcium in rats (Kishimoto *et al.*, 1993).

Calcium increases renal sodium excretion, without necessarily affecting renal perfusion or GFR. This appears to depend on increased renal production of prostaglandins. It has been suggested that part of the mechanism of pressure natriuresis is a rise in medullary hydrostatic pressure leading to the release of prostaglandins which reduce tubular reabsorption of sodium. Nitric oxide is also involved (see below). The increase in prostaglandin production may depend on the availability of sufficient calcium to trigger it (Ruilope *et al.*, 1994). PTH may also inhibit Na–K ATPase in the proximal tubule (see above).

Interactions between sodium and calcium which could lead to hypertension are not confined to excretion; one of the most eloquent hypotheses concerning the aetiology of hypertension hinges on their transcellular exchange and its effect on arteriolar responsiveness to constrictor stimuli (Haddy, 1980; De Wardener and Clarkson, 1985). Intracellular calcium has also been suggested as a key factor in the regulation of renin secretion (May and Peart, 1989). It is, therefore, appropriate to consider possible relationships between sodium and mechanisms which may underlie hypertension.

Sodium and Possible Mechanisms of Hypertension

"We must ask whether our kidneys are really able to eliminate such large quantities of salt" (Von Bunge, 1902). The most coherent hypothesis linking excess sodium intake with hypertension rests on concepts advanced by De Wardener (De Wardener and McGregor, 1980; De Wardener, 1990), Blaustein (1977, 1993), Haddy (Haddy, 1988; Haddy *et al.*, 1978) and Guyton (1992b); it links inadequate pressure natriuresis to vascular effects of sodium transport inhibiting natriuretic hormones (ASTI/EDLI; *Chapter 1*) and their secondary effects on intracellular calcium (Kaplan, 1990).

The Natriuretic Hormone (ASTI) Hypothesis

There are five main links in the logical chain which sustains this hypothesis:

1. The normal defence against retention of excess salt is the 'pressure natriuresis' induced by ECF volume expansion. At normal sensitivity, even small increases in pressure produce large increases in excretion, hence long term stability of pressure is ensured (Mizelle *et al.*, 1993). If, however, pressure natriuresis is defective, its sensitivity blunted, sodium retention will sustain ECF expansion. Increases in medullary interstitial pressure mediated by nitric oxide are an important factor in normal pressure natriuresis (Granger, 1992; Nakamura *et al.*, 1993; Ikenaga *et al.*, 1993; Majid *et al.*, 1993; Schultz and Tolins, 1993). Chronic nitric oxide inhibition in the renal medulla of conscious rats reduced medullary perfusion and caused sodium retention and increased arterial pressure (Mattson *et al.*, 1994). In humans, however, intravenous infusion of L-NMMA to inhibit nitric oxide synthesis caused natriuresis but still raised arterial pressure (Haynes *et al.*, 1993). In this case, the natriuresis could be secondary, i.e. pressure natriuresis; the direct effect of nitric oxide inhibition is to cause sodium retention (Lahera *et al.*, 1992). Stress produces changes in neural control of sodium excretion which predispose to excessive salt retention (Osborn, 1991; DiBona, 1992).

2. The sustained ECF expansion may still be corrected by increased secretion of natriuretic hormones, notably ANP and ASTI (Graves and Williams, 1987; Goto *et al.*, 1990). The latter, however, is the natural ligand for the cardiac glycoside receptor (EDLI, endogenous digitalis-like inhibitor), hence it is a natural 'active sodium transport inhibitor' (ASTI). Indeed, recent evidence suggests that ouabain may be a natural endogenous ASTI secreted by the adrenal cortex *(Chapter 1)*.

3. Suppression of Na–K ATPase activity raises intracellular sodium (Simon, 1990).
4. Outward calcium exchange for extracellular sodium is inhibited by increased intracellular sodium; as a result, intracellular calcium is increased (De Wardener and McGregor, 1980, 1982).
5. The increased intracellular calcium produces excessive arteriolar sensitivity to vasoconstrictor stimuli.
6. Cardiac glycosides may also increase the release of noradrenaline and reduce its reuptake into sympathetic nerve endings (Woolfson *et al.*, 1994).

It is important to note that, while increased cardiac output is the **initial** cause of hypertension when ECF volume expands, the **sustaining** cause is vasoconstriction, interpreted by some as an autoregulatory response protecting against excess perfusion (Krieger *et al.*, 1991).

This hypothesis does not exclude other important events, e.g. secondary changes in vascular structure (Korsgaard *et al.*, 1993), but it provides a precise explanation for the interacting influence of high salt and low calcium intake (which, as already noted, **raises** the normally low concentration of **intracellular** calcium). Importantly, there is evidence that excess sodium affects arterial size, compliance and pulse wave transmission, independent of changes in arterial pressure (Safar *et al.*, 1994). The hypothesis would also anticipate that the classical inhibitors of Na–K ATPase, cardiac glycosides, should increase arteriolar contractility and they do (Poston *et al.*, 1993). Chronic administration of ouabain causes hypertension in rats (Yuan *et al.*, 1993). There are other interactions between sodium and calcium which affect arteriolar tone; hydraulic stress causes conformational changes in the glycosaminoglycans of the vascular wall, which are influenced by both ions (Bevan, 1993). Endogenous inhibitors of sodium transport may also increase vascular tone by reducing endothelin-dependent relaxation (Poston *et al.*, 1993), whereas inhibition of nitric oxide occurs *in vitro*, but not *in vivo* (Woolfson *et al.*, 1994). Various Na–K ATPase inhibitors, including vanadate and cardiac glycosides increase arterial pressure when given acutely to anaesthetised dogs and intravenous cardiac glycosides have similar effects in conscious humans (Haddy, 1988). The hypothesis would also be consistent with lack of the stimulatory effect of insulin on Na-K ATPase, as a result of either defective secretion or resistance, predisposing to hypertension (Weder, 1994). Paradoxically, recent results in rats suggested that chronic salt loading stimulated Na–K ATPase in a number of tissues, rather than inhibiting it (Li *et al.*, 1994).

The data supporting the existence of changes in Na–K ATPase activity (or other markers such as ouabain inhibited transport of labelled Na, K or Rb) in hypertension are mainly derived from red and white blood cells (RBC and WBC)—the latter perhaps more persuasive as red cells are so atypical in their properties. Since changes could be secondary effects of hypertension, the most impressive evidence comes from normotensive offspring of hypertensive parents. An important subgroup of the evidence concerns differences in transport associated with salt-sensitive hypertension (Weder, 1991). The original studies by Morgan *et al.* (1981) did not show that the transport change caused hypertension, but they did demonstrate that salt loading affected the sodium pump of RBC, provided they were exposed to plasma from salt loaded subjects; these changes were characteristic of hypertensive rather than normotensive subjects. Subsequent experiments with RBC have produced both supportive and negative data; it is not clear whether the long life-span of the RBC may contribute to this conflict. Perhaps more important is the fact that many control patients have sodium intakes which arguably make them salt-loaded (Weder, 1991)—the price of an oblivious disregard for sodium requirement which has already been discussed (*Chapter 5*) and which will trouble us further. Thus, the original study, unlike many which followed, involved 'salt restriction' as the comparator for salt loading, but even this included intakes up to 100 mmol/day—low compared with **customary** intake but, at 1.4 mmol/kg/day, more than double the likely **upper** limit of requirement. Interestingly, Zemel *et al.* (1988) showed that excess calcium blocks the effect of salt loading on RBC Na–K ATPase (consistent with its effect on hypertension).

Data from white cell studies, fewer in number because they are not so simple to perform, consistently show a greater tendency to suppression of pump activity in hypertensives. Important though such evidence may be, what matters is not actually sodium transport in blood cells but a particular consequence (increased intracellular sodium) in particular cells (arteriolar smooth muscle). Moreover, since the sodium pump functions as a bilge-pump and increases sodium extrusion as intracellular sodium rises, even changes in its activity which are predictive of hypertension, preceding its emergence, are not necessarily primary; they could be responses to other factors affecting intracellular sodium, including other ion exchanges and membrane permeability. This may explain the fact that in some inbred rat models of hypertension, sodium pump activity is supranormal (Tuck and Corry, 1994). It is also possible that, as with clinical use of cardiac glycosides, there may be a compensatory regulation of the number of pump sites in the presence of a chronic inhibitor of pump activity (Rayson and Gilbert, 1992).

Evidence for increased intracellular sodium in hypertensives is, in fact, substantial, starting with simple studies of vascular sodium content in experimental animals (Tobian and Binion, 1952) and a growing wealth of supportive data in RBC (Hilton, 1986) and WBC. The latter include correlations between the increases of Na_i and of diastolic pressure and the presence of increased Na_i in the normotensive offspring of parents with essential hypertension (Tuck and Corry, 1994).

Changes in Na–K ATPase activity or intracellular sodium are not the only evidence for circulating transport inhibitors. Dahl, 25 years ago, showed that responses characteristic of his salt-sensitive hypertensive inbred rats could be transmitted to normotensive rats via blood in parabiotic experiments (Dahl *et al.*, 1967). Much of De Wardener's experimental evidence involves glucose-6 phosphate dehydrogenase (G6PD) (De Wardener, 1988), a very indirect marker of Na–K ATPase activity which is perhaps unfortunate, particularly when microassays are available (Doucet, 1988). Plasma from hypertensive patients not only depresses Na–K ATPase *in vitro*, but to an extent reflect the degree of hypertension (Hamlyn *et al.*, 1982). Among other functions related to sodium balance, the AV3V area of the brain *(Chapter 4)* has been suggested as a site of ASTI production (Kelly, 1987; Haddy, 1988; Tuck and Corry, 1994), alongside earlier evidence that the hypothalamus was an important source (De Wardener, 1988). Most recently, the adrenal cortex has been identified as a source of ouabain and the latter has been suggested as a physiological regulator of Na–K ATPase activity *(Chapter 1)*. Moreover, an adrenal cortical tumour, secreting ouabain, has been identified as a cause of human hypertension (Blaustein, 1993). It is also conceivable that withdrawal of this hormone contributes to the hypotensive effects of adrenal insufficiency.

Other systems involved in sodium transport have been suggested as contributing causes or genetic markers of essential hypertension, most notably increased Na^+–Li^+ exchange, an *in vitro* marker similar (but not identical) to *in vivo* Na^+–H^+ exchange (Weder, 1991; Tuck and Corry, 1994; Carr and Thomas, 1994). The latter removes from cells excess H^+ generated during metabolic activity; it does so by exchange for ECF sodium and is specifically inhibited by amiloride (Bobik *et al.*, 1994). Interpretation is complicated by the fact that, for example, Na^+–Li^+ countertransport shows similar abnormalities in conditions which also influence hypertension, e.g. obesity and elevation of plasma insulin or triglycerides (Tuck and Corry, 1994). Again, interpretation is complicated by the variety of factors affecting the Na^+–H^+ antiport in a wide range of tissues. Nevertheless, Bianchi *et al.* (1988) regard increased Na^+–Li^+ countertransport as the most consistent ionic transport alteration in essential hypertension, although the magnitude of the change in transport does not correlate with the severity of the hypertension (Carr and Thomas, 1994). Unlike Na^+-Li^+, the evidence for Na^+-H^+ as a genetic marker is poor (Tuck and Corry, 1994), although antiporter activity is enhanced in some patients with essential hypertension (Rosskopf *et al.*, 1993). There is also evidence for depressed Na–K–2Cl transport in hypertensives, particularly those who are salt-sensitive (Weder, 1991; Tuck and Corry, 1994), although in other hypertensives it is increased (Garay *et al.*, 1994). These conflicting data may reflect differences in the *in vitro* conditions used, e.g. composition of the medium and the intracellular sodium content (Carr and Thomas, 1994). Sharma and Distler (1994) have suggested that essential hypertension involves a fundamental disturbance of acid–base balance with increased generation and excretion of acid; if so, the resulting chronic metabolic acidosis would create an endogenous sodium load from bone (Michell, 1976b). It might also impair the function of magnesium-dependent pumps such as Na–K ATPase and Ca-ATPase (Newman and Amarasingham, 1993).

In assessing ion-transport correlates of hypertension it is important to demonstrate effects which precede the increase in pressure and to exclude the influence of other risk factors such as changes in plasma lipids which also effect cation transport systems (Linjen *et al.*, 1994). Many early studies were flawed by inappropriate control groups (Samani, 1994). As well as affecting vascular reactivity, changes in cation transport can also influence the development of vascular hypertrophy and proliferation of smooth muscle cells (Bobik *et al.*, 1994). Thus, changes in Na^+-H^+ antiporter activity may either mediate the mitogenic effects of angiotensin or provide a stable intracellular fluid (ICF) pH which allows mitogenesis to occur (*Chapter 1*).

While the Guyton and the De Wardener–Blaustein–Haddy hypotheses of the link between sodium and hypertension are synergistic in their main arguments, there is an important difference. In the latter, either excess consumption or defective excretion could underlie the problem, i.e. ASTI could induce hypertension even if the increase in sodium excretion matched the increased intake. Thus, while De Wardener (1990) emphasises the importance of defective excretion in various strains of hypertensive rat, and Brenner and Anderson (1989) have suggested an inborn deficit in the number of nephrons as a key predisposing factor in essential hypertension, it would not be a prerequisite for the effect of excess salt, if the ASTI hypothesis is correct. In the Guyton formulation, there has to be a defect in excretion or ECF volume expansion could not persist. It is, however, worth emphasising that the data supporting this concept are obtained from dogs on sodium intakes of 3–25 mmol/kg/day, i.e. 10–60 times their nutritional requirement and that dogs on chronic excess salt do show increases in both ECF volume and blood pressure (Michell, 1989c, 1993). Indeed, recent evidence from humans suggests that, as the magnitude of a salt load increases, so does the delay in excreting it (Brier and Luft, 1994).

Before returning finally to the question of whether there is a causal link between excess dietary salt and hypertension (and whether or not it is important), it is necessary to consider that hypertension also causes changes in electrolyte balance, notably sodium; these vary with the cause of the hypertension.

Effects of Hypertension on Fluids and Electrolytes

One of the most consistent features of patients with essential hypertension (and other forms of hypertension) is an exaggerated response to natriuretic stimuli (Williams and Dluhy, 1994). This seems strange when **defective** sodium excretion features prominently among theories linking salt with hypertension. Presumably the natriuresis, which seems to involve a prior increase in venous pressure but remains inadequately explained, is a consequence, rather than a cause of hypertension. Suppression of sodium reabsorption in the loop of Henlé, as a result of increased pressure transmitted to the medullary vasa recta, may contribute to the exaggerated natriuresis (Woods, 1983), however, in hypertensive rats the natriuresis precedes the onset of hypertension (De Wardener, 1980). It does not result from expansion of plasma volume which is usually normal, or even subnormal, in essential hypertension, perhaps because of increased transcapillary escape of albumin caused by increased capillary pressure (De Wardener, 1990). In contrast, sodium space (ECF volume) is related to arterial pressure in hypertensive but not normal subjects; there is, however, a subset of younger hypertensives, with subnormal ECF volume, but this seems to be a secondary change rather than a primary feature (Robertson, 1987; Beretta-Piccoli, 1990; Beretta-Piccoli *et al.*, 1982).

Excessive natriuresis is a feature of low-renin hypertensives, including those with primary hyperaldosteronism, rather than those with normal renin levels (Romero and Knox, 1988). It is possible that a contributory factor towards the exaggerated natriuresis is the rise in the circulating level of natriuretic peptides, both ANP and BNP (natriuretic peptide from the brain), in hypertension. This is generally associated with more severe elevations of pressure and the rise in BNP is particularly associated with left ventricular hypertrophy (Richards, 1994). Laragh and Seeley (1992) argue that exaggerated natriuresis also occurs in high renin hypertension through sensitisation of pressure natriuresis and, perhaps, as a result of increased heterogeneity of nephron function. Prognostically, the outlook for high renin hypertension is more gloomy, with a greater risk of secondary cardiovascular damage.

The characteristics of secondary hypertension depend on the underlying cause, thus hyperalderostonism *(Chapter 8)* is characterised by potassium depletion and metabolic alkalosis, not by oedema, thanks to 'renal escape' *(Chapter 7)*. Some have attributed 'low-renin' hypertension to excess of an unspecified mineralocorticoid, but this seems improbable since this form of hypertension is not associated with potassium disturbances. Aldosterone plays a key role in malignant hypertension and here the kidney does not escape from its salt-retaining effects because renal function is already compromised. The renal damage is also the cause of the enhanced aldosterone secretion because of the associated increase in renin release. Thus angiotensin-II closes the vicious cycle, further increasing aldosterone, vasoconstriction, arterial pressure and renal damage (Laragh and Seeley, 1992).

The main cause of secondary hypertension in humans is renal disease; it is reputedly the same situation in dogs, but recent evidence suggests that this is not the case (Michell and Bodey, 1994). There are a number of reasons why renal disease could potentially predispose to hypertension:

1. Excess sodium retention in chronic renal failure (CRF) as a result of nephron loss and a mismatch between the increased glomerular filtration rate (GFR) and tubular reabsorption in the surviving intact nephrons. To maintain sodium balance, they need to reduce their fractional reabsorption but increase their absolute reabsorption. Even a slight mismatch can potentially result in substantial wastage or retention of sodium *(Chapter 8)*.
2. Increased production of angiotensin II as a result of renovascular disease (causing reduced renal perfusion) or localised areas of reduced perfusion resulting from renal damage.
3. Reduced production of antihypertensive lipids including prostaglandins (Muirhead, 1983).

It is often suggested that glomerular disease is more likely to cause hypertension than medullary/interstitial disease (McGregor, 1977; Ledingham, 1985), but this view is far from being widely accepted (Brod, 1978).

There is a vicious cycle between hypertension and renal disease: the latter not only causes hypertension but can also result from it.

It is not simply systemic pressure which matters, but the exposure of glomerular capillaries to raised perfusion pressure. In diabetics, even modest elevations of systemic pressure, within the normal range, may be harmful because inappropriate relaxation of the afferent arteriole allows greater transmission of pressure to the glomeruli (Tuttle *et al.*, 1991; Miles and Friedman, 1993). There is a tendency in non-diabetics to assume that co-existing renal disease is caused by hypertension, rather than proving it, hence exaggerating its incidence as a consequence of hypertension (Weisstuch and Dwarkin, 1992; Schlessinger *et al.*, 1994).

The prevalence, and cause, of hypertension varies with the type of renal disease. In humans, where glomerular disease is the prevalent form, roughly half the patients are likely to be hypertensive at presentation (Kincaid-Smith and Whitworth, 1988). Obviously, this varies with the country and the diagnostic centre, but it also varies with the type of glomerular disease and is especially common when glomerular lesions are accompanied by systemic disease, notably vasculitis. In relating hypertension to glomerular disease, it is important to realise that 10–15% of patients with a diagnosis confirmed by biopsy may not yet have detectable proteinuria (Kincaid-Smith and Whitworth, 1988).

Prime examples of non-glomerular diseases which commonly cause hypertension are polycystic kidney disease and pyelonephritis due to reflux nephropathy (in which the hypertension may then cause renal failure); this is a potent cause of hypertension, even in children (Lieberman, 1983). The hypertension of polycystic kidney disease is probably renovascular (Chapman and Schrier, 1991). With the exception of renovascular disease (Robertson, 1988), the specific causes of hypertension in renal parenchymal disease (as opposed to general possibilities) remain to be defined (Kincaid-Smith and Whitworth, 1988). Thus, while sodium retention with expansion of plasma volume may contribute in many patients (Kotchen, 1983), it is neither consistent nor predictable, although glomerular disease is broadly regarded as 'salt retaining', whereas reflux nephropathy is 'salt losing'. Acute tubular necrosis, despite increased renin secretion and sodium retention, does not typically cause hypertension (Ledingham, 1985).

There may well be a neural (sympathetic) component to the salt retention of CRF (Feld *et al.*, 1990). On the other hand, uraemia is among the earliest examples of a stimulus to the production of a circulating inhibitor of sodium transport (Bricker *et al.*, 1968; Blaustein, 1977). Nevertheless, most patients with hypertension which results from renal disease will have it long before they become uraemic. The increased pressure results from increases in both peripheral resistance and cardiac output (Edmunds and Russell, 1994). Subsequently, structural changes contribute; these arise from hypertrophy and calcification.

Even where salt retention does not cause hypertension, sodium restriction may well reduce arterial pressure; the problem is that in doing so, it may well decompensate what remains of renal function. Recently, Muirhead (1994) has suggested that the renal medulla is the source of 'medullipin I' which is hepatically activated to medullipin II, a vasodilator which also causes natriuresis. Loss of the ability to produce such vasodepressor lipids would be a plausible reason for hypertension in renal medullary disease. It might also suggest that some forms of liver disease should cause systemic (as opposed to portal) hypertension. This does not seem to have been observed in humans, although it has in dogs (Michell and Bodey, 1994).

There is a real danger that the ancient triangle — salt, renal disease, hypertension — may be simplistically misinterpreted to the extent that patients are assaulted with hypotensive polypharmacy to protect them from the cardiovascular, cerebral and ocular effects of hypertension as well as its adverse effects on the kidney. Thus, there is good evidence that ACE-inhibitors probably protect the remaining renal function in the majority of patients with CRF, probably by reducing glomerular pressure and proteinuria (Tolins and Raij, 1991; Brown *et al.*, 1993; Lewis *et al.*, 1993; Michell, 1993, 1995). In combination with diuretics and salt restriction, especially with injudicious dosage or severity, however, the result in some patients will be nephrons which can no longer maintain their GFR in the face of an inadequate perfusion pressure, because this defence depends on efferent arteriolar constriction mediated by angiotensin II (Sullivan and Johnson, 1983; Kincaid-Smith and Whitworth, 1988; Eknoyan and Suki, 1991; Tuttle *et al.*, 1991; Weinberg, 1993). This is particularly true of patients with renovascular hypertension (Robertson, 1988), however, the earliest renal change in essential hypertension is a reduction in renal perfusion which is reinforced by subsequent sclerosis; GFR is sustained by efferent constriction (Woods, 1983). Similarly, polypharmacy may predispose to hyperkalaemia by additive effects. Reduction of angiotensin II reduces aldosterone secretion (despite the independent stimulatory effect of hyperkalaemia) and β-blockers impede cellular uptake of potassium (and also reduce renin release). This would be particularly likely in patients on potassium supplements (to counter the effects of diuretics) or with inadequate insulin levels (since insulin also stimulates cell uptake of potassium).

Special instances of hypertension and its effects on sodium and renal function occur in diabetes mellitus and in pregnancy. Various factors are involved in diabetic hypertension, including diabetic nephropathy, but also, as already noted, increased risk of renal damage in essential hypertension. Both insulin-dependent (IDDM) and non-insulin dependent (NIDDM) diabetes coexist with a high prevalence of hypertension (Corry and Tuck, 1991). In many patients, insulin resistance causes high circulating levels of insulin; this includes cases of obesity, essential hypertension and, surprisingly, IDDM. Here there is often a defective response, as well as defective secretion. While the rise in renal perfusion and GFR characteristic of early diabetes (and eventually the nephropathy) probably reflects effects of hyperglycaemia, hyperinsulinaemia restricts renal sodium excretion and increases body sodium (Corry and Tuck, 1991). It also raises the sensitivity to vasoconstrictor stimuli, including angiotensin II (Semplicini *et al.*, 1994). Reduced insulin dosage may increase sodium excretion and reduce both body weight and blood pressure (Corry and Tuck, 1991). Granted that insulin stimulates Na–K ATPase (De Fronzo, 1981; Williams and Epstein, 1989; Perrone and Alexander, 1994), the effect on sodium excretion is not perhaps surprising but, presumably, this effect does not apply to arterioles if the hypothesis linking hypertension to inhibition of arteriolar Na–K ATPase is correct. The answer may be that insulin raises intracellular sodium (which would then stimulate Na–K ATPase) by Na–H exchange, however, the direct effects of insulin on arterioles seem to be consistent with activation of Na–K ATPase, i.e. relaxation (Buchanan, 1991).

Pregnancy Hypertension

Pregnancy involves substantial changes in both renal function and sodium balance and there may be important species differences (*Chapter 7*). Basically, however, it is a hypotensive state with increased cardiac output, supranormal GFR and expanded ECF volume, including plasma volume. Vascular resistance is reduced; it is not simply a matter of increased uterine perfusion acting as an AV shunt (Paller and Ferris, 1994). Nevertheless, one of the most feared complications of pregnancy is 'eclampsia'—convulsions due to hypertension. Pregnancy hypertension is a condition developing during the second half of gestation, particularly the later stages. It is also known as pre-eclampsia or toxaemia of pregnancy. It involves both hypertension and impairment of regional blood flow and utero-placental perfusion in particular (Terragno and Terragno, 1988). Other causes of hypertension may be superimposed on normal or pre-eclamptic pregnancy (Terragno and Terragno, 1988).

Before considering it further, a diversion into pseudo-comparative medicine is essential. Eclampsia occurs in bitches, but the common factor is the literal meaning—convulsions; canine eclampsia is a result of lactational demand causing hypocalcaemia. A more formidable trap arises from pregnancy toxaemia in sheep; it is an acetonaemic/ketotic state caused by calorie demands of pregnancy, especially with twins ('twin lamb disease'). Regrettably, failure to grasp this fundamentally different pathogenesis has subjected sheep to deliberate induction of pregnancy toxaemia, e.g. by superimposing starvation, in an effort to 'model' the human condition (Thatcher and Keith, 1988; Michell *et al.*, 1988). Such models are all the more futile since sheep, like other species, respond to starvation with natriuresis (Michell, 1981).

One of the earliest changes in pre-eclampsia is a fall in GFR—but this is masked by the supranormal GFR of normal pregnancy (Goldberg and Schrier, 1991). The early changes in cardiac output are very variable but once the condition is established, cardiac output is usually reduced (Lindheimer, 1993). Oedema, rather than hypertension, is the first sign and proteinuria develops later, potentially leading to nephrotic syndrome. Plasma volume is below normal for pregnancy and haemoconcentration can be severe. Similar degrees of hypovolaemia occur in pre-eclamptic women, whether or not they also have oedema (Dal Canton and Andreucci, 1986). A range of other disturbances occur, including coagulation abnormalities, hepatic dysfunction and renovascular damage. The condition particularly affects first pregnancies and twin pregnancies; the incidence increases as parturition approaches.

The suggested mechanisms remain numerous and include an abnormal immune response to pregnancy, leading to immune-complex mediated damage; placental underperfusion (though some regard this as a consequence); failure to sufficiently enhance prostacyclin production (which normally attenuates vascular reactivity during pregnancy and also inhibits platelet aggregation). At the same time, production of leukotrienes rises, increasing capillary permeability. Some regard the fundamental lesion as a generalised state of endothelial injury induced by unknown cytotoxic factors from ischaemic placenta and enhancing release of vasoconstrictors such as endothelin, while impeding the release of endothelially-derived relaxing factors including nitric oxide and prostaglandins (Goldberg and Schrier, 1991). The spiral arteries of the uterus may be especially vulnerable to defective prostacyclin production since they depend on it, together with morphological changes, to sustain the 40–50-fold increase in blood flow demanded by pregnancy (Terragno and Terragno, 1988). Inhibition of Na–K ATPase and increases of cell Na^+ or Ca^{2+} have also been suggested, as with other forms of hypertension, but with severe data conflicts. Others have suggested a primary change in calcium balance with reduced excretion and increased cellular uptake (Taufield *et al.*, 1987). Changes in Na^+–Li^+ countertransport are unlikely to be important, since it increases during pregnancy whether or not there is pre-eclampsia (MacPhail *et al.*, 1993). Heilmann *et al.* (1993), find a reduction of sodium extrusion from RBC by Na–K cotransport in pre-eclampsia; normally this increases during pregnancy and failure to do so could reduce the deformability of red cells.

Aldosterone, renin activity and angiotensin in pre-eclampsia reach intermediate levels, above normal but below those for normal pregnancy. Sensitivity to pressor effects of angiotensin II precedes the onset of hypertension. Pre-eclamptics show more sodium retention than during normal pregnancy and poorer excretion of sodium loads, despite their subnormal gestational aldosterone levels. This is attributed to defective renal escape from the effects of aldosterone: all these changes are just as likely to result from pre-eclampsia as to contribute to its causes (Goldberg and Schrier, 1991; Whitworth and Brown, 1993). In pregnant sheep, salt loading combined with reduced uteroplacental perfusion caused hypertension, whereas neither alone did so (Leffler *et al.*, 1986). Despite periods in the last 80 years when it has been fashionable either to salt restrict or salt load pregnant women, there is insufficient evidence to conclude that restriction is prophylactic or contradict the idea that it may at least reduce the likelihood of convulsions (Steegers *et al.*, 1991).

Two issues remain to be addressed in this chapter; the role of salt intake, or its moderation, in human hypertension and the lessons which may be available from comparative medicine. First, however, it is necessary to point out that, whatever its role in hypertension, sodium may well be an independent risk factor predisposing to left ventricular hypertrophy (Daniels *et al.*, 1990; Harmsen and Leenen, 1992). While this is often taken simply as an adverse result of hypertension, even as evidence for its existence, the relationship is less simple (Michell, 1994a). Changes in wave reflection may alter the pressure load on the left ventricle without affecting pressure measurements from conventional sites (Yaginuma and O'Rourke, 1993). Investigation is obviously hindered by the technical problem of sensitive detection of elevated arterial pressure and of ventricular thickening, by ultrasound or especially by ECG. Nevertheless, there is reason to suspect that ventricular hypertrophy can anticipate the development of hypertension and, perhaps through increased cardiac output, be involved in its onset (Michell, 1994a; Schunkert *et al.*, 1994). Alternatively, the link may lie in a genetic predisposition to both, perhaps through the ACE gene. The adverse effects of excess salt may also include those on calcium balance, as already noted, but these are not restricted to hypertension—they may contribute to loss of bone mineral, e.g. in post-menopausal women (*Chapter 9*)

Salt and Human Hypertension

The extent to which salt contributes to human hypertension remains a moot point; there are many well-informed advocates at both extremes of the argument, as well as intermediate views. An involvement of salt as an exacerbating or initiating cause of many cases of human hypertension seems beyond reasonable doubt. The real disagreement seems to be what action, if any, is warranted (Brown *et al.*, 1984; Dustan and Kirk, 1989). Unfortunately, the discussion often seems animated by something more akin to religious fervour than dispassionate debate (Michell, 1984a).

There are two types of evidence linking salt to hypertension in studies of humans:

1. acute intervention studies—these examine the beneficial effect of salt restriction in hypertensives or seek hypertensive effects of short-term salt loads;
2. cross-cultural—these compare levels of arterial pressure, age-related changes and prevalence of hypertension in different population groups according to their sodium intake.

The problem with acute intervention studies is their brevity; no one would expect persuasive evidence of links between smoking and cancer, alcohol and hepatic damage from short-term binges of smoking or inebriation. As Kurtzman (1985) remarked, to really test the link between salt and hypertension "you would have to show that a low salt diet prevented the development of hypertension in predisposed individuals if instituted early enough". The nearest we can approach is by studying low salt cultures—they are not inherently resistant because once they move to Western diets they become susceptible (Shaper, 1967; Poulter and Sever, 1994).

Of course, the possible differences in cross-cultural studies go far beyond diet, let alone dietary sodium—but not always; it may be no more than salt water vs fresh water for cooking (Page *et al.*, 1974). Among Polynesian populations, for example, an age-related rise in blood pressure was absent in those consuming less than 1 mmol/kg/day, whereas it occurred in the populations on higher intakes (Norton, 1992). There appears to be no exception to the rule that in low salt cultures hypertension is rare and blood pressure is stable despite advancing age; there are a few instances of really high salt cultures which apparently escape hypertension, but most are dubious (Michell, 1989c).

A major problem with intervention studies, apart from their short duration, is the recurrent issue of arbitrary classification of 'high' or 'low' intake and the lack of perspective on requirement. Thus, in a study (McCarron *et al.*, 1984) which supposedly showed that "higher intakes of Na were associated with lower mean systolic blood pressure and lower absolute risk of hypertension", among other problems (Michell, 1989c), intakes were estimated rather than measured, and they were well above requirement in both groups (>1 mmol/kg/day), yet the difference in intake between groups was small (12%).

There are, however, two issues in the salt–hypertension debate and of these, only one can be addressed by acute intervention studies; **whether or not salt restriction can help those already showing a trend towards hypertension**, by avoiding or reducing the need for medication. Important though this is, it is overshadowed by the value of knowing **whether lower levels of salt intake throughout a population and throughout life (including, perhaps, maternal intake) would reduce the prevalence of hypertension**; a public health prophylactic effect as opposed to individual therapeutic benefit. Because the causes of hypertension are multifactorial, failure to restrict salt intake may allow the age-related rise of pressure to obliterate other favourable aspects of lifestyle such as low alcohol consumption (Stamler, 1993). It is possible that vegetarian diets, as well as low salt diets, minimise the age-related rise in blood pressure. Historically, 'vegetarian' implied low salt but in 'accultured' societies, vegetarians often have similar sodium intakes to other individuals; their diets differ in a number of other factors suspected of influencing hypertension (Rouse and Beilin, 1984).

The definitive evaluation of the evidence on salt intake and hypertension is found in three papers by Law and collaborators (Law *et al.*, 1991a,b; Frost *et al.*, 1991). They reassessed the statistics and performed meta-analysis of cross-cultural studies involving 47,000 people from 24 communities, as well as 78 trials of salt restriction and 14 of differences of blood pressure within populations. The latter are usually the weakest data, since they tend to compare levels which are all far above requirement. The conclusion is that the link between excess dietary salt and increased prevalence of hypertension is both statistically and clinically significant. Objections have been raised (Swales, 1991), but these mainly reflect excessive expectations from acute intervention studies in normotensives and exaggerated fears of sodium depletion; they are well rebutted by Law *et al.* (1991c) and by Stamler (1993); the Law analysis is also supported by the appraisal of published studies by Elliott (1991). Despite this, controversy continues, for example, in a recent flurry of correspondence in *The Lancet* (cited by Michell, 1994a). The grudging acceptance of benefit—in such a widely prevalent and unpleasant disease—is remarkable.

Moderate reduction in sodium intake may have a minor effect upon blood pressure, **perhaps roughly equal to the effect of diuretics.**

(cited in Michell, 1984a: emphasis added)

Diuretics are still front-line therapy, they are expensive and, for some, their side effects are unpleasant. Those opposing moderation of salt intake frequently comment on the 'small' reductions of arterial pressure which are likely to result. It is worth emphasising, however, the benefits which may flow from even small reductions in pressure, e.g. a 40% reduction in stroke for a 5 mmHg fall in diastolic pressure (COMA, 1994).

What is strange in this war of attrition is that many of the protagonists struggle towards the concept that there may be a 'threshold' for an effect on pressure in the range 50–100 mmol/day, i.e. 0.7–1.4 mmol/kg/day (Muntzel and Drueke, 1992). This has been obvious for more than a decade (Freis, 1981), yet such protagonists still agonise over the effectiveness or otherwise of reducing intake to double the requirement from quadruple the requirement (70 from 160 mmol/day), let alone higher levels. The conclusion of the NKF Foundation on non-pharmacological management of hypertension (Anderson, 1993) is that the data are "imperfect but nevertheless compelling"; the advice of the American Food and Nutrition Board is that a sodium intake of 5–22 mmol/day (0.1–0.3 mmol/kg/day) is reasonable (Ernst, 1991). If lower intakes become more prevalent, there may be a need to examine any interactions with antihypertensive drugs (Alderman *et al.*, 1993).

The most fascinating illustration of the pitfalls of interpretation which arise from failure to take account of sodium requirement is provided by 'Intersalt' (Anon, 1988; Michell, 1989d, 1994f). This study involved 10,000 people drawn from 52 populations and is separate from the evidence of Law *et al.* (1991 a,b) who did not include it in their meta-analysis. Four centres were excluded because their intakes (inferred from excretion) were regarded as abnormally low: this only left centres with median intakes above 100 mmol/day (except one, 96 mmol/day). As a result, a significant positive correlation between systolic pressure and sodium intake was lost, moreover, inclusion of all 52 populations showed an age-related rise of arterial pressure which was absent in the four centres arbitrarily excluded. In fact, populations with intakes above 100 mmol/day had a prevalence of hypertension of 18%, whereas it was 5% or less in populations whose intakes were below 0.8 mmol/kg/day (Michell, 1989d). How low is that? It is still some 25% above the likely high extreme of mammalian sodium requirement *(Chapter 5)*. Subsequent refinement of the statistical analysis, to allow for 'regression dilution bias' suggests that reduction of sodium intake by 100 mmol/day would be slightly more effective than originally predicted (Dyer *et al.*, 1994).

The burden of proof in the continuing controversy over the role of salt in human hypertension should shift from those seeking to moderate sodium intake to those who continue to assert that the traditional surplus is harmless. "We live on a diet that has no current justification and we know little about the consequences of such large salt intakes" (Bulpitt, 1986).

Comparative Aspects of Hypertension

This discussion concerns naturally occurring hypertension as opposed to development of inbred models, notably in rats. Mostly, it refers to dogs because not only are they classic experimental models for human hypertension, but more is known about them. Much less is known about feline hypertension (Michell, 1993), although the greater preponderance of glomerular disease and more frequent reports of ocular damage consistent with hypertension suggest that it might be more prevalent than in dogs. The problem is that indirect techniques which work well in dogs are much more difficult in cats (Michell, 1993). Equine blood pressure is seldom measured clinically, except for anaesthetic monitoring, but hypertension occurs in laminitis, perhaps as a result of the effect of vascular injury, and one survey suggests that 8% of horses may be hypertensive (Michell, 1993). Sheep seem resistant to hypertension caused by salt-loading, even at levels up to 46 mmol/kg/day (Michell, 1985a).

Despite the dog's role as the subject of the original Goldblatt model of hypertension, few veterinarians graduating before the 1980s would have heard of it as a canine disease. Even now, blood pressure measurement is not part of the normal veterinary consulting room routine. Yet it always seemed likely that canine hypertension was a missing rather than a missed disease because dogs with experimental hypertension develop classical retinal lesions; although veterinary ophthalmology is a well developed clinical specialty, the condition is seldom encountered clinically (Michell, 1988c, 1989c). This would suggest that hypertension is unusual or that retinal damage is a late development, perhaps requiring large or sustained elevations of pressure.

An early study using direct measurements in 1000 trained dogs (Katz *et al.*, 1957) found only 1% with mean pressures above the supposed normal value (which was rather high). Similarly, Spangler *et al.* (1977) found a less than 1% prevalence of hypertension, this time with indirect (oscillometric) measurements, but again in young dogs. Recent studies suggest that the Dinamap oscillometric monitor offers excellent precision in conscious dogs, provided it is used preferably with a tail cuff rather than a limb cuff (Vincent *et al.*, 1993; Bodey *et al.*, 1994). It has, therefore, been used for an epidemiological survey of some 2000 pet dogs in various regions of the U.K. (Michell and Bodey, 1994). This confirms that canine hypertension is unusual, that it is seldom primary and that the causes of secondary hypertension include diabetes mellitus, thyroid dysfunction and renal disease, with higher pressure in Cushing's syndrome and hepatic disease. The possible link between hepatic disease and medullipin has already been discussed in this chapter. The link with hypothyroidism (the usual cause of thyroid dysfunction in dogs) may not only reflect loss of the stimulatory effect of thyroid hormone on Na–K ATPase *(Chapter 1)*, but increases in the concentration of endogenous inhibitors of sodium transport (Blaustein, 1993).

The hypothesis underlying our survey was that the real interest of the dog was its resistance, rather than its susceptibility to hypertension. A low prevalence of essential hypertension could simply be a matter of genetics, but since many of our concepts of renal physiology and of the pathophysiology of CRF are based on dogs, the acid test would seem to be whether dogs with renal disease are resistant to hypertension (Michell, 1994a). The early data suggest that dogs with a measured loss of GFR have a much lower prevalence of hypertension than humans with a corresponding decline in GFR (Michell, 1994a). If this conclusion is sustained, it will be important to investigate why this species, with similar renal function to humans and experiencing a comparable chronic excess of dietary salt, should be resistant (Michell, 1989c, 1994a). One factor may be more effective excretion of excess sodium, since dogs are resistant to the hypertensive effects of salt loading (unless stressed) and also to oedema, whether caused by salt loading or glomerular disease (Michell, 1983a, 1989c). Stress also restricts sodium excretion in rats, especially those with spontaneous hypertension and this results from changes in renal sympathetic nerve activity (Koepke, 1989).

Even more intriguing is the finding that normal dogs include breeds with average systolic pressures in the range which would be borderline hypertensive in humans and, therefore, many ostensibly normal individuals with systolic pressures around 200 mmHg (Michell and Bodey, 1994). Are these high pressures adaptive or maladaptive? Many of these are athletic breeds, although the high pressures are inherent rather than resulting from competition or the associated feeding and training. Is this essential hypertension, in the absence of evidence of harmful effects on renal, ocular or cardiovascular function, simply because other breeds may have average systolic pressures below 120 mmHg? Some differences have been identified in greyhounds, e.g. differences in baroreceptor function, lower peripheral resistance (due perhaps to a lower collagen:elastin ratio) and a higher cardiac index (Fischer *et al.*, 1975; Cox *et al.*, 1985). It is disturbing that so much 'classical' circulatory physiology has utilised such an atypical breed. The main question, however, is why these breeds seem so oblivious to the adverse effects of pressures which would be regarded with alarm and urgently treated in human patients.

The natural history of canine hypertension, therefore, suggests that dogs may be resistant not only to the condition but to its effects; precisely because the dog has served as a model species for human hypertension, these matters deserve serious consideration and further research.

Conclusion: Salt and Hypertension—Whose Decision?

> . . . *the fact remains that most animals, including man, crave salt in their diet. Because of this salt appetite it is difficult to enforce a low-sodium diet and, when it is attempted, compliance generally is very poor.*
>
> (Freis, 1981)

> *If equal evidence had related salt to a similarly fatal but far less common disease, cancer, it would have evoked intense campaigns against it, long ago.*
>
> (Dahl, 1972)

> *At every crossway on the road that leads to the future, each progressive spirit is opposed by a thousand men appointed to guard the past.*
>
> (Maeterlinck, cited by Dahl, 1972)

More than a few will be guarding the real forces in the industrial market place; those promoting and protecting the interests of the major investors in the richest sources of profit. Often cited as 'wealth-creators', they undoubtedly create wealth for governments in the form of tax—whether on salt in former centuries or tobacco in this one—hence there is a common interest, provided that not too many of the electorate are upset, for example by a perceived threat to health:

> . . . *in public affairs as well as in private affairs, and for the same reasons, we are subject to contrivance that serves the industrial system.*
>
> (Galbraith, 1967)

Food fads are a menace. Even 'quality' bookshops have far more shelves devoted to alternative medicine and trendy diets than genuine information on medicine or public health. In part, the medical community is to blame; as hypotheses on fibre, lipids, antioxidants, wax and wane, with full exposure in the health columns and popular science programmes, even the morning news shows, the perception is that one idea is much as valid as another and that most have a short 'shelf-life'. This is inevitable when only a minority of the electorate and a scarcity of politicians have any genuine insights into the nature of biomedical research and, in particular, the character of scientific evidence.

Market forces are at their worst in health-care because the basis for informed consumer choice is infinitely more complex than with instant coffee and the consequences of error are far more serious than with cars or CD players.

> . . . *the selling of goods—the management of demand for particular products—requires well-considered mendacity.*
>
> (Galbraith, 1967)

Much of the food fad literature is an unsubstantiated source of distraction or confusion in areas where the realities are already complicated enough. It is simply not possible to do lifelong nutritional experiments on people and data from experimental animals always leave a question mark. So the evidence—as in much biomedical science—concerns the balance of probabilities, whereas politicians and non-scientists expect prescriptive advice based on scientific certainties. An added complication is that nutritional choices, unlike consumer durables, are inherently irrational, frequently having their origins in cultural beliefs far older than science (Michell, 1984a).

So who should decide? Is the salt controversy properly a matter of public health or *laissez faire?* It is both—as is tobacco. In both cases, there is a consensus that public education is needed, though differences as to how vigorous that should be. It is also increasingly acknowledged that exposure to risk should not be inflicted by passive smoking. The problem is with passive salt consumption. The majority of sodium intake in 'Western industrialised' countries is unsolicited: added salt may be as little as 15% (Bull and Buss, 1980; Larsen, 1982; Sanchez-Castillo *et al.*, 1987; Edwards *et al.*, 1989; Elmer *et al.*, 1991; Stamler, 1993). Moreover, of the sodium contained in food, the majority is added salt in processed foods, not naturally contained sodium. The justification is usually related to facilitation of manufacturing procedures, improvement of texture or palatability and—even at the end of the twentieth century—preservation (Bauman, 1982; Marsh, 1982). While salting for preservation may have cultivated our appetite for salt for many centuries, it is scarcely the best available option at the end of this one.

Consumer acceptability, of course, is a real problem. A healthier but less attractive product simply will not sell unless customers have strong views on the relevant health issues. The problem lies in the fact that, particularly with salt, our expectations are indelibly coloured by our experience, especially our early experience. Our taste for salt, our appetite for it and its effect on our arterial pressure are crucially dependent on early exposure and perhaps on that of our parents (as discussed in this chapter and *Chapter 4*). It is insufficient to allow adolescents or adults to reach their own conclusions—the damage may already be done (Grobbee, 1994). Mothers choose baby food that seems tasty—to them—without realising that adult tastes are highly developed by experience; as a result, food that contains sodium levels safely consistent with requirement seem insipid. This was exactly why it needed regulation rather than market forces to reduce the salt content of infant foods (Denton, 1982). Even during the first six months of life, salt intake significantly affects blood pressure (Hofman *et al.*, 1983). By the time we are adults, salt has, for many, become an addiction (Michell, 1984a) to the extent that they cannot reduce their intake, even when they have first class medical reasons for doing so.

Why is it so difficult to maintain a low sodium intake?

(Dahl, 1972)

The counterargument is that, whereas a reduction of tobacco use harms no one's health, a reduced salt intake may do so (Laragh and Pecker, 1983; Swales, 1990). If so, it is likely to be a small minority, in unusual circumstances (Michell, 1980b, 1984b; De Wardener and Kaplan, 1993), because significant reductions in arterial pressure are attainable through substantial reductions of intake, without sacrificing ECF volume, plasma volume or cardiac output (Fagerberg *et al.*, 1985). Fears have also been expressed that low salt diets may elevate blood lipids. If so, it happens only with extreme restriction (0.3–0.7 mmol/kg/day) and not at 1.2 mmol/kg/day (Del-Rio and Rodriguez-Villamil, 1993; Ruppert *et al.*, 1993). No one is advocating reductions below requirement, merely elimination of the vast casual excess intake (Michell, 1994a). Moreover, even this is left to choice—no one suggests a ban on salt cellars. The problem, at present, is that a consumer who believes that the balance of evidence leads them to reduce their salt intake faces great problems in doing so. Even when foods are labelled, the data are often uninterpretable. Worse, as we have seen in *Chapter 5*, many clinicians, even many research workers, have little idea of the actual requirement. To allow informed consumer choice, therefore, we need:

1. clearer understanding of sodium requirement among clinicians, nutritionists and health workers;

2. widespread, clear and intelligible labelling of foods, e.g. in mmol/100 g;
3. sounder public education, especially of prospective parents;
4. real pressure on manufacturers to reduce the sodium content of their products to a genuine practicable minimum.

If consumers prefer more—let them add it to taste, as long as they understand the probable consequences.

With considerable pleasure, almost the last reference incorporated into this book was COMA (1994). It is a report from the body appointed to advise the U.K. government on diet and health. The recommendations include a one-third reduction in salt intake and, in particular, less use of salt in processed foods. They appeared in the national newspapers of the eleventh day of the eleventh month; it is to be hoped that they will, therefore, become the subject of widespread remembrance.

7

Endocrine Effects on Normal and Abnormal Sodium Excretion

Introduction

It is traditional to see hormones in a regulatory role and to associate it with their concentration in plasma. More recently, for various hormones, it has been realised that local effects may be as important or more important than those associated with plasma concentration. It has also been realised that plasma concentration may change rapidly in response to pulsatile patterns of secretion. Plasma concentrations are affected by rates of metabolic clearance as well as secretion rates and the effects of a given concentration depend on the responsiveness of the target tissue, notably the receptor population (which may be up- or down-regulated) and the activity of second messengers (such as cAMP, cGMP and calcium). For example, high salt intake not only increases atrial natriuretic peptide (ANP) secretion but also the receptor density (Shapiro and Dirks, 1990). In the last decade, molecular biology has taught us much about the genetic factors and structural features which determine both the properties of the hormone and the responsiveness of its receptors.

Important though they are, these are perhaps matters of detail or refinement. What is often less clear than we assume is the regulatory importance of the hormonal system. With some it is obvious: the role of the renin–angiotensin–aldosterone axis in relation to regulation of sodium balance and arterial pressure, and that of aldosterone in relation to potassium balance, are characterised by their responsiveness to physiologically plausible stimuli and the appropriateness of their effects in the context of a negative feedback loop. Deficiency of the hormones is reflected in dysregulation or recognised clinical disturbances. With other hormones the position is unclear; beyond the period of active bone growth it is hard to be persuaded of an essential role for calcitonin in calcium regulation.

155

It also affects renal sodium excretion, but we are unaware of a feedback loop, i.e. calcitonin has not been shown to respond to aspects of sodium balance. It is, however, possible that the relevant stimuli have not yet been identified, or even the main target tissue. Are there, for example, effects on bone sodium or enteric sodium transport? Conversely, while the regulatory role of ANP has seemed attractively self-evident in allowing the heart to 'offload' excessive preload (by reducing plasma volume) or afterload (by reducing arterial pressure), the actual physiological importance of these properties has been questioned (see below).

While we should not assume, as was once the tendency, that almost any effect of a hormone is regulatory, however improbable the dosage required to elicit it, we should also be cautious about dismissing hormonal effects as merely incidental. The same might once have been said of the kaliuretic effects of aldosterone, readily viewed as a troublesome side effect of a salt-conserving hormone. It is also possible for hormones to reveal effects which are residual rather than currently important because evolution has tended to find new uses for ancient molecules. Prolactin, for example, became a major regulator of salt and water balance in non-mammalian vertebrates, long before such properties could appear relevant to the demands of lactation (Hirano, 1986). Indeed, the evidence for a major role in mammalian fluid balance is weakened by early experiments using preparations contaminated with ADH (Barron, 1987). Nevertheless, prolactin stimulates Na–K ATPase activity in the ascending thick limb of the loop of Henlé and it may have a particular role in maintaining the relatively high fluid space of newborn animals (*Chapter 2*). Otherwise, its role in water and electrolyte balance in mammals remains controversial (Brown and Brown, 1987).

While comparative medicine is often founded on the assumption that one species can serve as a more convenient model for another, the heart of the subject is truly species diversity. Comparisons of differences are as instructive as confirmation of similarities and it is all too readily assumed that what happens in rats, rabbits, dogs and people stands for all species. It is, for example, widely believed that oestrogens are consistently salt-retaining hormones, but this is untrue (see below).

Are the kidneys the main regulators of water balance? The question seems foolish until we consider dogs (O'Connor and Potts, 1988). They normally tend to produce concentrated urine and vary their intake rather than their output, according to their needs. This is undermined if they are offered tasty solutions rather than water, when they behave more like people—drinking for the flavour and excreting the excess. Sheep also provide evidence of regulating their water intake rather than urine output, although, being excellent urinary concentrators, they can manage with very little water indeed if their forage has an adequate moisture content. The significance of renal water regulation, therefore, depends on the level of fluid intake.

The same is true, in herbivores, of renal sodium excretion (*Chapter 3*); its importance depends on sodium intake. More important, in all species, the characteristics of renal function depend on sodium intake and so does the endocrine 'climate' represented by a considerable range of hormones. Since there is no self-evident normal set point for body sodium, rather a dependence on sodium intake (*Chapter 5*), and since sodium intake in many experiments is quite arbitrary or even unstated, there are particular problems in relating the significance of hormonal changes to sodium regulation. The answer to many questions related to sodium depends on the intake at which the question is put. For example, sodium loss increases late in pregnancy in sheep, but only if their intake is excessive (Michell *et al.*, 1988).

It is easy to assume that hormonal changes in pregnancy relate to the physiological demands. If so, they need not relate to the demands which we perceive. For example, is the expansion of extracellular fluid (ECF) volume in normal pregnancy a physiological need or a reflection of generous sodium intake? Is the increase in glomerular filtration rate (GFR) physiological? At the very least, it increases renal energy requirements by necessitating reabsorption of the increased filtered load of sodium. We have already seen (*Chapter 4*) that it is reassuring to our homeostatic instincts that salt appetite increases during pregnancy in rats, but it increases far more than is required for the actual demands of pregnancy and in sheep it does not increase at all. Pregnancy creates many demands, particularly on the cardiovascular and endocrine system, and these may secondarily affect sodium regulation. There are also changes in sodium balance during the reproductive cycle and a reproductive role for these is even less clear except, perhaps, in species such as rats and dogs where pseudopregnancy is often a feature of the cycle.

With these caveats, it is necessary to consider the effects of hormones which either regulate or alter sodium balance. Among the latter are **parathyroid hormone (PTH)** and **calcitonin**, both of which increase sodium excretion despite their opposite effect on calcium excretion; the renal effects of calcitonin may be pharmacological rather than physiological (Koeppen, 1990; Bourdeau and Attie, 1994). Those of PTH probably reflect inhibition of both Na–proton exchange and Na–K ATPase in the proximal tubule (Bertorello and Katz, 1993). **Insulin** stimulates Na–K ATPase and may therefore cause sodium retention (*Chapters 1 and 6*) as well as providing a defence against hyperkalaemia; the latter directly stimulates insulin secretion (Williams and Epstein, 1989). Endothelins have a biphasic effect on sodium reabsorption in the proximal tubule; initially increasing it but inhibiting at higher concentration (Garcia and Garvin, 1994). **Epidermal growth factor (EGF)** is produced by the thick ascending limb of the loop and can induce natriuresis as well as inhibiting the effect of ADH on the collecting duct (Bankir *et al.*, 1989). It also causes secondary polydipsia. At cellular level it has a range of effects, including promotion of Na–H exchange and mitogenesis *(Chapter 10)*. It is possible that the diuretic effects of EGF are indirect and mediated by prostaglandins (Gow and Phillips, 1994). EGF also promotes enteric sodium uptake *(Chapter 3)*. The role of **prostaglandins** in renal water and electrolyte excretion is also uncertain, however, and while they clearly affect it, their precise role in regulatory mechanisms is far from clear (Knox and Granger, 1992). The main hormones involved in sodium regulation are angiotensin II, aldosterone, atrial natriuretic peptide and endogenous inhibitors of sodium transport originating from the adrenal cortex or the hypothalamus; the latter have already been discussed in *Chapters 1 and 6*. **Kinins, prostaglandins and catecholamines**, including **dopamine**, are also important in the modulation of renal sodium excretion, although their role is less clear (Kirchner and Stein, 1994). It is likely that their effects depend on local, intrarenal generation rather than circulating levels (Knox and Granger, 1992). A possible role for **erythropoietin (EPO)** has recently emerged; volume expansion caused by salt loading reduces packed cell volume (PCV) and EPO production increases; it causes further salt retention as a result of intrarenal generation of angiotensin II. This vicious cycle could contribute to its hypertensive effects (Brier *et al.*, 1993; Naomi *et al.*, 1993; Bunke *et al.*, 1994). In evolution, however, the salt retention might have been advantageous in conditions combining hypovolaemia with anaemia, e.g. chronic parasitic diseases. Anaemia causes sodium retention despite a high cardiac output (which maintains oxygen delivery) and a reduced peripheral resistance—some have ascribed the sodium retention to a response to vasodilatation (Anand *et al.*, 1993). There is another potential interaction between EPO and renal function; PCV is the main determinant of blood viscosity and, while colloids have received extensive attention as physical factors affecting tubular absorption, the importance of viscosity has been neglected (Cowley and Roman, 1989), at least in recent years. Glomerular filtration inevitably increases PCV and blood viscosity within the proximal tubule.

Adrenal Steroids

The adrenal cortex produces a range of steroids, including some which are intermediates rather than functional hormones. Apart from androgenic and oestrogenic steroids, the main hormones are the glucocorticoids (cortisol, corticosterone, cortisone, 11 dehydro-corticosterone) and the mineralocorticoids (aldosterone, deoxycorticosterone [DOC], and hydroxydeoxycorticosterone). Progesterone is not an adrenal hormone but a vital precursor with sufficient structural similarity to compete with aldosterone and, by having less mineralocorticoid effect, to antagonise its effects. Mineralcorticoids promote sodium retention and potassium secretion in a variety of tissues (Fig. 7.1)

At one point, it seemed that the specificity of responses to adrenal steroids rested on the ability of mineralocorticoids and glucocorticoids to preferentially (but not solely) occupy Type I and Type II receptors, respectively. It was long known that glucocorticoids at high dosage produced mineralocorticoid effects, but this could be attributed to occupancy of the alternative receptor.

The question of receptors is especially important with steroid

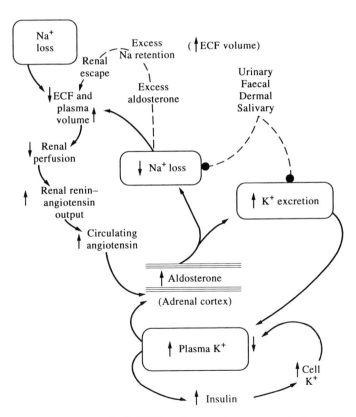

FIG. 7.1. Extracellular fluid (ECF) volume, aldosterone and plasma K^+.

hormones, since they have a distinctive mode of action which does not rely on second messengers. Lipid solubility allows them to enter cells and it is the hormone–receptor complex which enters the nucleus and activates the response; this results in the synthesis of new protein. This pattern was first revealed for aldosterone and one of its activation products is Na–K ATPase (Verrey, 1990), although this is not its only effect on sodium transport *(Chapter 1)*.

During the 1980s, it became clear that mineralocorticoid receptors were less specific than had initially been assumed. In fact, since they had no differential affinity for mineralocorticoids or glucocorticoids and the latter circulate at higher concentration, it was hard to see how Type I receptors could function as mineralocorticoid receptors (Funder, 1990). So far, the search for further mechanisms conferring specificity has yielded clues rather than definitive answers. The initial idea that specificity was conferred, at least in kidney, by a plasma protein (transcortin) was incorrect. The idea that affinity was similar but the subsequent effects different also proved hard to sustain. It appears more likely that *in vivo* there are mechanisms which exclude glucocorticoids from mineralocorticoid receptors, at least in some tissues, notably kidney. In particular, in tissues dependent on mineralocorticoid responsiveness for their function (e.g. kidney, salivary gland, colon), glucocorticoids are converted to metabolites which cannot occupy the Type I receptor; this is done by an enzyme, 11β-hydroxysteroid dehydrogenase, 11-HSD (Funder, 1990).

This sequence of events is worth recounting because the discovery which provided the key was a rare clinical syndrome combining symptoms of apparent mineralocorticoid excess with measurable suppression of aldosterone secretion. The answer was not, as first thought, a novel mineralocorticoid and attention turned to mechanisms which might enable cortisol to act as a mineralocorticoid. The decisive clue was, again, clinical in origin—the ab.!ity of liquorice to produce symptoms of mineralocorticoid excess. Originally attributed to weak agonist activity, this effect of liquorice rested on its ability to block 11-HSD and the absence of this enzyme provided an explanation for the syndrome of apparent mineralocorticoid excess (Funder, 1990). The anti-ulcer drug, carbenoxolone, has similar effects *(Chapter 11)*. The explanation may be incomplete, since some patients do have evidence of 11-HSD and it is possible that other enzymes are involved in metabolic exclusion of glucocorticoid from Type I receptors (White, 1994).

Further clues regarding receptor specificity came from another clinical condition, pseudo-hypoaldosteronism, in which Type I receptors are lacking. Several lines of evidence show that in these patients glucocorticoids can elicit mineralocorticoid effects, but from

Type II receptors (Funder, 1990). This suggests that the effects of the hormone–receptor complex on transcription may involve a pathway with components shared not only by glucocorticoids and mineralocorticoids but potentially by other steroids.

The boundaries of receptor specificity still remain blurred, with differences between tissues and, for example, the existence of a much wider distribution of Type I receptor sites in the nephron than anticipated on functional grounds. Whether they are all true Type I receptors or Type II receptors 'illicitly' occupied by aldosterone remains a moot point (Hulter, 1994). Moreover, the presence of the receptor needs the coexistent presence of 11-HSD to allow it to function specifically as a mineralocorticoid receptor. It is also important to consider that the receptor population is not static and, for example, it remains possible that the reduced population of Type I receptors in pseudohypoaldosteronism is a response (downregulation) rather than a cause (White, 1994). Evidence has also begun to emerge for further types of receptor and for the ability of aldosterone metabolites to regulate receptor sensitivity, including differences between sexes (Hulter, 1994).

We therefore understand better what we have long understood— that the specificity between mineralocorticoids and glucocorticoids is relative rather than absolute—but we still cannot provide a complete explanation. Further specificity is conferred by differences between cells concerning which genes are regulated by the hormone–receptor complex and this may also change during cell differentiation (Verrey, 1990).

Having considered what enables a mineralocorticoid to elicit its effects, it is necessary to discuss the outcomes of receptor activation. Since aldosterone acted as a prototype for general concepts of the steroids' mode of action, the focus of the discussion becomes the regulation and effects of aldosterone secretion.

Aldosterone: Regulation and Effects

Na–K ATPase is an aldosterone-induced gene product which is also activated by aldosterone *(Chapter 1)*. In fact, aldosterone affects the number of pump sites and their activity by increased leakage of sodium into cells and by induction of citrate synthase, a key Krebs cycle enzyme, potentially influencing the supply of ATP to the pump (Hulter, 1994). While the latter may contribute to the effects of aldosterone, it is not a prerequisite, the response to increased leakage being more important (Verrey, 1990; Marver, 1992). It is merely one illustration of the complexity of these regulatory mechanisms; not only are they numerous but interdependent and therefore likely to produce atypical responses in simplified experimental systems. Moreover, while there are excellent reasons to believe that the renal effects of aldosterone are well represented by those on amphibian epithelia, the most rigorous evidence still comes from the models (Verrey, 1990).

Glucocorticoids also stimulate Na–K ATPase; apart from the direct effects they are liable to have secondary effects in the nephron because they increase GFR and hence the reabsorptive load of sodium (Katz, 1990). They also stimulate Na–H exchange in the proximal tubule, hence increasing intracellular sodium (Kinsella, 1990). While increased entry of sodium into cells is also almost certainly one component of the renal Na–K ATPase response to aldosterone, evidence for a direct effect is substantial (Katz, 1990). The evidence is complicated by the use of amiloride [to restrict inward leakage of sodium into intracellular fluid (ICF)] at high dosage, when it also inhibits Na–K ATPase. The outcome of increased Na–K ATPase activity, particularly in the cortical collecting duct, is to promote the reabsorption of sodium and the secretion of potassium. The latter may be augmented (at least in some species) by a secondary hyperpolarisation resulting from increased pump activity. This hyperpolarisation allows increased entry of potassium into cells, hence facilitating its secretion (O'Neil, 1990). The latter is further facilitated by increased electronegativity of the lumen resulting from the stimulation of sodium reabsorption (Delmez *et al.*, 1990).

Apart from these 'classical' effects, and those on urinary acidification, including enhanced secretion of both protons and ammonia buffer (Stone *et al.*, 1990), aldosterone also binds to a membrane receptor and adrenal steroids exert a tonic or permissive role in activation of the adenyl cyclase system (Hulter, 1994; White, 1994). Some of these effects appear to be independent of gene transcription.

Because the kidney has multiple sites of sodium reabsorption, it can escape from salt-retaining effects of excess aldosterone, simply by reducing sodium reabsorption from other nephron sites, in response to ECF volume expansion (Fig. 7.1): other tissues do not escape. The kidney **cannot escape from the kaliuretic effects**; indeed, they are liable to be enhanced by increased delivery of sodium to the distal nephron, thus facilitating excretion of potassium. Excessive hydrogen ion secretion also continues, hence the usual clinical picture of hypokalaemia and potassium depletion with metabolic alkalosis (conditions which reinforce one another; Michell *et al.*, 1989; Hulter, 1994), but the absence of oedema. Hypertension results probably from increased ECF volume, but also other factors (Delmez *et al.*, 1990).

The nephron segments involved in escape include both the loop and, as expected, the proximal tubule, hence there is the likelihood of increased losses of other ions, such as calcium. The distal nephron does not escape, but continues to absorb sodium (Gonzales-Campoy *et al.*, 1989). Apart from ECF volume, the stimuli for escape include an increase in ANP secretion, increased renal perfusion pressure, suppression of renin and decreased renal sympathetic nerve activity (Hulter, 1994). Both the effects of ANP and perfusion pressure on escape depend on responses in deep rather than superficial nephrons. As well as renal perfusion pressure, increased renal interstitial pressure contributes to escape (Haas and Knox, 1990). This results from reduced intrarenal generation of angiotensin II and renomedullary vasodilation (Ruilope *et al.*, 1994). Dopamine is not involved in the mechanism of renal escape (Coruzzi, 1994), but in view of the similarity to pressure natriuresis, nitric oxide and prostaglandins probably are (Ruilope *et al.*, 1994). Changes in renal nerve activity are also likely to contribute (Gonzales-Campoy *et al.*, 1989).

The stimuli to increased aldosterone secretion (Fig. 7.1) are relatively simple:

1. A direct effect of hyperkalaemia on the adrenal cortex (glomerulosa cells). This does not require increased angiotensin but complete suppression of angiotensin, e.g. by angiotensin-converting enzyme (ACE)-inhibitors, blocks the response (Hulter, 1994).
2. Increased plasma angiotensin secretion.

The importance of aldosterone as a potassium-regulating hormone is often under-emphasised in the traditional focus on sodium conservation. Not only is secretion directly stimulated by hyperkalaemia, but potassium is excreted into faeces, sweat and saliva, as well as urine, depending on species. In addition, it is transferred into cells (Williams and Epstein, 1989; Krishna, 1990): these responses add up to a powerful defence against hyperkalaemia (Fig. 7.1). The importance of this role for potassium is more obvious in herbivores since their diets are rich in potassium, whereas most 'Western' human diets are low in potassium (Michell, 1978a). In various diseases, however, hyperkalaemia is a potent threat. The idea that potassium—rather than sodium—regulation may be the main role for aldosterone is strengthened by the observation (in dogs) that blockage of angiotensin II prevented them from conserving sodium during sodium depletion, without affecting aldosterone (Knox and Granger, 1992).

The importance of adrenocorticotrophic hormone (ACTH) is as an acute stimulus rather than one which chronically alters aldosterone output. It is involved in the increase of aldosterone secretion caused by stress and perhaps in diurnal changes (Seeley and Laragh,

1990; Hulter, 1994; White, 1994). Other subsidiary stimuli to aldosterone secretion include metabolic acidosis, PTH and reduced ECF osmolality; secretion is reduced by hyperosmolality and by dopamine. The effect of osmolality on aldosterone secretion is insensitive compared with the effect on ADH secretion (*Chapter 2*), thus changes equivalent to more than 5 mmol/l of Na^+ are required (Taylor *et al.*, 1987; Espiner and Nicholls, 1993). It is also insensitive compared with the effect of potassium; a fall of plasma sodium from 149 to 136 mmol/l had only a quarter of the stimulatory effect on aldosterone excretion compared with a rise in potassium of 0.7 mmol/l (Blair West *et al.*, 1963). In as far as plasma sodium concentration has only a limited effect on aldosterone secretion, increases have more influence than decreases, although sensitivity to sodium is increased by angiotensin II (Merrill *et al.*, 1987, 1989). In the absence of angiotensin II, sodium depletion fails to enhance aldosterone secretion, emphasising the minor role played by plasma sodium as a stimulus (Mitchell and Navar, 1989). Hypernatraemia may also suppress release of renin (Hollenberg *et al.*, 1994). The renin–angiotensin system is not only a regulator of aldosterone but also, independently, of renal sodium excretion.

Renin–Angiotensin System (RAS)

While many tissues have localised renin–angiotensin systems, the only significant source of circulating renin is normally the kidneys and its only known physiological effect is to promote the formation of angiotensin II, hence the two together act as if they were a single hormone. It is conceivable that the renin precursor molecule (prorenin) could have regulatory effects within the kidney, but this remains a matter for future research (Laragh and Seeley, 1992). Some of the ambiguities concerning the factors which regulate renin release may arise precisely from the fact that, in the case of the kidneys, the systemic as well as the local effects of the renin–angiotensin system are important.

Apart from regulating aldosterone secretion, the RAS, through angiotensin II, stimulates thirst, sustains arterial pressure in hypo-volaemia, and has direct effects on renal sodium excretion; these direct sodium-conserving effects may be at least as important as those mediated by aldosterone (Hall and Granger, 1994). They result from reduced peritubular capillary pressure (efferent arteriolar constriction) and enhanced sodium transport, particularly but not exclusively in the proximal tubule. By stimulating salt transport at the macula densa, angiotensin may sensitise tubuloglomerular feedback. The net effect is that suppression of angiotensin II, by removing a powerful stimulus for salt retention, greatly sensitises pressure natriuresis (Hall and Granger, 1994). Some have argued that angiotensin II also promotes antidiuretic hormone (ADH) secretion; the evidence for the physiological importance of such an effect is dubious (Chowdrey and Lightman, 1989), though it may potentiate the response to osmotic stimulation (Michell, 1991; Robertson, 1993). On the other hand, ADH suppresses renin release, both by raising intracellular calcium and, eventually, by the expansion of plasma volume (Laragh and Seeley, 1992). Thus, whether or not it is important, there is the basis for a negative feedback loop between renin and ADH, via angiotensin II.

Angiotensin II is one of the most potent vasoconstrictors known and also constricts glomeruli (Hollenberg *et al.*, 1993). It is clear that the RAS plays a pivotal role in the defence against hypovolaemia and hypotension through its effects on vascular resistance, sodium excretion and water intake, although, curiously, the pressor effects of angiotensin are least potent in these conditions, where they seem to be most needed (Laragh and Seeley, 1992). The vasoconstrictor effects of angiotensin II, like those of ADH, are important for survival of severe haemorrhage (Schadt, 1994). The RAS is also involved in some forms of hypertension (Laragh and Seeley, 1992), notably those involving compromised renal perfusion. It has additional metabolic effects (Mitchell and Navar, 1989) and ACE-inhibitors have effects which are not attributable to those on blood pressure or fluid and electrolyte balance (Michell, 1995b). Caution is required, however, in interpreting such side effects. Not only might they arise from properties other than ACE-inhibition (which also inhibits breakdown of kinins,) but they could also reflect changes in potassium balance; although potassium regulates aldosterone secretion, independent of angiotensin, extreme RAS suppression does curtail potassium excretion (see above). Cellular potassium depletion has wide-ranging effects on metabolism and as aldosterone levels fall, the amount of exchangeable potassium at a constant plasma potassium concentration also falls, i.e. almost certainly intracellular potassium is reduced (de Fronzo and Smith, 1994). While the original research on the RAS concerned the systemic consequences of increased renin release from the kidney, there has recently been a rapid growth in the awareness of the presence of the RAS system in a variety of tissues, including blood vessels, heart and brain, and of its potential for local as well as systemic effects (Ganten *et al.*, 1989).

The renal origin of renin, from the juxtaglomerular cells of the afferent arteriole, makes it well placed to respond to changes in renal perfusion pressure and, via feedback from the macula densa, distal delivery of solute. Hence changes in renin secretion contribute to autoregulation of glomerular perfusion, maintenance of GFR and glomerulotubular balance, as well as responses to hypotension or hypovolaemia. These responses are actually to angiotensin II: renin accelerates the generation of angiotensin I from its hepatically synthesised precursor, renin substrate. The lungs, with their high rate of perfusion, are the main site of conversion of angiotensin I to angiotensin II (A-II) by ACE. Angiotensin II has its own direct effects on sodium excretion (i.e. independent of aldosterone), decreasing it at low dose and increasing it at high dose (Hall *et al.*, 1990). The sodium conserving effect is mainly due to stimulation of Na^+-H^+ exchange in the proximal tubule (Rouse and Suki, 1994). In some pathological states, e.g. chronic renal disease or cirrhosis, the natriuretic effect may predominate and it is not caused by increased arterial pressure but by reduced distal reabsorption of sodium (Seeley and Laragh, 1990). Apart from effects on blood pressure and water and electrolyte balance, angiotensin II is also a growth factor for renal cells and may, therefore, be involved in tubular hypertrophy (Wolf and Neilson, 1993).

Renin release is also enhanced by β-adrenergic or dopaminergic stimulation and a reduced intracellular calcium concentration in juxtaglomerular cells; this contrasts with many other hormones which are secreted in response to increased ICF calcium (Mitchell and Navar, 1989; May and Peart, 1989; Seeley and Laragh, 1990; Hollenberg *et al.*, 1994). α-Adrenergic stimulation reduces renin secretion (Mitchell and Navar, 1989). It is suppressed by ANP (see below) and by angiotensin II, a negative feedback response. Although we take it for granted, it is a curious mechanism because it depends merely on angiotensin itself, not its ability to correct the initiating disturbance (Laragh and Seeley, 1992). Protein intake increases renin secretion and, although the exact mechanism remains uncertain, this may contribute to the characteristic increase in GFR. ACE inhibition, while reducing A-II, increases renin secretion as a result of reduced feedback. Strangely, renin suppresses hepatic production of renin substrate whereas A-II increases it; renin substrate (angiotensinogen) may be rate limiting in A-II production (Hollenberg *et al.*, 1994). Both neural and non-neural stimulation of renin release is substantially mediated by prostaglandins (Osborn and Johns, 1989). The neural control of renin release is a key factor in mediating responses to signals from cardiopulmonary receptors and also psychological stimuli such as stress or anticipation of exercise (Laragh and Seeley, 1992).

Apart from A-II, an octapeptide, the subfragment A-III (a heptapeptide) also inhibits renin release, stimulates aldosterone secretion and causes renal vasoconstriction, but its physiological importance is uncertain (Hollenberg *et al.*, 1994). Sodium intake not only influences the plasma concentration of A-II, but also the receptor population in vascular smooth muscle and the adrenal cortex. Sodium depletion sensitises the adrenal cortex, whereas vascular sensitivity increases with salt intake (Hollenberg *et al.*, 1993). This means that following acute hypovolaemia, arterial pressure can be sustained by less and less angiotensin as its effects on renal sodium conservation become obvious (Laragh and Seeley, 1992). The effect of s..lt intake on renin secretion almost certainly involves increased ECF volume and its effects on renal perfusion and renal nerves (Benabe and Martinez-Maldonado, 1993; Kopp and Di Bona, 1993), but could also involve sodium and chloride delivery to the macula densa. Plasma sodium concentration is an unimportant influence in the physiological range (Hollenberg *et al.*, 1993). The separate effects of potassium and the RAS system are emphasised by the fact that hypokalaemia, rather than hyperkalaemia, stimulates release of renin. Very high levels of potassium intake suppress renin secretion but the detailed relationship between potassium and the release of renin is complicated (Laragh and Seeley, 1992).

The relationship between macula densa stimulation and renin release is also complex, the linkage to the afferent arterioles involving the extraglomerular mesangium and possible mediators such as adenosine, prostaglandins and nitric oxide. Despite this complexity, it seems clear that the basic stimulus to the macula densa is decreased luminal concentration of chloride, rather than sodium, and associated changes in chloride transport (Mitchell and Navar, 1989; Lorenz *et al.*, 1993). There is increasing evidence for the importance of locally formed angiotensin II, as opposed to circulating angiotensin II, in the control of sodium excretion (Hall, 1986).

ANP

The discovery of ANP converted our perception of the heart from being merely a very finely regulated pump to a self-regulating endocrine gland. There is no other example of such rapid progress from discovery of a radically different hormone to its identification, sequencing and synthesis within three years (Genest, 1986; Cantin and Genest, 1987). Like calcitonin, however, doubt has begun to emerge whether its theoretical role is matched by its actual importance (Goetz, 1990; Richards, 1990; Michell, 1991).

ANP secretion is mainly from the atria in response to stretch, not necessarily pressure (Sagnella and MacGregor, 1994). There is a lower secretion rate from the ventricles, but this can increase in heart failure

(Kenyon and Jardine, 1989). Data from cardiac transplant patients show that cardiac denervation does not prevent ANP responses to changes in sodium intake (Singer *et al.*, 1994). Atrial tachycardia increases ANP secretion, but natriuresis is not always marked and clinical sodium depletion does not result (Briggs *et al.*, 1990; Espiner, 1994). Right-sided overload reduces both preload and afterload through the effects of ANP. Many of these present ANP as the functional antagonist of aldosterone, opposing its effects (sodium retention, increased blood pressure) and suppressing its secretion. ANP does not, however, suppress the direct effect of aldosterone on sodium transport mediated by Na–K ATPase (De Bold *et al.*, 1988), i.e. it produces responses which oppose its actions rather than blocking them. Moreover, it does not underlie the phenomenon which alerted De Wardener and others to the likely existence of natriuretic hormones, the natriuretic response to saline infusion (Cowley and Roman, 1989; Michell, 1991). In fact, ANP receptors have not been found in the proximal tubule (Shapiro and Dirks, 1990), though ANP could still affect sodium reabsorption there, via interactions with angiotensin II (Hall and Granger, 1994).

The focus of these responses is to reduce circulating volume and arterial pressure by both renal and cardiovascular effects, including transfer of ECF from plasma to interstitial fluid (Laragh and Atlas, 1988). Recently it has become clear that, like angiotensin, ANP comes from various sites and may be merely one of a number of natriuretic peptides. Thus, some have argued that urodilatin, from the kidney itself, may be a more important influence on sodium excretion (Goetz, 1993; Gerzer and Drummer, 1993). Others have questioned whether ANP, for all the logic of its effects, is actually important as a physiological regulator of sodium excretion (Richards, 1990).

In heart failure, atrial biosynthesis and secretion of ANP increases and expression of the ANP gene in the ventricles, normally suppressed, is reactivated (Imura and Nakao, 1989). Circulating ANP (α-ANP) is a 28-amino acid peptide consisting of the terminal sequence of γ-ANP, the usual 126-amino acid storage form of the hormone. Its beneficial cardiovascular effects in heart failure are to reduce circulating volume through natriuresis and increased capillary permeability and to reduce arterial pressure, thus achieving a reduction of preload and afterload, but at the price of a potential tendency to sustain oedema (Cantin and Genest, 1985; Atlas and Laragh, 1986; Maack, 1988; Lassiter, 1990). The natriuretic effect involves both increased GFR and reduced sodium reabsorption in the collecting duct (Maack *et al.*, 1984; Atlas and Laragh, 1986; Zeidel and Brenner, 1987), hence the antagonism to the effects of aldosterone. ANP also reduces its secretion directly and by suppressing renin production. This, in turn, depends on the increased filtered load of sodium resulting from the raised GFR (Seeley and Laragh, 1990). The increased GFR involves a rise in filtration fraction, but is not essential for the natriuretic effect of ANP (Espiner, 1994). The reduction of peripheral resistance occurs mainly when this is abnormally high. Normally hypotension causes a compensatory increase in heart rate mediated by baroreceptor reflexes, but ANP opposes this response (Volpe, 1992).

The enhancement of GFR by ANP may contribute to the increased single-nephron GFR in the surviving nephrons of patients with chronic renal failure (Tuso *et al.*, 1994). The importance of the increased GFR cannot be over-emphasised; it means that, as well as increasing sodium excretion, ANP also necessitates an increase in sodium reabsorption (Maack, 1988). It also means that ANP is likely to have its greatest importance in conditions such as heart failure where circulating levels are very high and GFR is subnormal. The basis of the increased GFR is afferent arteriolar dilation; additional effects on capillary permeability and efferent constriction may also contribute (Kirchner and Stein, 1994). The potential for interaction with the effects of nitric oxide seems obvious but, as yet, scarcely investigated. Thus, Shultz and Tolins (1993) found that salt loading increased nitric oxide production in rats and that nitric oxide increases both GFR and sodium excretion, while inhibition of nitric oxide synthesis has the opposite effects, provided its hypertensive effect (which causes pressure natriuresis) is excluded. Yet the possible involvement of ANP was not discussed. There is evidence to suggest that ANP-induced natriuresis is independent of nitric oxide (Romero *et al.*, 1992). Arterial relaxation induced by ANP probably is mediated by nitric oxide (Morikoti *et al.*, 1992) and so is pressure natriuresis (Romero *et al.*, 1992; Ikenaga *et al.*, 1993; Majid *et al.*, 1993).

ANP may also impede water conservation (and hence facilitate the development of hypernatraemia) because it suppresses ADH secretion and promotes 'medullary washout' (Seldin, 1990), hence undermining the maintenance of highly concentrated interstitial fluid around the collecting ducts, on which final removal of water from urine depends. Since high levels of aldosterone also tend to raise plasma sodium concentration, this is an exception to the antagonism between the effects of the two hormones. ANP also impedes the effect of ADH on collecting duct permeability (Bankir *et al.*, 1989) and opposes the dipsogenic effect of angiotensin II (Kenyon and Jardine, 1989). As with sodium excretion, doubts exist about the importance or reproducibility of these effects of ANP on ADH secretion and water excretion, at least in patients with cardiac failure and increased levels of plasma ANP (Bichet, 1989).

The problems concerning the importance of ANP as a regulator of sodium excretion (or plasma volume) arise from early experiments using large doses and later experiments, showing that while physiological levels are effective, they are inconsistent with the time course or magnitude of responses to natriuretic stimuli (Kirchner and Stein, 1994). Problems also arise from the fact that the wide distribution of receptors includes those involved in its breakdown, as well as its effects (Maack, 1988). There is also the inevitable question of whether hormone assays are detecting the active sequence or breakdown products. Less convincing is the evidence that cardiac denervation abolishes the natriuretic response to atrial distension in experimental dogs, whereas ANP secretion still increases; cardiac denervation has numerous effects and it would be more disturbing if the natriuresis persisted without an increase in ANP (Richards, 1990). There are other circumstances in which ANP increases but sodium excretion does not, however no one suggests that it is the **only** factor affecting sodium excretion. Finally, we should not necessarily expect the same factors to be involved in natriuretic responses to cardiac overload, increased ECF volume (and salt intake) or hypernatraemia. Urodilatin, for example, appears to be involved in the natriuretic response to acute hypernatraemia and its central nervous system (CNS) effects (Gerzer and Drummer, 1993). Thus, while the exact importance of ANP in a variety of physiological circumstances remains to be defined, there is little reason to doubt its contribution, and perhaps that of other natriuretic peptides, in states combining high circulating levels and a pathophysiological drive to salt retention, especially where GFR is reduced. Not unexpectedly, heart failure *(Chapter 8)* offers the best example. In experimentally induced cardiac failure in dogs, a model which increased ANP (ventricular tachycardia) caused less salt retention than one which did not. In calves, complete heart replacement by a mechanical pump caused sodium retention, whereas if the ventricles but not the atria were replaced, ANP was secreted and sodium retention was avoided (Espiner, 1994).

Although hypernatraemia suppresses ANP secretion (De Bold *et al.*, 1988), the importance of ANP in clinical disturbances of ECF sodium concentration, rather than plasma volume, remains undefined. There is, however, evidence of its importance in the adaptation of brain cells to hypernatraemia (Fraser and Arieff, 1994). Moreover, in a number of conditions, such as chronic renal failure and mineralocorticoid escape, it is becoming clear that plasma levels of brain natriuretic peptide (BNP), as well as ANP, are increased

(Ishizaka *et al.*, 1993; Westenfelder *et al.*, 1993; Yokota *et al.*, 1993). BNP can be produced by the heart as well as the brain, indeed the cardiac ventricles may be the main source, and circulating levels increase in heart failure (Espiner, 1994; Nicholls, 1994). ANP may also be involved in potassium regulation *(Chapter 4)*.

Apart from its physiological interest, ANP's therapeutic potential has attracted considerable interest, particularly as an antihypertensive agent and in the treatment of cardiac disease and acute renal failure *(Chapter 8)*. Because of its rapid metabolic clearance, long acting analogues are more promising (Tan *et al.*, 1993). It has recently been suggested that ANP stimulates the release of endogenous sodium transport inhibitor, which is regulated by the AV3V area of brain (Songu-Mize and Bealer, 1993). While this would reinforce the natriuresis, it might also undermine the antihypertensive effect of ANP, assuming that depression of sodium transport is a key link in the pathogenesis of hypertension *(Chapter 6)*. As well as the therapeutic use of exogenous ANP, there is the possibility of manipulating endogenous levels by controlling the rate of breakdown. In acute renal failure, ANP's ability to promote renal perfusion and increase GFR has particular potential. In all these instances, it is too soon to say whether pharmacological potential will be converted into clinical efficacy (Lieberthal and Levinsky, 1990; Sagnella and MacGregor, 1994).

MSH

MSH causes natriuresis without altering arterial pressure or renal haemodynamics, although it also increases ANP secretion. MSH receptors have not been found in the kidney and it remains unclear whether this hormone has any independent effects on the regulation or pathophysiology of sodium excretion (Valentin *et al.*, 1993).

Kallikrein–Kinin System

Kallikrein within the cells of the distal nephron releases kinins into its lumen. As in other tissues, they may affect sodium and water excretion directly or via interactions with prostaglandins (Cowley and Roman, 1989; Padfield, 1989; Carretero and Scicli, 1994). It is far from clear that they are involved in the renal regulation of salt and water balance (Briggs *et al.*, 1990), but their effects can be enhanced in the presence of ACE-inhibitors which also inhibit the degradation of kinins (Carretero and Scicli, 1994). Obviously, A-II-receptor antagonists do not have this drawback (Laragh and Seeley, 1992).

Pregnancy

It is well known that pregnancy alters renal function and electrolyte balance, but the degree to which such alterations are 'normal' remains uncertain (Lindheimer and Katz, 1980). Pregnancy also affects renal structure, increasing both the vascular and interstitial space and dilating the renal calyces, pelvis and ureters (Davison, 1983). While pregnancy often causes sufficient sodium retention to cause oedema, and some would consider this normal (Paller and Ferris, 1990), the significance of this retention, which exceeds the needs of the conceptus, remains uncertain (Gallery, 1984; Aber, 1985; Davison, 1986). At present it is not clear whether sodium retention during pregnancy is primary or compensatory, i.e. whether pregnancy is an 'overfill' or 'underfill' state (Schrier and Durr, 1987). Accordingly, advice on sodium intake during human pregnancy remains fraught with uncertainty (Mickelsen, 1982; Barron and Lindheimer, 1984). Worse, there is 'a lack of basic understanding of the place of disordered sodium metabolism in hypertensive pregnancy' (Gallery, 1984).

Pregnancy potentially presents multiple challenges to sodium regulation. Firstly, there is the accumulation of sodium in the foetus and foetal fluids. As a daily requirement, the quantity involved is small; it increases towards the end of pregnancy but, even then, it is probably less than the initial demands of lactation (*see also Chapter 5*). Fascinatingly, the mother can compensate for an artificially increased foetal demand created by drainage of foetal urine (Gibson and Lumbers, 1992). Moreover, maternal ANP may have important regulatory effects on foetal blood pressure and urinary sodium turnover (Cheung, 1991). Inevitably, there are likely to be substantial species differences associated with the duration and endocrine control of pregnancy and the number of offspring. There are also likely to be differences between humans and quadrupeds due to the differing mechanical effects of the dilated uterus on circulation (Lindheimer *et al.*, 1986).

As well as the intra-uterine and intramammary accumulation of sodium, there is a more general increase in body sodium, particularly associated with expansion of ECF volume; oedema is not unusual in otherwise normal pregnancy (Paller and Ferris, 1983). The degree to which this is a physiological necessity or a reflection of high salt intake remains a matter of debate until specific data can be obtained. Cramer, in 1906, was the first to advocate salt restriction during pregnancy, to limit the development of oedema, and de Snoo from 1913 onwards strongly advocated the benefits of salt restriction for all pregnant women, including prevention of pre-eclampsia (Steegers *et al.*, 1991). Steegers *et al.* (1991) placed women on sodium intakes of 0.3 mmol/kg/day, although urinary data suggest that intake was closer to 0.6 mmol/kg/day. Compared with women on unrestricted diets, they had similar arterial pressures during pregnancy, with lower cardiac output offset by higher peripheral resistance. The lower cardiac output was attributed to a lower plasma volume and, since PCV fell, a lower blood volume than women on control diets. The study is limited by inability to make direct measurements of fluid spaces and by both incomplete dietary compliance and differences in food intake as well as sodium, thus the salt-restricted women gained less fat during pregnancy. Perhaps the most interesting outcome was that, despite these changes, the women on the low sodium intakes produced babies of normal birth weight and placental weights were not significantly affected (though both were 6% lower). Certainly, pregnant women on even lower sodium intakes (10 mmol/day, 0.15 mmol/kg/day) adjust their renal sodium excretion as rapidly as normal and remain in sodium balance (Paller and Ferris, 1994).

Pregnancy is accompanied by profound changes in cardiovascular function and, despite increased cardiac output and expanded plasma volume, is essentially a hypotensive condition. There are also changes in capillary permeability. A key determinant of the cardiovascular changes is the additional vascular volume imposed by the placental circulation and, to a lesser extent, the increased mammary circulation. The combination of expanded intravascular volume and reduced arterial pressure would lead us to anticipate that pregnancy is an 'underfill' state and, consonant with this view, the renin–angiotensin–aldosterone system is activated and sodium is retained.

It would be naive, however, to expect that the impact of pregnancy on sodium regulation could be viewed purely in terms of appropriate responses to foetal sodium accumulation and changes in maternal circulation. It is also a period of radical change in the endocrine environment and, therefore, in the levels of hormones which, whether or not they regulate sodium balance, can certainly affect it. There are also changes in the regulation of water balance which affect plasma sodium concentration. Changes in salt appetite have already been discussed (*Chapter 4*). Just as the changes in renin and angiotensin suggest an 'underfill' state, the early changes in ANP are compatible with 'overfill' (see below). Whether pregnancy is associated with 'underfill' or 'overfill', the key determinant of the appropriate ECF volume during gestation is probably the plasma volume which best sustains placental perfusion (Davison and Lindheimer, 1989).

Since the chicken and the egg is essentially a reproductive conundrum, it is appropriate to begin with the central question of gestational sodium balance: which comes first, the sodium retention or the expansion of ECF volume? This discussion should be related to the very different changes associated with pre-eclampsia (*Chapter 6*). It is not only ECF volume which changes, but also the hydration of structural components of connective tissue, probably caused by oestrogens (Paller and Ferris, 1983).

Haemodynamic Changes

There is an early increase in cardiac output, probably in response to reduced peripheral resistance and ahead of changes in plasma volume (Abraham and Schrier, 1994). This vasodilation is not attributable to increased maternal or foetal metabolic demand and precedes major changes in uterine blood flow (Paller and Ferris, 1994). The initial increase in cardiac output results from increased stroke volume, but subsequently, although cardiac output remains high, it is sustained by an elevated heart rate and stroke volume returns to normal (Paller and Ferris, 1990). The volume of red blood cells (RBC) in circulation increase but plasma volume increases more, hence PCV falls; this *per se* would reduce blood viscosity and would potentially reduce arterial pressure. Alongside these early changes is a marked increase in renal perfusion and GFR, although filtration fraction falls because of reduced efferent arteriolar resistance. The reduced renal vascular resistance has been attributed to elevated prostaglandin concentration (Paller and Ferris, 1994). The rise in GFR precedes the increase in plasma volume (Schrier, 1988). The latter precedes any increase in fluid intake, at least in rats (Atherton *et al.*, 1982). An important clinical implication of the supranormal GFR is that plasma creatinine concentration should be below normal (Davison, 1986). The rise in GFR does not depend on foetal or placental factors since it also occurs in pseudopregnant rats (Baylis, 1982) and it may precede the increase in renal blood flow (dal Canton and Andreucci, 1986).

Unless there is perfect adjustment of glomerulotubular balance, an increased GFR promotes loss rather than retention of sodium. A reduced filtration fraction and a fall in plasma albumin would also predispose to a rise rather than a fall in renal sodium excretion (Davison and Lindheimer, 1989). Moreover, increased GFR is often an accompaniment of increased ECF volume, though the rise in GFR during pregnancy is particularly striking—perhaps 50% or more. These considerations therefore suggest an 'overfill' rather than an 'underfill' state, i.e. an expansion of ECF volume driven by sodium retention. On the other hand, the reduced vascular resistance, mediated in part perhaps by increased prostaglandin levels and blunted responsiveness to the vasoconstrictor effects of angiotensin (Paller, 1984), seems more compatible with 'underfill'. Vascular synthesis of prostacyclin and PGE_2 makes an important contribution to the maintenance of foetal circulation and the necessary adjustments of maternal circulation (Terragno and Terragno, 1988).

Schrier and Briner (1991) argue that pregnancy is an 'underfill' state due to arterial vasodilation, but that it has two atypical features, supranormal renal function and unimpaired ability of the kidney to escape from the sodium retaining effects of aldosterone; the former may explain the latter. The key factor, therefore, is a potent vasodilator, as yet unidentified, to which the kidney is exceptionally sensitive. Evidence from rats suggests that prostaglandins or kallikrein could fulfil this role (Oddo *et al.*, 1993). Some evidence in rats, but not all, indicates that nitric oxide may be the vasodilator and, in addition, that relaxin, the uterine hormone involved in parturition, also increases cardiac output (Abraham and Schrier, 1994). The placenta produces nitric oxide and data from experimental animals suggest that defective production may contribute to pregnancy hypertension *(Chapter 8)*, but data from women are conflicting (Podjarny *et al.*, 1994).

As in oedema *(Chapter 7)*, ANP is normal or increased in human pregnancy, and in rats its renal effectiveness remains normal (Paller and Ferris, 1994); on balance, therefore, pregnancy sodium retention is more suggestive of an 'overfill' phenomenon. Nevertheless, pregnant women have an undiminished ability to excrete acute sodium loads and their ANP response to volume expansion is enhanced (Lowe *et al.*, 1992a). They also increase their ANP secretion in response to dietary salt loads even when plasma volume is not significantly increased (Lowe *et al.*, 1992b). Not all studies agree that ANP levels are normal during gestation; Thomsen *et al.* (1993, 1994) regard a reduced secretion of ANP as a factor in the increased blood volume of pregnancy. Moreover, intracellular sodium of RBC is reduced, suggesting increased rather than reduced activity of Na–K ATPase (MacPhail *et al.*, 1992, 1993; Webb *et al.*, 1993). In the last trimester of pregnancy, ANP falls as aldosterone rises (Thomsen *et al.*, 1994). ANP does not explain the vasodilation of early pregnancy.

Before considering the hormonal changes of pregnancy—and why the renin–angiotensin–aldosterone axis should be activated despite 'overfill', it is helpful to discuss changes in water balance, since these also relate to the question of 'underfill' vs 'overfill'.

Water Balance

Here we encounter a different chicken with its egg; there is a profound increase in water turnover during pregnancy with higher urine output and enhanced fluid intake—which comes first, and why? Neither the extremes of urine concentration or dilution are impaired (Paller and Ferris, 1994), despite the increased prostaglandin levels, which oppose the effects of ADH on the collecting duct. The key change seems to be a reset 'osmostat', such that both thirst and ADH are activated to retain water in defence of a lower 'normal' plasma sodium concentration. A small part of the fall in plasma osmolality results from the fall in blood urea associated with supranormal GFR (Lindheimer and Katz, 1986; Lindheimer et al., 1986). Increased thirst precedes the change in the threshold for ADH secretion, both in women and in sheep (Michell et al., 1988; Paller and Ferris, 1990). In pregnant sheep, ADH is still responsive to hypernatraemia but at a given arterial pressure, plasma ADH is lower (Keller-Wood, 1994). The ADH response to increased osmolality appears normal in pregnant sheep whose blood pressures were not measured (Bell et al., 1986). Hyponatraemia often arises from hypovolaemia (Chapter 8), but this is not the case in pregnancy because plasma volume is increased. Moreover, ADH secretion does not seem to be responding to an illusory 'underfill', because imposition of actual hypovolaemia produces the normal increase of ADH secretion in pregnant rats (Barron and Lindheimer, 1984). The osmotic threshold for oxytocin release is also reduced during pregnancy in rats (Koehler et al., 1993). It is unlikely that the changes in ADH release result from the effects of oestrogen, alone or in combination with progesterone or prolactin, or the effects of the renin–angiotensin system (Lindheimer et al., 1986). Oxytocin may have mild natriuretic effects in non-primate mammals (Conrad et al.,1993)

If ADH secretion is supranormal at normal plasma osmolality, it is all the more strange that urine output is dramatically increased in late pregnancy, both in women and in sheep (Michell et al., 1988; Gines et al., 1994). While ADH breakdown is increased, hence women with diabetes insipidus need a higher dose during pregnancy, the fact remains that ADH levels in normal pregnant women are not reduced (Gines et al., 1994). In a sense, they have a form of nephrogenic diabetes insipidus and we need to identify a cause of renal resistance to ADH. The changes in renal perfusion and the possibility of 'medullary washout' would be worth considering but for the fact that the extremes of urine concentration are still attainable. It is possible that human chorionic gonadotrophin contributes to polyuria and polydipsia during pregnancy (Davison et al., 1988).

Hormonal Changes

Aldosterone secretion increases during pregnancy and to a greater degree than expected from the concurrent increase in renin activity or angiotensin II (Brown et al., 1992; Paller and Ferris, 1994); the latter occurs despite the increase in ECF volume. The suggestion that the underlying cause is competitive antagonism of aldosterone by progesterone is not supported by the fact that pregnant women have no problem with renal sodium conservation, even on intakes as low as 10 mmol/day, and sheep increase their aldosterone levels during pregnancy, whereas progesterone, in this species (as in dogs), is not natriuretic (Johnson *et al.*, 1970; Michell *et al.*, 1988). There is renal resistance to the kaliuretic effects of aldosterone during pregnancy (Nolten and Ehrlich, 1980; Gallery, 1984), for example, patients with primary aldosteronism become normokalaemic instead of remaining hypokalaemic; progesterone may attenuate the kaliuretic response in humans, but it does not provide a complete explanation (Brown *et al.*, 1986; Paller and Ferris, 1994). It is clear that pregnant sheep can conserve sodium from urine even on sodium intakes which suppress aldosterone and some component of the renal sodium retention in rats and humans is also independent of aldosterone (Churchill *et al.*, 1981; Atherton *et al.*,1982; Michell *et al.*, 1988). Presumably this reflects enhanced proximal reabsorption which could impede potassium excretion by reducing distal delivery of sodium. In women, unlike sheep, aldosterone rises even during pregnancy on high sodium intakes (Nolten and Ehrlich, 1986).

Non-renal sources of renin may contribute to the increase seen in pregnancy. The uterus produces renin which may sustain uterine blood flow; angiotensin II reduces uterine vascular resistance, perhaps through stimulation of prostaglandin synthesis (Paller and Ferris, 1994). The rise in circulating prostaglandins during pregnancy therefore results from increases in both renal and uterine synthesis. Angiotensin does not underlie the increased thirst in pregnancy, at least in sheep, as polydipsia occurs in animals whose aldosterone is suppressed by high sodium intake (Michell *et al.*, 1988).

Oestrogens have been implicated in the salt retention of pregnancy; apart from renal effects, they may underlie some of the characteristic changes in the gel matrix of interstitial fluid. Oestrogen receptors are found in the kidney, mainly in the proximal tubule (Katz and Lindheimer, 1977; Scheibler and Danner, 1978), but they may also potentiate the effects of aldosterone in the distal nephron (Berl and Better, 1979). This probably explains the observation, over 65 years ago, that pregnancy alleviated the effects of adrenalectomy in bitches (Rogoff and Stewart, 1927). Subsequently, the sodium retaining effects of oestrogens have been confirmed and assumed to contribute to their hypertensive effects (Thorn *et al.*, 1938; Christy and Shaver, 1974; Crane and Harris, 1978). Oestrogens increase renin substrate levels substantially (Seeley and Laragh, 1990). Nevertheless, Johnson *et al.* (1970, 1972) showed that in dogs, oestrogens caused sodium retention independently of aldosterone and without increase in arterial pressure. Since neither reduction of GFR nor the renal 'escape' mechanism were involved, they speculated that oestrogens might have a direct tubular effect (Johnson and Davis, 1976). Oestrogen has been reported to stimulate sodium transport in frog skin (Tomlinson, 1971).

Oestrogens do not cause sodium retention in all species, however, and in sheep they have a biphasic effect, initially increasing and then decreasing sodium excretion in both urine and faeces (Michell and Noakes, 1985). In any case, sodium retention begins early in ovine pregnancy whereas the rise in oestrogen occurs later (Michell et al., 1988). A biphasic effect of oestrogen, similar to that in sheep, was seen in one study of humans (Katz and Kappas, 1967). Renal sodium retention in pregnant sheep, unlike other sheep, is particularly dependent on intact innervation (Aberdeen et al., 1992). Pregnant sheep are also more resistant to suppression of angiotensin II by high sodium intake (Quillen and Nuwayhid, 1992); nevertheless, even on sodium intakes of 1 mmol/kg/day, aldosterone secretion remains low throughout pregnancy (Michell et al., 1988).

Sadly, pregnancy further illustrates the problem that answers to questions about the physiology of sodium depend on the level of intake at which the question is put. In sheep on high intakes (1.0 mmol/kg/day), sodium excretion increases towards the end of pregnancy as a result of greatly increased urine volume; on lower intakes (0.1 mmol/kg/day) renal sodium retention is sustained throughout pregnancy (Michell et al., 1988). Very similar trends are seen in rats (Bird and Contreras, 1986). The conclusion which might readily be reached, that animals are unable to retain sodium in late pregnancy because of an inadequate increase in aldosterone secretion, would therefore be incorrect. Yet there is often little consideration or justification of the sodium intake on which experiments are conducted. For example, Cha et al. (1993) examined changes in renal perfusion, GFR, filtration fraction, fractional proximal tubular reabsorption, urinary sodium excretion and osmolar and free water clearance, both in pregnant and control sheep. Unfortunately, they did so on sodium intakes of 120 mmol/day, probably 2 mmol/kg/day (bodyweights were not stated), i.e. in animals subject to chronic sodium loading and its various consequences, on top of pregnancy. Standard diets for rats in experiments on pregnancy (Barron, 1987) often provide around 15 mmol/kg/day, enough to underwrite the total sodium retention of pregnancy every 2–3 days.

The electrolyte physiology of pregnancy will remain susceptible to considerable confusion until a much more disciplined distinction between a generous excess of sodium and a 'normal' intake, i.e. one modestly above requirement, is generally adopted.

(Michell et al., 1988)

8
Clinical Aspects of Extracellular Volume and Sodium Concentration

The range of clinical disturbances affecting extracellular volume and sodium concentration is huge. This chapter therefore begins with a general introduction and is then divided into the following sections:

1. Volume excess (oedema).
2. Sodium concentration (hypo- and hypernatraemia).
3. Volume depletion.
4. Renal dysfunction.

Introduction

Sodium is an element; it cannot be generated, broken down or converted in the body so, in essence, clinical problems should be simple and arise from deficit, excess or maldistribution. While it cannot be interconverted, the interaction of sodium regulation with various physiological mechanisms adds complexity. For example, as already noted, the consequences of excess sodium can include calcium depletion, while sodium depletion can predispose to either hyper- or hypokalaemia; such interactions are considered further in *Chapter 9*.

Sodium Depletion, Dehydration and Plasma Sodium

The primary consequence of sodium depletion is a fall in extra-cellular fluid (ECF) volume, including circulating volume, with sodium concentration unaffected *(Chapter 1)*. As volume depletion increases, thirst and antidiuretic hormone (ADH) respond to hypo-volaemia rather than plasma sodium concentration and hyponatraemia (which would normally suppress water retention) develops (Fig. 8.1). Correction does not, therefore, require hypertonic solutions

179

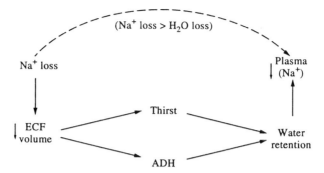

FIG. 8.1. Sodium loss and hyponatraemia.

of sodium, but restoration of ECF volume. This necessitates the use of solutions with similar sodium concentration to plasma: hypotonic solutions of sodium will simply allow water to escape into ECF, which is futile when the priority is repair of ECF volume (Fig. 8.2). There is no escape from the obligation to provide roughly 145 mmol of sodium for every litre of ECF volume needing to be replaced.

Secondary consequences of sodium depletion include interference

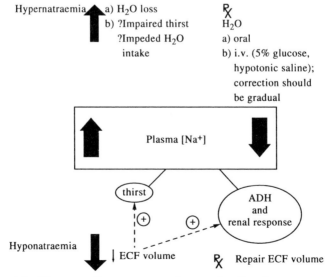

FIG. 8.2. Plasma sodium concentration: main disturbances and their correction.

with renal regulation of water and potassium excretion and plasma pH *(Chapter 2)*.

Just as hyponatraemia is not the immediate outcome of sodium depletion, hypernatraemia is not the result of excess sodium, unless it is given parenterally or taken in very large quantities without access to water. Hypernatraemia indicates water loss combined with factors which prevent adequate compensatory drinking; these may include pain, weakness, dysphagia, vomiting, or simple water deprivation. Water loss alone is unlikely to cause hypernatraemia, thus, with diabetes insipidus, the outcome is compensatory polydipsia rather than hypernatraemia.

The clinical consequences of both hypernatraemia and hyponatraemia include a wide range of symptoms, many of them behavioural, none of them pathognomonic. There is substantial overlap between these conditions or with other electrolyte disturbances, notably hypercalcaemia and hypomagnesaemia (Michell, 1979a). Many result from redistribution of water and cell swelling or shrinkage, especially in the central nervous system (CNS) *(Chapter 1)* and the intensity of the effects depends substantially on the speed as well as the magnitude of the change. Partly, this is because of the ability of brain cells, given time, to offset the osmotic effects of hypernatraemia by generating additional intracellular organic solutes ('idiogenic osmoles'), which may include taurine.

One of the most extreme manifestations of hypernatraemia is salt poisoning in pigs, characterised by CNS disturbances including circling, blindness and convulsions. While high salt intake may predispose, the usual cause is water deprivation, often due to freezing temperatures. Frequently, the symptoms are precipitated not by the deprivation but by suddenly regaining access to water; the extreme thirst associated with the severe hypernatraemia is sufficient to induce a sudden reduction of the elevated plasma sodium concentration (Taylor, 1983; Blood *et al.*, 1994).

Sodium Excess: Hypertension and Oedema

The primary consequence of excess sodium intake or retention (apart from enhancement of thirst and ADH) is expansion of ECF volume. There is also likely to be a rise in glomerular filtration rate (GFR) (Katz and Genant, 1971; Pitts, 1974; Bell and Navar, 1982; Roos *et al.*, 1985): partly, this results from atrial natriuretic peptide (ANP) (Maack, 1988). It is, however, unlikely that hypervolaemia will be clinically significant, let alone that oedema will result, because of endocrine and renal mechanisms for excretion of excess sodium. Thus, the clinical consequences of excessive mineralocorticoid (aldosterone) secretion involve hypokalaemia rather than oedema; hypertension is also common (Delmez *et al.*, 1990).

The kidney is the long-term regulator of arterial pressure and, in theory, the natriuresis caused by even slight rises in arterial pressure is sufficient to eliminate excess sodium without a lasting rise in pressure (Fig. 8.3). The development of hypertension, therefore, requires a defect in this mechanism, i.e. something that makes the slope of the relationship between pressure natriuresis and arterial pressure less steep (Guyton, 1992b).

The cause of **hypertension** is not usually expansion of ECF volume *per se*, however, and may be a side-effect of activation of mechanisms for the excretion of excess sodium; for example, increases in the circulating concentration of active sodium transport inhibitors (ASTI/EDLI: *Chapter 1*) may raise the intracellular sodium concentration of arteriolar muscle and, through exchange with calcium, increase its sensitivity to vasoconstriction (Haddy, 1988; De Wardener, 1990). Even in dogs, the classic experimental species for hypertension research, and despite their low natural prevalence of hypertension, there is evidence to suggest that while, indeed, a chronic excess of sodium is efficiently excreted, the price is a measurable increase in both ECF volume and arterial pressure

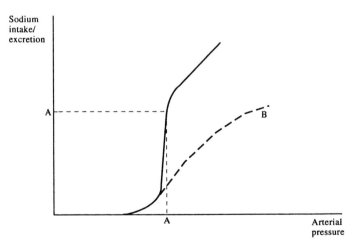

A. There is a range of sodium intake within which any resulting rise of arterial pressure is promptly corrected by the powerful natriuretic response reflected in the steepness of the pressure natriuresis curve

B. Defective renal sodium excretion flattens the pressure natriuresis curve i.e. increased sodium intake is excreted at the price of a substantial increase in arterial pressure

FIG. 8.3. Renal regulation of arterial pressure: pressure natriuresis.

(Michell, 1993). In humans, the evidence suggests that chronic ingestion of excess salt, rather than age, underlies the lifelong rise in arterial pressure which is usually regarded as normal. These issues have been discussed in more detail in *Chapter 6*, but the conclusion is that, at least in humans, the most prevalent clinical condition arising from excess sodium is hypertension.

The most important clinical condition associated with maldistribution of sodium is **oedema**. The classical explanation is easily understood; oedema involves the expansion of the major component of ECF, i.e. interstitial fluid, from the minor component, i.e. plasma *(Chapter 1)*. Whatever the underlying cause, this can only occur if plasma volume is constantly replenished by the normal renal defence against hypovolaemia. Thus, whatever the primary cause of oedema, the kidney is always the enabling cause (Michell, 1991). To that extent, the use of diuretics is not simply symptomatic treatment *(Chapter 11)*. Their effectiveness is enhanced by both direct and indirect vascular effects, e.g. if natriuresis exceeds the rate of mobilisation of oedema fluid, the resulting hypovolaemia will lead to vasoconstriction and further uptake of interstitial fluid *(Chapter 1)*.

If the dosage of the diuretic is sufficiently excessive (and these are very potent drugs with considerable variation in their effect on individual patients), the rate of the natriuresis may far outstrip the speed of mobilisation of oedema fluid and result in acute hypovolaemia. The point is that, while diuresis (natriuresis) is the visible outcome, it is not the primary objective; that is mobilisation of oedema fluid. Sustained rather than intense diuresis is therefore important and awareness that, as the dosage exceeds that needed for adequate diuresis, the likelihood of various side-effects increases (Michell, 1988d).

The effectiveness of a given dose of any diuretic tends to decrease as the amount of fluid mobilised increases and the patient returns to their 'dry weight'. This reflects the fact that most diuretics have their main action on a single nephron segment (except osmotic diuretics). Thus, as ECF volume falls, other nephron segments can compensate for, i.e. oppose, the effects of the diuretic. Osmotic diuretics are useful for restoring urine output, e.g. in acute renal failure, or reducing cell swelling, but they are inappropriate for the treatment of oedema because their effects include expansion of ECF volume.

Although this approach to the role of sodium retention in oedema is physiologically satisfying, it is uncomfortably at variance with a growing variety of clinical observations. Essentially, these cluster around two foci; evidence that inappropriate sodium retention precedes rather than follows the onset of significant oedema and evidence that circulating volume is either normal or even increased rather than subnormal. The 'solution' to the latter has been to concoct the concept of 'effective plasma volume', i.e. that although measurable plasma volume is not reduced, it falls in some crucial site associated with receptors monitoring circulatory volume. Whether this is a useful interim concept highlighting our inadequate awareness of the receptor mechanisms responsible for the monitoring of ECF volume (and perhaps interstitial, as well as plasma volume) or whether it is simply defence of a hallowed concept against besetting data, remains a matter of fierce debate. Interestingly, this characterises the literature concerning the oedema associated with cardiac, hepatic or glomerular disease, respectively. A further layer of complexity is added by considering not only what 'fools' the sodium-retaining mechanisms, but what prevents the salt-excreting mechanisms (notably ANP) from adequately correcting the excessive retention of sodium. These questions are explored further (see below) but, as yet, there is no definitive resolution of the conflict between the 'underfill' and 'overfill' concepts of oedema, either in disease or in pregnancy (*Chapter 7*).

Other Aspects of Abnormal Sodium Balance

Important disturbances of sodium balance are associated with the effects of both surgery and anaesthesia. These include endocrine and renal effects of either anaesthesia or surgery (through reflexes which remain active during surgery or as a result of loss or redistribution of fluid). Removal of distended gut may subtract substantial volumes of fluid; obstructive colic in horses provides perhaps the extreme example. On the other hand, sequestration of fluid, e.g. in obstructed or twisted intestine or damaged tissues, causes depletion, notably of ECF volume, without external loss.

In contrast, thirst, ADH and the renin–angiotensin–aldosterone axis may all be activated by surgery and its aftermath, in the absence of demonstrable deficits. As with the 'underfill' vs 'overfill' controversy in explanations of oedema, there is, therefore, considerable debate as to whether postoperative salt and water retention is a primary problem or a response to deficits, some of which may be difficult to measure (e.g. sequestration of fluid in abnormal extravascular sites, i.e. 'third space' formation). It seems clear that the tendency towards excess water retention is greater than that for excess sodium retention, so that hyponatraemia becomes likely. For this reason, and because the main reason for the use of parenteral fluids during surgery (or immediately before or after) should be defence of circulating volume, they should normally have plasma-like concentrations of sodium. Low sodium solutions are mainly intended for maintenance, i.e. substitution for oral intake of water and electrolytes; this is a less pressing priority which can usually await recovery and often, therefore, be achieved by the normal route: orally.

Apart from their impact on ECF volume, renal function and the central nervous system (CNS), disturbances of sodium balance also affect the prevailing concentration of a variety of hormones. The consequences of acute disturbances, or of chronic changes in sodium intake, therefore include changes in the endocrine environment with potential effects well beyond those on fluids, electrolytes or arterial pressure. Interactions with catecholamines and insulin, for example, will greatly affect metabolic pathways, as will changes in intracellular potassium.

Hypertonic, Hypotonic and Isotonic Dehydration

These terms, correctly used, refer to the effect on the patient, not the supposed composition of their losses. Thus, hypertonic dehydration generally equates with hypernatraemia, hypotonic dehydration with hyponatraemia. Losses do not have to be hypertonic to cause hypotonic dehydration; it would be rare for faecal water to contain a higher concentration of sodium than plasma, yet hyponatraemia is a common consequence of diarrhoea. That is because the associated hypovolaemia enhances thirst and renal water retention while intake of sodium is usually reduced by inappetence, clinical intervention or vomiting. When diarrhoea does cause hypernatraemia, it may reflect not so much a lesser content of sodium in stool water, but greater dermal or respiratory water loss, e.g. in warm, dry surroundings or during fever.

Two important conclusions emerge:

1. in assessing the impact of pathological losses of sodium, water or electrolytes, the combined impact of **all** losses, not just the obvious abnormal losses, needs to be considered;
2. unlike an *in vitro* system, the ultimate effect of the losses depends not only on their composition but on compensatory responses, notably those involving thirst and renal function.

Isotonic Dehydration

Since there is no change in the osmolality of the interstitial fluid bathing cells, there is no redistribution of water between intracellular (ICF) and ECF, no change in cell volume (Fig. 8.4). There may be transfer of interstitial fluid to plasma as a result of a rise in plasma albumin (haemoconcentration caused by fluid loss) or vasoconstriction

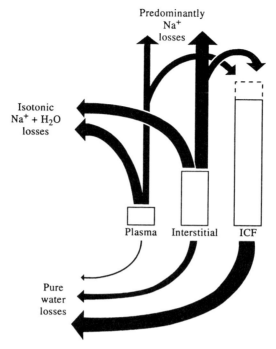

Fig. 8.4. Fluid compartments and the losses they sustain as a result of salt or water depletion. Note the loss of extracellular fluid (ECF) Na^+ to intracellular fluid (ICF) as well as externally in predominant sodium loss.

(response to hypovolaemia). The impact of the loss is solely on ECF and 25% of the loss, therefore, is sustained by plasma (in the absence of adaptive responses).

Hypertonic Dehydration

The extracellular hypertonicity extracts water from ICF so that the increase in osmotic concentration is the same in ICF and ECF. Each compartment, therefore, comes to equilibrium when it has suffered an identical percentage loss of volume. But since plasma volume is 5% of body weight, whereas total body water is 60% or more, less than 10% of a pure water deficit is drawn from plasma. This reinforces the importance of hypernatraemia, rather than hypovolaemia, as the immediate danger caused by water loss. It also illustrates the appropriateness of 'dehydration natriuresis' as a defence against hypernatraemia, despite the inevitable sacrifice of ECF volume. In effect, the hypertonic deficit is converted to a greater but isotonic deficit; a particular benefit is the correction of cell shrinkage caused by hypertonic dehydration.

It is instructive to calculate just how much water loss will cause a dangerous rise in ECF sodium concentration (say 15 mmol/l) in the absence of other responses. Since the intracellular solute must increase its concentration in parallel, it is simplest, in making this calculation, to pretend that sodium is dissolved in all the body water, not just ECF.

Therefore, in a 70 kg animal (with 60% body water):

Normal total solute is equivalent to 42×145 mmol (60% of 70). If the sodium concentration rises from 145 to 160 mmol/l due to hypovolaemia, this solute in a new volume (v) is at a concentration of $(42 \times 145)/v = 160$ mmol/l. The new volume is therefore 38.1 l, and the loss is therefore 3.9 l (i.e. $42.0 - 38.1$).

The percentage loss in plasma volume (and body water) is thus less than 10% at this point: if normal blood volume were 5 l, the loss of circulating volume (0.33 l) would be modest even for a blood donor. In contrast, isotonic loss of 4.5 l would remove approximately one-third of ECF, including plasma (equivalent to 1.5 l of haemorrhage).

It is often said that dehydration equivalent to 5% loss of bodyweight is scarcely detectable, 10% is obvious and 15% or more is severe. Clearly, assuming an otherwise healthy patient (i.e. no underlying chronic disease) the impact depends greatly on the relative loss of sodium and water. Dehydration to the extent of 5% bodyweight in the above example would be 3.5 l, and if it were pure water loss its effects, even on plasma sodium concentration, would be unlikely to cause symptoms. A loss of 10% would probably cause symptomatic hypernatraemia (a rise exceeding 15 mmol/l). On the other hand, an isotonic loss of 3.5 l would remove 25% of ECF volume and would almost certainly produce clinical signs.

These 'guesstimates' relating to bodyweight are even more suspect. Apart from the fact that gut fill will also affect them and that normal bodyweight is not usually known, dramatic variations can occur, even when it is measured. Thus, some cases may survive surprising degrees of weight loss, while others may die when losses add up to little more than 5% (Groutides and Michell, 1990). The latter may include sudden, severe diarrhoea when large volumes of fluid may be transferred to the intestinal lumen before being lost in diarrhoeic faeces and thus lowering bodyweight.

Hypotonic Dehydration

Hypotonic dehydration implies a shift of water from ECF to ICF and thus cell swelling. In addition, it implies that the loss of ECF volume is even greater than that resulting from the external deficits. At the same percentage loss of bodyweight, this is, therefore, the most extreme of the three forms of dehydration in its impact on ECF volume and, therefore, on circulatory function.

Having outlined the scope and principles of this discussion of clinical disturbances, it is now appropriate to consider a number of clinical contexts in which these disturbances commonly arise. Oedema provides an appropriate starting point because it poses the severest challenge to our current understanding of the regulation of sodium balance. Sodium depletion, and hypovolaemia are much more prevalent causes of direct threats to survival, both in humans and animals, than sodium retention but the pathophysiological principles underlying their management rest on much sounder foundations.

1. Volume Excess

Oedema

There are three main clinical conditions which are attributed to excess salt retention: oedema, hypertension and hypernatraemia. Hypertension can be 'volume dependent', i.e. sustained by expanded ECF volume, but usually it is not, certainly in the prevalent form of human hypertension (essential hypertension). Nevertheless, excessive salt intake certainly predisposes to increased arterial pressure, perhaps because of changes in the intracellular electrolytes of arterioles initiated by increased secretion of natriuretic hormones which inhibit Na–K ATPase (*Chapter 6*). Hypernatraemia is an unlikely consequence of excess salt intake, unless thirst and either ADH secretion or the renal response to it are impeded. There will, however, be expansion of ECF volume, though not necessarily to an extent detectable as oedema. Hypertension *per se* does not predispose to oedema and it is an unusual finding in uncomplicated essential hypertension. Firstly, the immediate outcome of excessive arteriolar constriction is to drop the perfusion pressure in the downstream capillaries and secondly, plasma volume is not usually increased.

Whether or not plasma volume increases in the chain of events leading to the common presentations of oedema remains a focus of debate. The only unambiguous context in which oedema and hypervolaemia usually coexist is during pregnancy. This is discussed further in *Chapter 7*, along with other endocrine aspects of sodium balance. The role of plasma volume and the mechanisms defending it in the development of oedema remains an area of appalling uncertainty. The reasons are partly historic; before the 1980s, very few physiologists saw much reason to accept the existence of specific mechanisms for the excretion of excess salt, beyond perhaps the effect on peritubular physical factors controlling fluid reabsorption in the proximal tubule *(Chapter 2)*. In nephrotic, hepatic and cardiac oedema the sodium retention does not result from a failure of ANP secretion; generally ANP levels increase (Cody, 1990).

As has already been emphasised, the regulation of sodium was conceived almost entirely in terms of the renal–adrenal defence against depletion and the defence against excess lay simply in suppressing it. The last 15 years have refined the central question of oedema, 'Why is excess sodium retained?' by adding 'despite the specific mechanisms available for its excretion?' They have not answered it. The 'underfill' and 'overfill' hypotheses have shifted in their balance of support but both remain viable. For a clinician, this may seem like a miscount of the number of angels on the head of a pin—a matter of detail—when the undisputed fact is that excess sodium is retained. For a physiologist, it is absolutely central to our understanding of sodium regulation. We have acquired a tolerance over the decades to the fact that, for all our knowledge of renal physiology, we cannot explain to an intelligent teenager whether the kidney causes the oedema by retaining too much sodium or whether the oedema causes the sodium retention.

Oedema is not abstruse or academic; it is common, and has been readily recognisable by ordinary people throughout time. We can tell them how the kidneys of dogs, rats and people respond to saline infusions, describe changes in gross urine output and composition, or in the microvoltages and microconcentrations of single nephrons. We can specify the genetic control and the amino acid sequences involved in both the molecules and receptors underlying many of these responses. But unlike diarrhoea, haemorrhage or adrenal insufficiency, where we can give convincing, not necessarily comprehensive, reasons for renal sodium retention, with oedema we cannot. It is reminiscent, but on a larger scale, of the difficulties which have existed in reconciling identical changes in either GFR or tubular reabsorption with opposite changes in the delivery of solute to the macula densa *(Chapter 2)*. But it is not a minor gap in our knowledge, it is a chasm—as if we were uncertain whether apnoea caused hypoxia or resulted from an excess of 'effective' oxygen. This account, therefore, emphasises reasons rather than management (though diuretics are considered further in *Chapter 11*).

The difficulty in identifying the precise causes of cardiac, hepatic and nephrotic oedema arises substantially from the validity and sensitivity of the techniques involved. It requires only a small mismatch between GFR and tubular reabsorption, which are both large, to cause substantial fluid retention. Measurements of plasma or ECF volume depend on the 'marker dilution' principle, which assumes that the marker is distributed uniformly, completely, and exclusively in the compartment which it measures. The ability of ECF markers to escape into cells or gut fluids is increased in any condition which affects the permeability of cell membranes or enteric epithelia. No marker is perfect, even in normal subjects, hence the convention of expressing ECF volume as a 'space' which is consistent for the chosen marker, i.e. inulin space, thiocyanate space, etc.

With plasma, the marker is assumed to distribute within the same space as albumin because it is either attached to it when injected or it binds subsequently. Increased capillary permeability allows the marker to follow albumin to abnormal extravascular distribution sites. The reduction in albumin concentration caused by leakage is attributed to dilution by an increased plasma volume. Hence, measured plasma volume is potentially increased when the actual volume is depleted. Similarly, escape of the marker will exaggerate the apparent extent of any expansion of ECF. It could, for example, happen that with a mildly subnormal ECF volume but substantial escape of the marker, there would be a physiological response to hypovolaemia, which was reflected in raised concentrations of salt-retaining hormones, when the measured ECF volume was spuriously increased. False reductions in the measured compartment are less likely, since they require false increases in marker concentration and there is no endogenous source of redistribution or production. The most plausible cause of a misleading increment of marker concentration would be areas of extreme underperfusion with which the marker could not equilibrate, or possibly perivascular loss around the injection site, allowing a delayed peak of entry into circulation.

The main risk of marker dilution techniques in clinical conditions, therefore, is that ECF may appear to be expanded and plasma volume overfilled when either is actually normal or even contracted.

In the case of plasma it is obvious (but sometimes neglected) that the regulatory system responds to changes in blood volume, hence haematocrit must also be measured—samples from peripheral veins do not necessarily represent the overall ratio of red blood cells (RBC):plasma in circulation. The result is again an underestimate of blood volume. This heterogeneity also affects direct measurements of blood volume with labelled red cells. Speculatively, if a sensory mechanism responded to perfusion as well as pressure, PCV would also have important effects via blood viscosity.

It is convenient to begin a detailed consideration of causes of oedema with nephrotic syndrome, since there is no doubt that the kidney is not only the enabling cause but the initiating cause, though not necessarily through primary retention of sodium.

Nephrotic Oedema

Twenty years ago, the simple explanation that nephrotic oedema resulted from a fall in plasma albumin and that renal sodium retention was a response to hypovolaemia seemed largely satisfactory (Swales, 1975; Schrier, 1976). This contrasted with cardiac failure or cirrhosis, where there was already a need to invoke 'effective' as opposed to measured blood volume. During the 1980s, however, it became clear that even in nephrotic syndrome, a straightforward 'underfill' explanation was not compatible with data from many patients, especially adults (Bailie, 1983; Schrier, 1988). Instead, the renal sodium retention appeared to be primary rather than compensatory, i.e. causing the oedema rather than responding to it (Czerwinski and Lach, 1982; Brown *et al.*, 1985). The direct effect of chronic hypoalbuminaemia on renal function is to reduce GFR, the more favourable filtration gradient being outweighed by reduced renal perfusion; this reduction in GFR predisposes to sodium retention in experimental dogs (Manning, 1987).

The strength of the 'underfill' hypothesis of nephrotic oedema rests on evidence for the importance of the following components (Humphreys, 1994):

1. reduced plasma albumin;
2. reduced plasma volume;
3. increased secretion of renin, angiotensin, aldosterone;
4. increased sympathetic activity.

Whatever their importance in fuelling the oedema, items 3 and 4 are indicators of a physiological response to hypovolaemia; their presence could have other explanations, but their absence would be hard to reconcile with hypovolaemia.

The main problems with the evidence are as follows:

1. Albumin concentration does not consistently fall far enough to explain the oedema. Not least, humans or rats with hereditary analbuminaemia are not oedematous. Humphreys (1994) attributes this to dilution of interstitial fluid protein. There is thus an automatic protection against oedema in that its formation reduces the concentration gradient which sustains it. This seems plausible in areas such as the lung where capillary permeability to protein is substantial (Michell, 1989b; Geheb *et al.*, 1994), but neglects the importance of subatmospheric pressure in interstitial fluid and the associated efficiency of lymphatic clearance of spilled protein. These provide a 'safety factor' *(Chapter 1)* whereby substantial falls in plasma albumin concentration do not cause oedema (Guyton, 1992a). Where plasma albumin falls convincingly, the

urinary loss cannot be the sole cause, since hepatic synthesis should normally be able to keep pace (Bailie, 1983; Goodwin, 1985; De Wardener, 1985). Whatever the importance of the reduced plasma albumin there is no doubt that retention of sodium and water, as well as restoring plasma volume initially, will tend to favour the development of renewed oedema and hypovolaemia by further dilution of albumin (Epstein and Perez, 1994). This is likely to be more important in oedema dependent on a reduced transcapillary oncotic gradient (glomerulopathy, enteropathy, lymphatic blockage, capillary damage), rather than raised capillary pressure. The sodium retention in nephrotic syndrome is unusual in that proximal reabsorption is reduced (Seldin, 1990) and it is possible that a reduction in the albumin concentration in peritubular capillaries could contribute to this.

2. Plasma volume is indeed normal or increased in many nephrotic patients and there is no reason to attribute this to technical error, especially with techniques based on blood volume and the use of radiolabelled RBC, thus avoiding problems associated with leakage of albumin, to which Evans Blue, for example, is bound (Dorhout Mees and Koomans, 1990). It could, however, remain true that the 'underfill' was adequately or even over-corrected by the renal response; the fact that the bottom of a boat is only slightly wet does not mean there is no leak, merely that the bilge pump is efficient.

What, then, of expanded plasma volume? Although fluid space measurements take some hours, they relate to a specific moment, the time of injection, once the result is calculated. We are again confronted by the problem of a snapshot of a 24-h process. Perhaps nephrotics 'overfill' during hours of activity because of 'underfill' during the majority of the day. Perhaps a central component is a mismatch between the diurnal rhythm of sodium excretion and that of plasma volume. This could explain Humphreys' suggestion (1994) that nephrotic oedema involves **both** 'underfill' and 'overfill' components. With hepatic cirrhosis there is clear disruption of the circadian pattern of sodium excretion with a nocturnal natriuresis mediated by increased secretion of ANP and loss of the nocturnal increase in renin and aldosterone secretion (Warner *et al.*, 1993).

The problem is that, not only can we not measure plasma volume reliably, we cannot do it repeatedly, let alone continuously, in clinical cases. A hypothesis invoking the importance of 24-h rhythms in plasma volume seems no more than a convenient escape so long as this remains unmeasurable. Once it becomes measurable, its importance may become obvious. Such rhythm differences are already acknowledged as important in the development of hypertension (Harshfield *et al.*, 1993) and in the response to diuretics (Cambar *et al.*, 1992).

The existence of a non-osmotic drive to ADH release provides strong evidence for the underfill concept. Most nephrotics have increased plasma ADH concentrations despite having a reduced plasma sodium concentration (and osmolality). Moreover, despite the controversy as to whether fluid space measurements should be expressed per 'dry'-weight (non-oedematous) or 'wet'-weight (which would tend to reduce plasma volume simply by increasing bodyweight), the plasma ADH is highest in those with the lowest blood volume, even when it is expressed per wet weight (Gines *et al.*, 1994). The possible role of the interstitial environment of osmoreceptors in abnormalities of ADH secretion seems to have had little consideration. If this fluid space, small though it is, became expanded it could reduce the transmural pressure gradient, i.e. offload the tonic tension in the receptor walls. This would mimic the effect of cell shrinkage, i.e. of the hyperosmotic stimulus to ADH secretion.

A subset of the evidence concerning plasma volume relates to its therapeutic expansion. If nephrotics are hypovolaemic, volume expansion should less readily cause natriuresis than in normal subjects. There is substantial evidence for blunted natriuresis in nephrotic patients (Humphreys, 1994), but it is not unequivocal. More important, such evidence could also support an overfill hypothesis which is clearly only viable if normal natriuretic mechanisms are impaired. The reduced natriuretic response occurs despite the direct effect of a fall in colloid concentration, which is to depress sodium reabsorption in the proximal tubule (Epstein and Perez, 1994). Clearly, therefore, it is essential to distinguish between experiments where plasma volume is expanded with saline (which reduces oncotic pressure) as opposed to colloids. The 'plasma volume expansion' produced by Koomans *et al.* (1984) and the 'volume expansion' produced by Valentin *et al.* (1992), in both cases to look at natriuretic responses in nephrotic syndrome, are really quite different, the former using hyperoncotic albumin and the latter saline. By comparison, the fact that the former was in people, the latter in rats, is a minor detail.

3. Physiological markers of underfilling, such as increased sympathetic activity or raised plasma concentrations of aldosterone or components of the renin–angiotensin system all fail to provide clear support for the existence of a real or perceived hypovolaemia in nephrotic oedema (Humphreys, 1994).

Renal denervation corrects the blunted natriuretic response of nephrotic rats which is associated with increased renal sympathetic nerve activity. In nephrotic people, evidence can be found for a heightened plasma noradrenaline concentration and its reduction by volume expansion; however, these are not consistent findings.

Aldosterone alone cannot 'explain' oedema, though it can reinforce it; aldosterone excess, *per se*, causes hypertension, hypervolaemia and potassium depletion, but not usually oedema. The kidneys (though not other organs) are known to escape its sodium-retaining but not its kaliuretic effects *(Chapter 7)*. The key indicators are data suggesting suppression of the renin–angiotensin–aldosterone axis, which normally implies hypervolaemia. Data for renin suppression are substantial and not explicable via inadequate renin substrate (Humphreys, 1994). Moreover, patients with high renin continued to retain salt even if angiotensin-converting enzyme (ACE)-inhibitors were used to block its effects. Of course, the interpretation of such data is complicated by the fact that sodium excretion can be affected by the underlying glomerular disease, not just the resulting nephrotic underfill. Such effects could include impairment of renin secretion (Schrier, 1988). Evidence from rats showed that hypoalbuminaemia *per se* did not reduce plasma volume but did so if the rats had pre-existing renal damage (Epstein and Perez, 1994).

Apart from the data which contradict the underfill hypothesis, the overfill hypothesis is supported by specific evidence (Humphreys, 1994). For example, proteinuria and glomerular injury leading to reduced GFR both cause renal sodium retention; the reason, in the case of proteinuria, is uncertain. In experimental unilateral glomerulonephritis in dogs, the altered sodium excretion is confined to the abnormal kidney; it cannot, therefore, result from changes in plasma volume. It may reflect insensitivity to ANP, since even when nephrotic patients produce similar amounts of ANP to normal people they show less response to it. Similar data have come from rats and suggested explanations include increased sympathetic activity and defective production of second messenger, i.e. cGMP (Humphreys, 1994).

A wealth of conflicting data contribute to the difficulty in judging between the 'overfill' and 'underfill' explanations of nephrotic oedema. Some of the arguments and pitfalls are complex, but it bears repeating that one factor contributing to the confusion is depressingly simple—sodium intake. This is often not comparable in the experiments under comparison, not least because it is sometimes unknown, sometimes unspecified and seldom at a justified level. This problem is prevalent in many areas of research on sodium, as has already been emphasised *(Chapter 5)*.

A simple 'underfill' concept would anticipate considerable parallels between nephrotic oedema and that associated with hepatic disease, since the latter potentially has an even greater influence on plasma albumin concentration. The main parallel, however, turns out to be similar if not greater uncertainty as to the mechanisms. Hepatic oedema perhaps provides the toughest challenge to the plausibility of fluid space measurements since, as Peters (1948) pointed out, on the one hand, leakage of a protein-bound marker would give a false impression of expanded plasma volume, whereas expansion of the portal capillaries would certainly increase **plasma** volume, but not **circulating** volume. Here lay the seeds of the concept of effective circulating volume which evolved from Peters' concept that the determinant of fluid retention might be "not the actual volume of circulating plasma but some function usually related to the volume of circulating plasma—such as renal blood flow. . . .".

Hepatic Oedema

The underfill hypothesis of ascitic fluid accumulation attributes sodium retention to a reduced plasma volume. This results from loss of excessive quantities of fluid from hepatic sinusoids and splanchnic capillaries because of hypoalbuminaemia and portal hypertension (Epstein and Perez, 1994). The latter causes excessive lymph production because of the permeability of hepatic sinusoids and the excess lymph leaks into the peritoneal cavity (Badalamenti *et al.*, 1993); lymph production in cirrhotics may increase 20-fold (Levy, 1993). Renal vasoconstriction and salt retention in cirrhosis can, however, occur while blood volume is normal (Amend, 1993) and recently there has been growing emphasis on the role of excessive vasodilation (especially splanchnic), as well as reduced intravascular fluid, in reducing the filling of the circulatory system (Arroyo *et al.*, 1993; Abraham and Schrier, 1994). This mismatch between the accessible vascular volume and the volume of blood available to fill it constitutes a fall in 'effective' circulating volume. The nature of the vasodilator, whether it is produced by the liver or whether hepatic inactivation is impaired, remains uncertain, though endothelium-derived relaxing factor (EDRF), which is probably nitric oxide, has been suggested (Badalamenti *et al.*, 1993; Lang *et al.*, 1993; Abraham and Schrier, 1994). Endogenous opiates may also contribute (Abraham and Schrier, 1994).

One component of the reduction in effective arterial blood volume may be increased arteriovenous shunting in the splanchnic, pulmonary, dermal and muscle vasculature (Schrier, 1988). Factors causing abnormal venous pooling can readily deplete the volume of blood on the arterial side of the circulation while expanding the total circulating volume (Schrier, 1988), because the veins represent the storage or capacitance side of the system and normally contain about four times as much blood.

Obstruction to hepatic outflow seems to be a particularly important factor in ascites and not simply through changes in actual or effective intravascular volume (Brenner and Rector, 1986). Hepatic sodium receptors may be involved. Portal hypertension eventually leads to a reduced cardiac output with a disproportionately large decline in GFR and sodium excretion. This is exacerbated by aldosterone; secretion is increased in response to activation of the renin–angiotensin system, but hepatic breakdown is also impaired. There is little relationship between changes in aldosterone and either worsening or improvement of ascites (Epstein and Perez, 1994). Nevertheless, the aldosterone antagonist, spironolactone, often induces substantial natriuresis and improvement of the ascites, suggesting that failure of renal escape from excess aldosterone, rather than the level of aldosterone *per se*, may be more important (Caramelo and Schrier, 1990). The rise in aldosterone in ascitic patients follows a decrease in arterial pressure and attempts to block the systemic rise in angiotensin may be counterproductive, since it seems to be an important protection against hypotension in this circumstance (Brenner and Rector, 1986; Badalamenti *et al.*, 1993). Thus, ACE-inhibitors may cause hypotension in cirrhotic patients (Caramelo and Schrier, 1990). The intrarenal effects of angiotensin may, however, contribute to sodium retention independently of aldosterone (Epstein and Perez, 1994).

ANP concentration is increased rather than depressed (Gerbes *et al.*, 1987; Epstein and Perez, 1994), i.e. the response to sodium retention is appropriate but insufficient. The increased ANP levels result from enhanced secretion, not impaired hepatic breakdown (Warner *et al.*, 1993). Similarly, there is evidence to suggest that ASTI/EDLI secretion is also increased in patients with cirrhosis (La Villa *et al.*, 1990). The response to endogenous or exogenous ANP is suppressed (Badalamenti *et al.*, 1993; Warner *et al.*, 1993). This resistance may result from increased proximal tubular reabsorption restricting the delivery of sodium to the distal nephron, where ANP acts, and also the opposing effects of angiotensin II (Wong and Blendis, 1994). Warner *et al.* (1993) suggest that in the early stages of liver disease, increased ANP prevents overt ascites, but that subsequently genuine loss of fluid from the vascular compartment attenuates the rise in ANP and activates salt-retaining mechanisms. It is important to reiterate that, while ANP is an appropriate response to volume expansion, it not only corrects the increased cardiac preload by increasing sodium excretion, but also by increasing the transfer of fluid from plasma to interstitial fluid—thus worsening oedema (Warner *et al.*, 1993). Its vasodilator effects will also worsen oedema (Warner *et al.*, 1993) and may undermine 'effective' circulating volume.

Cirrhotics are generally less able to excrete sodium loads, whether or not they are ascitic (Gentilini and Laffi, 1992). Central to the overfill hypothesis of ascites is the observation that renal salt retention precedes the accumulation of fluid and occurs despite an expanded plasma volume (Wilkinson, 1981; Schrier, 1988). This can also be reconciled with the 'underfill' concept if vasodilation rather than capillary leakage is seen as the key factor undermining 'effective' circulating volume. One problem with distinguishing between these alternatives is the fact that patients may differ or even show characteristics of both at different stages of their disease, e.g. whether their plasma levels of renin, angiotensin, aldosterone are normal or increased (Epstein, 1983; Wilkinson, 1984; Yanover *et al.*, 1986). Calculation of peripheral resistance from measurements of cardiac output and mean arterial pressure does not clarify the importance of vasodilation because of compensatory changes, e.g. in cardiac output or vascular tone (Schrier, 1988). Experimental evidence in dogs does not support the importance of peripheral arterial vasodilation as a key factor in the early sodium retention of cirrhosis and suggests that factors other than an increased splanchnic vascular space are involved (Levy, 1994). Intrahepatic hypertension may be an important stimulus. Against the 'overfill' hypothesis is evidence that human patients improve their sodium and water excretion in response to expansion of plasma volume by intravenous fluids (Epstein and Perez, 1994).

The renal mechanisms underlying sodium retention in cirrhosis are still not well defined. GFR is often but not invariably reduced, but tubular reabsorption of sodium is increased both in the proximal and distal nephron (Anderson, 1994). GFR may be bolstered by efferent constriction and increased filtration fraction, thus reducing hydrostatic pressure and raising oncotic pressure in peritubular capillaries; this favours sodium reabsorption at a given GFR. Increased sympathetic activity may contribute to the sodium retention. Renal prostaglandin production is increased, but this may be a response to the vasoconstrictor effects of angiotensin and vasopressin (*Chapter 7*). In advanced cirrhosis, prostaglandin production falls; this may contribute to the initiation of hepatorenal syndrome (Caramelo and Schrier, 1990).

As with cardiac failure, water retention may be even more pronounced than sodium retention, with a resulting hyponatraemia; this reflects excessive secretion or impaired breakdown of ADH (Bichet and Schrier, 1984). Water excretion is also impaired by increased proximal tubular reabsorption of sodium and reduced delivery of sodium to the diluting segments of the nephron (Arroyo et al., 1994). Again, the severity of such problems varies greatly between patients (Gines et al., 1994). Some may resist the effects of excess ADH by increasing their renal production of prostaglandins (Arroyo et al., 1994). It is important to distinguish between true hyponatraemia (due to water retention) and depression of plasma sodium concentration by increases in glucose or lipids in blood (Anderson, 1993). The stimulus to hypersecretion of ADH may well be peripheral vasodilation and the resulting fall in 'effective' circulating volume, indeed, the vasoconstrictor effects of ADH may help to correct this defect and to maintain arterial pressure (Gines et al., 1994). They may also, however, impede renal perfusion and restrict renal function. It is clear, nevertheless, that renal sodium retention in hepatic disease is caused by increased reabsorption, not reduced GFR (Epstein and Perez, 1994).

Important evidence concerning mechanisms of ascites comes from experiments using 'head-out water immersion' (HWI) of humans. This redistributes blood and increases 'central' blood volume without directly altering plasma composition, moreover, unlike volume expansion with intravenous fluids, the stimulus is rapidly reversible. HWI improves sodium and water excretion in cirrhotic patients, supporting the idea of an underlying deficit in this 'effective' 'central' blood volume (Gentilini and Laffi, 1992). It does so, at least in part, through increased renal prostaglandin production—a response seen in various circumstances where renal perfusion is compromised (Humes and Weinberg, 1986; Badr and Brenner, 1994; Epstein and Perez, 1994). Other factors possibly contributing to renal sodium retention in cirrhosis include increases in sympathetic activity, circulating catecholamines and defective kinin formation (Badalamenti et al., 1993).

Despite evidence that sodium retention precedes ascites, the overfill theory seems implausible, since the renin–angiotensin–aldosterone and sympathic nervous systems are both activated (Henriksen and Larsen, 1994). Studies in experimental animals suggest that the rise in aldosterone is an early change which, if blocked, prevents renal sodium retention (Bernardi et al., 1994). More important, patients with cirrhosis but without ascites still show normal renal escape from the effects of excess mineralocorticoid (Badalamenti et al., 1993). Moreover, when mineralocorticoids were given to cirrhotics, only a minority developed ascites because the majority showed 'escape', especially those with the greatest increase in plasma volume (Schrier *et al.*, 1994a). Whether plasma volume is actually expanded or contracted, patients with oedema as well as ascites respond clinically, e.g. to diuretic therapy, as if they were volume expanded, whereas those without oedema respond as if they are 'underfilled' (Anderson, 1993).

Profound changes in renal function occur in hepatorenal syndrome, a form of acute renal failure (ARF) with a functional basis and no primary renal damage (thus hepatic transplantation restores normal renal function). There is a drastic decline in renal perfusion and GFR, but there is renal sodium and water conservation and the urine composition therefore resembles that associated with prerenal failure (e.g. due to dehydration) rather than ARF (Michell, 1983a; Amend, 1993). The reasons remain unclear; suggestions have included the accumulation in liver disease of 'false transmitters', i.e. vasoactive substances which override the normal control of intrarenal blood flow (Michell, 1983a; Levy, 1993). Thromboxane production is increased but does not seem to play a causal role as similar increases are seen in patients with severe hepatic disease but free of renal failure. Moreover, thromboxane inhibitors fail to alleviate hepatorenal syndrome (Moore *et al.*, 1992; Badalamenti *et al.*, 1993). Attention has also been directed to the role of endotoxins as cirrhotics have increased enteric colonisation by bacteria and have less hepatic protection against the absorption of endotoxin (Gentilini and Laffi, 1992). Against this idea, patients with hepatorenal syndrome may improve their renal function following a portocaval anastomosis (Levy, 1993). There is potentially a pathogenic role for both vasoconstrictors (e.g. angiotensin, renal nerves), which directly reduce renal perfusion, and vasodilators, which may do so indirectly, via their systemic effects. There is no satisfactory animal model of this condition. ANP levels do not necessarily increase in hepatorenal syndrome but levels of VIP and substance P, vasodilator peptide hormones normally cleared by the liver, are raised. Unfortunately mortality remains almost 100% and the key to survival is correction of the underlying hepatic problem (Schelling and Linas, 1990).

Among the possible causes of renal vasoconstriction in hepatorenal syndrome, attention has focused on leukotrienes, platelet activating factor (PAF), endothelins, as well as angiotensin II and sympathetic activity together with impaired production of kinins and prostaglandins (Moore *et al.*, 1992; Arroyo *et al.*, 1993; Badalamenti *et al.*, 1993; Laffi *et al.*, 1994). The concept is that cirrhotics who maintain their renal function do so by enhancing their renal prostaglandin production, whereas this capacity is exhausted in hepatorenal syndrome (Gentilini and Laffi, 1992). Diminished intrarenal prostaglandin production has also been implicated in the sodium retention and resistance to diuretics which are characteristic of this syndrome (Badr and Brenner, 1993). There is not convincing evidence, however, for beneficial therapeutic effects of prostaglandins in this condition (Clewell and Walker-Renard, 1994). Excessive use of diuretics is liable to precipitate hepatorenal syndrome in patients with cirrhosis; they are especially susceptible to the effects of hypovolaemia (Levy, 1993). Obviously, the risks are also increased by the use of non-steroidal anti-inflammatory drugs, since the normal defence of renal perfusion against excessive vasoconstriction depends on prostaglandins (Badalamenti *et al.*, 1993; Chonko and Grantham, 1994). The importance of angiotensin II is unclear; increased secretion could equally be a cause or a result of impaired renal perfusion (Levy, 1993).

Among the unanswered questions raised by hepatorenal syndrome are whether the intense renal vasoconstriction is a primary or a compensatory change and what causes the profound oliguria; similar reductions in perfusion occur in chronic renal failure (CRF) without reducing urine output (Levy, 1993). This form of pre-renal failure is certainly unique in that it is independent of reductions of cardiac output and involves specific factors causing renal vasoconstriction.

Cardiac Oedema

The basic mechanisms of cardiac failure remain surprisingly resistant to rigorous explanation (Francis, 1994). Among them, the reasons for salt retention are central to current concepts of therapy *(Chapter 11)*. Here the debate on 'underfill' vs 'overfill' has parallels with that on the effects of 'forward' and 'backward' failure (Michell, 1991). Cardiac failure is defined by the inability of the heart to maintain a normal cardiac output **at a normal filling pressure** (Cannon and Martinez-Maldonado, 1983; Brozena and Jessup, 1990). Early suggestions that reduced cardiac output was a key determinant of sodium retention were clearly not compatible with various salt retaining states in which it increases (Abraham and Schrier, 1994). Moreover, in myocardial infarction, sodium retention precedes changes in resting cardiac output or venous pressure (Brenner and Rector, 1986). In experimental animals with induced heart failure, e.g. due to vascular lesions or myocardial infarction, renal sodium retention is detected before changes in venous pressure or ventricular performance (Chonko and Grantham, 1994). As the concept of heart failure focuses increasingly on its earlier stages, the cardinal feature is no longer perceived as pump failure, but the neuroendocrine responses, compensatory or otherwise, notably those leading to retention of salt and water (Young and Pratt, 1994).

Renal perfusion falls early in cardiac failure and this is greatly exacerbated by exercise, hence the beneficial effect of reduced activity (Hollenberg and Schulman, 1985). The renal response to inadequate perfusion is an increased filtration fraction; this maintains GFR but, by raising the protein concentration of blood continuing from the glomerulus to the peritubular capillaries of the proximal tubule, it facilitates reabsorption of tubular fluid. There is thus more retention of salt and water at a given GFR (Brenner and Rector, 1986). Increased filtration fraction does not occur universally, however, so other factors are involved (Cody and Pickworth, 1994). The renin–angiotensin system is also activated (Cannon and Martinez-Maldonado, 1983). The increased sympathetic activity in response to a reduced cardiac output also promotes renal sodium retention, through effects on both renal perfusion and sodium reabsorption in the proximal tubule (Abraham and Schrier, 1994). Apart from its effects on sodium balance, the net effect on renal function is to reduce GFR despite an increased filtration fraction (Oster *et al.*, 1994a). Nevertheless, the combination of renal and hypotensive effects means that patients with severe heart failure may experience deterioration of renal function with ACE-inhibitors.

The effect of salt retention and expanded blood volume (Anand *et al.*, 1989) is increased by the fact that cardiac failure prevents a normal protective response, increased capillary leakage at increased intravascular volume (Humphreys and Rector, 1985). As in other states of 'pre-renal' failure, prostaglandins are important in sparing renal perfusion from the vasoconstrictor response to diminished cardiac output (Shapiro and Dirks, 1990). Although heart failure is usually associated with reduced cardiac output, patients can have congestive heart failure with an increased output, e.g. due to coexisting anaemia or thyrotoxicosis; they still retain sodium and water (Cannon and Martinez-Maldonado, 1983). This is attributed to arterial 'underfill' due to peripheral vasodilation (Schrier, 1988).

Water retention in cardiac failure results from enhanced secretion of ADH; this occurs in the absence of an osmotic stimulus and is attributed to arterial 'underfill' or, in patients with 'high-output' cardiac failure, to arterial vasodilation (Gines *et al.*, 1994). The rises in ADH are sufficient to further impair cardiac function through increased peripheral resistance. There is also increased thirst due to angiotensin (Ramsay, 1989) and reduced delivery of sodium to the diluting segments of the nephron also predisposes to hyponatraemia (Oster *et al.*, 1994a). These trends are reinforced by the response to diuretics *(Chapter 11)*. Paradoxically, ACE-inhibitors can cause hyponatraemia. This is because the loss of feedback suppression by angiotensin elevates renin levels; these also increase in CNS and, since ACE-inhibitors do not cross the blood–brain barrier (BBB), rises in brain angiotensin drive an increased thirst. As well as changes in renal function, heart failure can cause renal lesions, e.g. tubular nephropathy. The medulla, which normally verges on hypoxia, is especially vulnerable to reduced perfusion. Some lesions are caused not by the disease but the drugs used to treat it (Lajoie *et al.*, 1994). These include diuretics *(Chapter 11)*; as always, adverse drug reactions involve a small but unfortunate minority of patients.

The central problem of cardiac oedema, regardless of 'overfill' or 'underfill', is why natriuresis fails to prevent it once right atrial pressure increases; this appears to be the obvious homeostatic role for ANP. Thus, apart from reducing ECF (and plasma) volume through natriuresis, it also reduces 'afterload' on the left ventricle through its hypotensive effects (Cantin and Genest, 1985; Atlas and Laragh, 1986). These include suppression of renin secretion and the renal response to ADH. But plasma volume is also reduced by increased capillary permeability, allowing interstitial fluid to act as an overflow; obviously this effect counteracts the beneficial effect of natriuresis on mobilisation of oedema fluid (Maack, 1988).

Evidence on the role of ANP needs to be interpreted with caution; changes in cardiac content reflect release as well as synthesis and, while ANP receptors are widely distributed, some may be concerned in its metabolism rather than its effects (Michell, 1991). Nevertheless, as in cirrhosis, it is clear that ANP secretion is increased in heart failure but insufficiently to prevent salt and water retention; renal responsiveness to ANP is reduced (Shapiro and Dirks, 1990) and this is exacerbated by concurrent increases in the secretion of renin, angiotensin, aldosterone and ADH. Neither a decrease in renal receptors nor an increase in renal inactivation of ANP account for its failure to prevent oedema (Abraham and Schrier, 1994). Nevertheless, its effectiveness is blunted by an increase in the renal endopeptidase which degrades it in the proximal tubule; inhibition of this enzyme in experimental heart failure increases the urinary excretion of ANP and also produces haemodynamic improvement (Young and Pratt, 1994). Secretion of the endogenous sodium transport inhibitor (EDLI; *Chapter 1*) is also increased in congestive heart failure. As with ANP, the question is why it fails to prevent excessive sodium retention. If, however, it is a key determinant of vascular tone *(Chapter 6)*, it would cause additional vasoconstriction, helping to maintain arterial pressure in the face of reduced cardiac output (Blaustein, 1993).

As well as effects on plasma volume and vascular tone, ADH predisposes to hyponatraemia and this is reinforced by impaired renal diluting capacity associated with reduced distal delivery of fluid (Cannon and Martinez-Maldonado, 1983; Bichet and Schrier, 1984). It is important to emphasise that, although the later effects of salt retention and vasoconstriction are counterproductive (Johnston *et al.*, 1987), the initial effects are compensatory, i.e. they sustain cardiac output (Schrier, 1988; Michell, 1991). Considering the extensive use of diuretics in the treatment of cardiac failure, it is remarkable how little is understood, even now, about the associated changes in sodium excretion or the balance between their compensatory or maladaptive effects (Cody *et al.*, 1986; Michell, 1991). Indeed, progress seems to proceed in a spiral—large circles which provide a measurable but small advance. The possibility that renal sodium retention is a primary rather than a secondary feature of cardiac oedema has gained support, yet it is precisely the concept which Peters (1948) set out to challenge—with the idea that it was a secondary response to an actual or perceived fall in circulating volume. The problem of 'effective blood volume' is discussed further in Michell (1995a).

Chronic Obstructive Pulmonary Disease

Chronic lung disease can cause severe retention of salt and water and, as in cardiac failure, there is reduced renal perfusion and GFR, together with similar neurohumoral changes (Anand *et al.*, 1993). Clinicians often attribute the fluid retention to right-sided heart failure. Cardiac output, however, is normal and both right and left ventricular function are normal, at rest and during exercise. The trigger factor is probably vasodilation and hypotension caused by hypercapnia, which also stimulates non-osmotic release of ADH; plasma ANP is also increased (Anand *et al.*, 1992; Sterns *et al.*, 1993). GFR falls far less than renal perfusion and the increased filtration fraction will favour sodium retention. The expansion of blood volume is far greater than expected, considering the increase in plasma and ECF volume. It reflects the rise in PCV stimulated by hypoxia. Acute hypoxia *per se* increases renal perfusion but causes sodium retention. Chronic hypoxia also causes sodium retention. Hypercapnia predisposes to sodium retention, probably through increased renal exchange of hydrogen ions for sodium (Anand *et al.*, 1993). Mechanical ventilation, combined with positive expiratory pressure, impedes venous return and superimposes a reduction of cardiac output as a further stimulus to sodium and water retention. Increased pressure in the posterior vena cava, reduced plasma levels of natriuretic peptides (ANP and brain natriuretic peptide, BNP) and increased secretion of renin and aldosterone contribute to the sodium retention (Rossaint *et al.*, 1993; Shirakami *et al.*, 1993). Oxygen therapy causes natriuresis in patients with chronic obstructive respiratory disease (De Siati *et al.*, 1993; De Angelis *et al.*, 1993) and this may be caused by an increase in endogenous digitalis-like inhibitors of sodium transport *(Chapter 1)*.

2. Sodium Concentration: Hypernatraemia and Hyponatraemia

Because plasma sodium concentration is sensitively regulated by the mechanisms controlling water balance, thirst and ADH, clinical abnormalities imply interference with both. Thus, water loss only causes hypernatraemia if weakness, pain, behavioural disturbances, vomiting or deprivation prevent ingestion and retention of increased amounts of water (Star, 1993). Excessive drinking (e.g. in psychogenic polydipsia) does not depress plasma sodium concentration unless renal factors interfere with dilution of urine and increased excretion of water. Excessive salt consumption causes thirst rather than hypernatraemia and sodium depletion causes hyponatraemia only belatedly, and indirectly, by impairing delivery of sodium to the diluting segments of the nephron (enhanced proximal reabsorption) and enhancing ADH secretion and thirst via hypovolaemia; the immediate cause of hyponatraemia, therefore, is usually an interference with water regulation caused by hypovolaemia. The ability of sodium-regulating mechanisms to correct hypo- or hypernatraemia is minimal, as will be seen. This is scarcely surprising if we are clear that essentially, though not exclusively, changes in sodium balance are reflected in changes of ECF volume, whereas changes in ECF sodium concentration result from abnormal water balance. Indeed, a recent book on sodium regulation (Seldin and Giebisch, 1990) does not index 'hypernatraemia'. Yet too many clinicians still persist in interpreting plasma sodium concentration as an index of sodium balance—a legacy of the 1940s when, for example, evidence for sodium retention in cardiac oedema was sought in the form of hypernatraemia (Peters, 1948).

Hypernatraemia

Hypernatraemia can arise regardless of increased or decreased body sodium. Where sodium is retained, it is likely to be a renal response to hypovolaemia and the cause of the hypernatraemia is likely to be excessive water loss. The latter reflects **all** losses. Thus, for example, it is misleading to expect faecal sodium concentration to predict whether plasma sodium will fall or rise. The losses only determine the change in concentration in an inert system; in a patient the renal response, induced by the relevant hormones, is the key determinant, hence hyponatraemia is the usual outcome of diarrhoea, despite the fact that a faecal sodium concentration approaching, let alone exceeding, plasma sodium is most unusual. Moreover, the losses are not just those associated with the primary pathology; dermal and respiratory water loss are also extremely important and susceptible to changes in body temperature (e.g. fever) as well as ambient temperature and humidity. Warm, dry surroundings inevitably increase the likelihood of hypernatraemia. Sweating is likely to reinforce hypernatraemia in primates because sweat usually contains less sodium than plasma, especially under the influence of aldosterone. In horses, however, there are substantial amounts of sodium and this may serve as a defence against hypernatraemia, e.g. during sustained exertion.

An iatrogenic cause of hypernatraemia is the use of osmotic diuretics *(Chapter 11)* since, unlike receptor-active diuretics, they increase water loss even more than sodium loss (Howard *et al.*, 1993). Obviously, injudicious use of parenteral fluids can also cause hypernatraemia, especially in unconscious patients, as they cannot drink. Less obviously, misguided use of salt tablets by sedentary individuals in hot, humid climates not only postpones their adaptive aldosterone-induced reduction of sweat sodium but may also cause hypernatraemia (Howard *et al.*, 1993). This may be reinforced by the tendency for humans to drink according to taste, habit or social context, rather than the demands of regulatory thirst (Michell, 1991; Fitzsimons, 1993; Zerbe and Robertson, 1994).

Once initiated, hypernatraemia is potentially self-accelerating because, by the time the condition is symptomatic, the early compensatory increase in thirst is blunted by lethargy and an increasing range of behavioural disturbances. The impact of a given degree of hypernatraemia is intensified by the speed at which it develops. This is because, although brain cells can regulate their sodium and potassium content swiftly in response to changes in tonicity (Schielke and Betz, 1992), they need time to generate additional idiogenic osmoles such as taurine to offset the increased osmolality of their surroundings and minimise the shrinkage which causes much of the damage, even meningeal haemorrhage (Oh and Carroll, 1992; Star, 1993). Other idiogenic osmoles probably include, apart from amino acids (e.g. glutamine), myoinositol, choline and sorbitol, as well as increased intracellular ions (Macknight *et al.*, 1993; Strange, 1993; Gullans and Verbalis, 1993; Lee *et al.*, 1994). It follows that if hypernatraemia has developed sufficiently slowly for such compensatory responses to occur, it must also be corrected slowly, otherwise the idiogenic osmoles will immediately draw excess water into cells and cause the problems of cell swelling usually associated with hyponatraemia. The generation of idiogenic osmoles is a response to hypertonicity, not hypernatraemia *per se*, but there are two provisos:

1. it may not occur in response to artificial solutes such as mannitol;
2. physiological solutes must create a gradient, not simply rise in parallel both in ECF and ICF.

Thanks to idiogenic osmole responses, humans have been known to survive plasma sodium concentrations as extreme as 85–272 mmol/l (Sterns *et al.*, 1993). Nevertheless, the majority of patients with a plasma sodium concentration outside the range 120–160 mmol/l will have neurological symptoms and among those with severe hypernatraemia, mortality may exceed 50% (Lee *et al.*, 1994).

It would easily be imagined that the hypernatraemia caused by hyperaldosteronism is an example caused by excess sodium but, once again, the immediate cause is disturbed water balance, specifically depression of ADH by a marked increase in plasma volume (Palmer and Alpern, 1993). Diabetes mellitus initially depresses plasma sodium concentration (but not osmolality) simply because additional water is committed to plasma by the osmotic effect of the increased blood glucose. Once glycosuria occurs, however, the osmotic diuresis raises plasma sodium, often above normal, and it is important to realise that in this case the degree of hypernatraemia is less than expected for the severity of the water loss, because of the direct effect of hyperglycaemia on plasma sodium concentration (Palmer and Alpern, 1993). Hyponatraemia is therefore more probable in mild or well controlled diabetes mellitus with hypernatraemia becoming more probable in severe or poorly controlled diabetes, especially those with ketoacidosis or hyperglycaemic, hyperosmolar, non-ketotic coma (Michell *et al.*, 1989; Oster *et al.*, 1994b). The latter superimposes impairment of water intake on water losses caused by osmotic diuresis. Quite apart from the fact that diabetes insipidus is infinitely more common in textbooks than patients, those with diabetes **mellitus** are actually more likely to have hypernatraemia than those with diabetes **insipidus**, except where the latter have impaired water intake.

Diagnosis of polyuric states should be simple, in theory, using criteria such as urine osmolality and its response to dehydration, but is complicated by the fact that patients with diabetes insipidus may still produce some ADH, while psychogenic polydipsia can impair urinary concentrating capacity by causing medullary washout *(Chapter 2)*. Moreover, changes in renal perfusion and flow rates along the nephron can cause slight changes of urine concentration without increases in ADH (Howard *et al.*, 1993; Palmer and Alpern, 1993;

Morrison and Singer, 1994). This results from the 'distal-trickle' effect (Oh and Carroll, 1992), i.e. restriction of distal delivery of water and solute by increased proximal reabsorption, thus allowing greater time for osmotic equilibration even when ADH levels and collecting duct permeability are very low. Electrolyte disturbances can also cause nephrogenic diabetes insipidus, e.g. potassium depletion, which may also cause renal lesions *(Chapter 2)*. Hypercalcaemia blunts the cAMP response to ADH, impairs the generation of concentrated medullary interstitial fluid (through effects on sodium transport); it also reduces renal perfusion and GFR, especially when it results from hyperparathyroidism (Morrison and Singer, 1994; Benabe and Martinez-Maldonado, 1993).

While water regulation provides the main defence against hypernatraemia, it has recently become clear that an important additional defence arises from 'dehydration natriuresis' (McKinley *et al.*, 1983; Ramsey and Thrasher, 1984, 1991; Michell, 1991). Partly this involves the ability of hypernatraemia to suppress aldosterone secretion (Merrill *et al.*, 1989), but dehydration natriuresis is particularly important on high salt intakes when aldosterone secretion is suppressed. Thus, sheep on moderate or high sodium intakes (0.4, 1.2 mmol/kg/day) showed increased urinary and faecal sodium excretion in response to partial water deprivation, whereas those on lower intakes (0.05 mmol/kg/day) did not (Michell and Moss, 1994). They also corrected their water deficits via thirst, rather than enhanced water conservation in response to ADH. Additional protection against hypernatraemia probably arose from movement of water into ECF from gut fluids. The natriuresis did not involve ASTI/EDLI or ANP, indeed, the latter tended to fall, perhaps because of hypovolaemia. Oxytocin levels increased, consistent with its response to dehydration, hypovolaemia or salt loading in rats (Balment *et al.*, 1980; Stricker *et al.*, 1987; Wells *et al.*, 1990). In rats, water deprivation enhances ANP secretion, whereas hypovolaemia does not (Zongazo *et al.*, 1992), suggesting that ANP could be involved in dehydration natriuresis, at least on the high salt intakes to which they are usually exposed. In humans, dehydration does increase the plasma concentration of ASTI/EDLI (Wellard and Adam, 1987). ANP can affect the release of ADH, but only at high concentration and this affects the ADH response to hypovolaemia rather than to hypernatraemia (Thrasher and Ramsay, 1993). The central nervous system is probably involved in the modulation of dehydration natriuresis (McKinley, 1992), though the efferent pathway does not appear to involve real nerves (Park *et al.*, 1989). In dogs with access to water, drinking and dehydration natriuresis are more important than ADH in the control of plasma sodium concentration (Cowley and Roman, 1989).

Hyponatraemia

Acute, severe hyponatraemia can be lethal; usually the patient's fate is in the balance by the time it is recognised but it remains within the clinician's power to seal it, as well as to save it (Sterns, 1990). Whether or not water intake is increased, the ultimate cause of hyponatraemia is inadequate dilution of the urine. This may result from higher levels of ADH than normal for the prevailing plasma sodium concentration or impaired dilution mechanisms within the kidney. Exceptionally, massive overdrinking combined with minimal solute intake can cause hyponatraemia (Sterns *et al.*, 1994). Impairment of dilution can involve reduction of sodium removal in the ascending thick limb of the loop of Henlé, or from the early distal nephron, or subnormal delivery of fluid to these segments from the proximal tubule. All of these can arise from hypovolaemia. Thus, while hypovolaemia is not the invariable cause of hyponatraemia, clinically it is the first to exclude or treat. In adrenal insufficiency, hyponatraemia results from both ADH secretion and impaired dilution as a result of hypovolaemia, together with the direct effect of glucocorticoid deficiency (Oh and Carroll, 1992; Bichet *et al.*, 1993; Robertson, 1994).

Hypovolaemia reduces the osmotic threshold for both thirst and ADH secretion. Moreover, modest hypotension or hypovolaemia have relatively little effect on ADH secretion, in contrast with its extreme sensitivity to changes in osmolality (Robertson, 1994). It is also very responsive to nausea, whether or not there is vomiting, whereas the supposed effect of emotional stress is not detected by measurements of plasma ADH; the effect of unsuspected nausea is potentially an important problem in experimental animals (Robertson, 1994). Changes in ADH response can be characterised as changes in the threshold at which rising osmolality accelerates the release of ADH and the slope of the response, i.e. the increase in plasma ADH as osmolality rises. In addition, the speed of increase of osmolality affects the ADH response (Robertson, 1994). Since hypertonic mannitol produces the same effects as saline, despite depressing plasma sodium concentration, the crucial change in plasma is osmolality rather than sodium (Robertson, 1994). Unfortunately, from the point of view of diagnosis, responses to alterations in osmolality span a considerable range of normal variation (Robertson, 1993).

Inappropriate secretion of ADH (SIADH) can cause hyponatraemia. It may result from a 'reset osmostat', or without any detectable change in the osmoregulation of ADH (perhaps due to an enhanced renal response). Osmotically inappropriate secretion of ADH also contributes to the hyponatraemia in oedematous conditions such as congestive heart failure, hepatic cirrhosis and nephrotic syndrome (see above). Various drugs and tumours can also cause SIADH (Oh and Carroll, 1992; Sterns *et al.*, 1993).

Hypothyroidism causes hyponatraemia even without reductions in cardiac output (Meyer-Lehnert and Schrier, 1990). This may result from reduced GFR and enhanced proximal tubular reabsorption of sodium (both of which impair diluting capacity) and enhanced secretion of ADH. The reabsorptive changes may reflect a loss of the stimulatory effect of thyroid hormone on Na–K ATPase (*Chapter 1*). This would also cause migration of sodium into ICF, reinforcing the hyponatraemia (Meyer-Lehnert and Schrier, 1990).

Hyponatraemia can also result from migration of sodium into cells (sick cell syndrome), e.g. in potassium depletion or ischaemic impairment of sodium pump activity (Michell, 1979e; Flear *et al.*, 1980; Oh and Carroll, 1992; Geheb *et al.*, 1994). Pseudohyponatraemia is caused by the presence of an abnormally large non-aqueous component in plasma, e.g. lipids or proteins; these contribute to the measured volume but not the aqueous volume containing sodium. Ion-specific electrodes avoid this pitfall because they directly measure sodium activity rather than concentration, but their validity is undermined if the sample is diluted because the same volume artefact now affects the calculated dilution factors (Oh and Carroll, 1992; Sterns *et al.*, 1993).

It is a mistake to believe that impaired urinary dilution will be sufficiently obvious to be detected by abnormally concentrated urine (Oh and Carroll, 1992). Even attempts to quantitate the osmotic impact of renal water excretion are not free of ambiguity. Thus, the conventional approach is to measure free water clearance, the difference between osmolar clearance and urine output. Osmolar clearance, like any other clearance, is the *U/P* ratio of concentration (in this case, osmolality) times the urine output:

$$C = \frac{U}{P} \times V$$

where C=osmolar clearance, U=UOsm, P=POsm, V=urine vol./time.

It represents, conceptually, the volume of plasma which could be totally cleared of **all** solute (at plasma concentration) in unit time. Additional water, in effect distilled water, represents dilution of urine, concentration of plasma.

Thus, $U_{vol} - C = FWC$ (free water clearance).

This (solute) 'free water clearance' is normally negative because urine is concentrated in terrestrial animals. The hallmark of a renal response to ADH, therefore, is an increased negative free water clearance. Rose (1986) argues that this concept is flawed by the fact that urea is a major contributor to the osmolality of urine but not plasma. In his view, since ICF and ECF are at osmotic equilibrium and the osmolality of ECF is mainly determined by sodium and potassium, the traditional osmolar clearance formula should be replaced by using the ratio of U/P Na+K, rather than U/P osmolality; simplifying, this approximates to $U_{Na}+K/P_{Na}$. As a result, if both Na and K in urine are low, the calculation shows that excess water is being excreted, whereas the urinary osmolality could still be high due to urea.

Unfortunately, while it has been realised for some time that hyponatraemia can not only cause reversible behavioural disturbances but brain lesions (Michell, 1979a), it has recently emerged that it may precipitate a serious demyelinating disorder of the pons, leading to dysphagia, dysarthria and quadriplegia. It is a matter of intense controversy whether this risk is heightened by over-rapid correction of hyponatraemia (Sterns, 1990; Oh and Carroll, 1992; Sterns *et al.*, 1993). Controlled and monitored restriction of water intake is a useful adjunct or alternative to the use of isotonic or hypertonic saline parenterally. The elderly are especially vulnerable to disturbances of plasma sodium concentration. This has less to do with changes in regulatory mechanisms, e.g. decline in renal function, much more to do with intercurrent disease and especially exposure to the side effects of drugs (Shannon *et al.*, 1984). Above all, there is the tendency to attribute the associated behavioural disturbances to age or coexisting disease, rather than specific electrolyte abnormalities, especially those caused by clinical misadventure.

3. Volume Depletion

Anaesthesia and Surgery

Anaesthesia and surgery can present some of the most ambivalent problems in fluid and electrolyte balance. Both with sodium and water there is the possibility of actual deficits incurred before surgery, because of the underlying condition, or because of the surgery itself. Deficits may be overt—evaporation from wounds, haemorrhage or the removal of fluid-engorged tissue, or covert, e.g. by sequestration in damaged tissues ('third space'). These should be predictable, even if they are not readily quantifiable. The problem is that anaesthesia and surgery involve drugs and stimuli which affect renal function and the hormones which control it, notably aldosterone and ADH. Some stimuli, e.g. traction of viscera, are not consciously perceived but still affect hormonal systems. As a result, there may be 'irrelevant' retention of sodium and water. It is not surprising, therefore, that in various decades, fashion has swung between 'pushing fluids' and using them much more selectively (Michell *et al.*, 1989). The uncertainties have been compounded by the potential unreliability of fluid space determinations when both capillary permeability and perfusion are abnormal, as already discussed.

Additional complications are superimposed by the effects of hypotension, should it occur, on the kidneys and the ability of some drugs to impede autoregulation so that even mild hypotension becomes a threat to renal function. It is also possible for patients to be anaesthetised when, unknown to those responsible, there is already substantial loss of renal function due to CRF but the condition is still asymptomatic and plasma creatinine (let alone urea) within the normal range. Hypovolaemia and hypotension readily decompensate renal function, especially when superimposed on the effects of anaesthesia. The direct effect of anaesthetics is usually to stimulate sodium transport at low dose and inhibit it at high dose and to impede water conservation; these effects are outweighed by those mediated by endocrine or neural effects (Mujais, 1986).

It is not only the drugs associated with premedication and anaesthesia which matter; patients on chronic non-steroidal anti-inflammatory therapy, for example, may be more susceptible to ARF because the normal renal protection against excessive vasoconstriction (in response to hypotension or hypovolaemia) is prostaglandin dependent.

Intercurrent disease is also important; patients with renal disease are more susceptible to metabolic acidosis or alkalosis (e.g. caused by diarrhoea or vomiting) and less able to correct the effects of injudicious fluid therapy. They are also less able to compensate for chronic changes in pCO_2. Where intraoperative hypokalaemia occurs, clinicians are sometimes too anxious to correct it. Unlike hypokalaemia caused by diuretics (*Chapter 11*), there is little evidence that it causes ventricular dysrhythmias and a risk that potassium infusions will cause dangerous complications (Wong *et al.*, 1993). The combination of compromised renal function, metabolic acidosis and extensive tissue damage heightens the risk of hyperkalaemia, whether or not there are underlying deficits of cell potassium (due to inappetance, gastrointestinal losses or urinary losses as a result of increased aldosterone secretion triggered by a mild hypovolaemia caused, for example, by vomiting or diarrhoea). Compromised cardiac function may have activated renal compensatory responses, including angiotensin-dependent constriction of the efferent arteriole in defence of GFR. This response may be impaired in patients on ACE-inhibitors.

The challenge, therefore, is to know whether fluid and electrolyte abnormalities, especially in the postoperative period, are responses to actual deficits or whether they are primary disturbances. These problems are reduced if, so far as possible (depending on the urgency of the underlying conditions), patients come to surgery with pre-existing deficits and disturbances already corrected (Michell *et al.*, 1989; Faber *et al.*, 1994).

It follows that when fluids are given **during** surgery, rather than before or after, the purpose should be clearly defined. Maintenance fluids (e.g. 1/5 N glucose–saline, i.e. Na 30 mmol/l, glucose 4% approx.) are designed to replace the electrolytes and water normally taken orally if, for some considerable period, oral intake is precluded. They are low in sodium because the ratio of sodium:water in the nutritional requirement is much less than that in plasma. The 4% glucose is merely a means of making the solution isotonic; its calorie content is negligible compared with daily requirement, let alone the increased requirements sometimes encountered in surgical patients. The question is why should such a solution be used **during** surgery? Before, perhaps, or after, because of inadequate oral intake but **during** surgery, in my view, the reason for giving intravenous fluids is to protect plasma volume and composition. ECF volume can only be restored by solutions with adequate 'osmotic skeleton', i.e. a sodium concentration around 140 mmol/l *(Chapter 1)*. Anything less allows water to migrate into ICF instead of staying where it is needed, in ECF. A subsidiary reason for giving fluids during surgery (in small quantities) is simply to maintain an intravenous access in case it is needed for an emergency.

Central to the dispute over fluid loss during surgery is the question of plasma sodium concentration. It may fall, as usual, in response to hypovolaemia. But it may also fall because, in general, the stimuli for irrelevant retention of water, notably excess ADH for the prevailing osmolality, outweigh those for retention of excess sodium. There should, therefore, be the greatest caution in using low-sodium fluids during surgery or in the immediate aftermath. That is not to say they should not be used, e.g. to repair water deficits, but that the reasons should be clear and it should not be a mindless routine ('give a litre of fluids', as in 'give them a kg of books'). The point is that low sodium solutions are inefficient in repairing ECF volume and may reinforce an existing trend towards hyponatraemia. They may have a role prior to surgery to offset the effects of preoperative food and fluid deprivation.

The effects of fasting are not simply those of reduced intake of electrolytes and water. Fasting has been seen to cause natriuresis in humans, rabbits, horses, sheep and rats (though not when salt intake is excessive). Most studies involve complete restriction of food intake for several days, but similar effects are seen in sheep with merely a 30% reduction of food intake for 48 h, similar to the natural fall at oestrus (Michell, 1981). The natriuresis is not simply a continuing excretion of excess dietary salt, since it also occurs when sodium intake is low. Among the suggested causes are cation loss to cover loss of anions generated by catabolism, metabolic acidosis and glucagon. Data concerning the latter are conflicting and restoration of protein intake interrupts the natriuresis despite persistent acidosis. Aldosterone levels are increased but the reason is unknown since plasma potassium is not consistently increased and renin is usually suppressed. This suggests that the excreted sodium is a temporary endogenous excess, either from gut or bone (Michell, 1981).

The actual effects of anaesthesia and surgery depend on the patient, the species, the anaesthetic, hydration and cardiovascular function, the extent, duration and severity of the surgery. Generalisations can therefore only draw attention to possibilities which need to be considered, to potential rather than consistent hazards. Most common anaesthetics depress myocardial contractility and cardiac output and are therefore liable to reduce renal perfusion and GFR (Faber *et al.*, 1994). Horses subject to 30–60 min of halothane anaesthesia raised their plasma creatinine by approximately 33% (Steffey *et al.*, 1993). GFR is more consistently reduced than renal blood flow (RBF) (Mujais, 1986), suggesting that filtration fraction falls and, therefore, that the direct effect would decrease sodium retention. The use of positive pressure to support respiratory function further depresses renal perfusion, GFR and sodium excretion (Sladen, 1994).

The impact of anaesthesia on autoregulation varies with the agent and, for example, halothane leaves it virtually unaffected. Ketamine is unusual in increasing renal perfusion (Sladen, 1994). The effects of anaesthesia are not restricted to general anaesthetics. Spinal anaesthesia can also cause hypotension. The most striking effect of surgery is a post-operative reduction in urine output, noted in dogs as long ago as 1905; this results from changes in GFR as well as ADH (Fieldman, 1988; Fieldman *et al.*, 1985). Thus, the increase in ADH which frequently accompanies surgery, whether or not there are osmotic or hypovolaemic stimuli, means that the patient will less readily excrete excess parenteral fluids, especially if they are hypotonic (Fieldman *et al.*, 1985). A variety of stimuli can increase ADH in anaesthetised patients, including hypoxia and visceral traction (Haas and Glick, 1978; Robertson, 1993). It is not clear whether the effect of stimuli such as pain, hypoxia or stress is direct or whether it is mediated by haemodynamic effects or nausea, a potent stimulus to ADH, but whatever the reason, ADH secretion increases (Robertson, 1993, 1994). The increase often persists for several days and is not prevented by deliberate expansion of plasma volume with colloids; it may represent a defence against hypotension (Fieldman, 1988).

Aldosterone secretion also increases during surgery; this rise only lasts a few hours if hypovolaemia is prevented (Le Quesne *et al.*, 1985). Sodium retention may nevertheless last for several days. Anaesthesia also activates the renin–angiotensin system and this is especially marked in hypovolaemia and contributes to a further decline in renal function (Mujais, 1986). Not surprisingly, the adverse effects of anaesthesia on renal function can be greatly increased by the depth of anaesthesia or by concurrent dehydration (Bruce, 1980; Gray *et al.*, 1989). A number of volatile anaesthetics can be nephrotoxic because their metabolites include fluoride which causes polyuria and tubular damage; halothane is not metabolised to fluoride (Sladen, 1994).

While there are reasons for retention of excess water and sodium, the trend is for over-retention of water to prevail and for plasma sodium concentration to fall (Michell *et al.*, 1989; Gann and Kenney, 1990). Thus, post-operative hypernatraemia is unusual, whereas hyponatraemia is common (Greco and Jacobson, 1990). This is reinforced in hypovolaemic patients by reduced delivery of sodium from the proximal tubule to the diluting segments of the nephron (Shapiro and Anderson, 1987). Sodium may also be lost into cells in underperfused tissues as a result of 'sick cell syndrome', i.e. failure of cells to maintain normal sodium transport (Flear *et al.*, 1980; Illner, 1984). Too many clinicians fail to grasp that the key to expansion of plasma volume is the presence of sodium in parenteral fluids at something close to its ECF concentration; there is a vague feeling that somehow a glucose–saline mixture must be more physiological, more nutritive, despite the actual triviality of the energy content. "One often sees patients in intensive care units receiving intravenous solutions too dilute for them to excrete, i.e. because. . . . they have excessive amounts of circulating ADH" (Faber *et al.*, 1994). Moreover, as discussed earlier, the danger is that if hyponatraemia is detected after sufficient time for brain cells to adapt (by retaining additional solute), rapid correction may not only cause additional symptoms but lasting damage, i.e. central pontine myelinolysis (Faber *et al.*, 1994). If low sodium solutions are used during surgery or the immediate recovery, two questions need to be asked:

1. Has blood volume been adequately protected?
2. What perceived risk of hypernatraemia or what actual losses of electrolyte-free water is the solution intended to correct?

Post-operative water intoxication remains one of the common causes of acute hyponatraemia: it is potentially lethal and is entirely avoidable by avoiding excessive administration of low sodium parenteral fluids (Sterns, 1990).

Finally, even among those who see the need to use a sodium-containing solution, there remains an excessive tendency to choose 'normal' or 'physiological saline'. It is neither. It is considerably higher in sodium than normal plasma and has neither bicarbonate nor a precursor. It will, therefore, cause dilutional acidosis unless renal function is sufficiently good to prevent it. The fear is frequently that 'excess' or 'over-rapid' bicarbonate will cause metabolic alkalosis and adverse effects on cerebrospinal fluid (CSF). Bicarbonate at around 25 mmol/l (or precursors to give the equivalent yield) is at higher concentration than acidotic plasma, but no higher than normal plasma; it merely avoids dilution of bicarbonate. It therefore seems rational that any acidosis should improve as a result of the selected parenteral fluid, rather than despite it, moreover healthy kidneys give excellent protection against metabolic alkalosis unless hypovolaemia or potassium depletion are superimposed (Michell *et al.*, 1989).

Hypovolaemia and Shock

Shock is a form of peripheral circulatory failure associated almost invariably with hypovolaemia and usually with a reduced cardiac output, but not always. The hypovolaemia may be relative, i.e. it can result from excessive vasodilation as well as actual loss of fluid.

Since sodium is the osmotic skeleton of ECF, hypovolaemia frequently results from sodium depletion caused, for example, by diarrhoea or adrenal insufficiency (*see Chapter 3 and p.221*). **Hypovolaemic shock** could result from primary sodium loss, e.g. in severe diarrhoea, but more usually it results from haemorrhage, leakage from plasma (e.g. with burns) or maldistribution of ECF. Indeed, shock is a caricature of the physiological response to haemorrhage; the features are recognisable but exaggerated to an absurd and damaging degree (Fig. 8.5). Thus, vasoconstriction and increased coagulability of blood contribute to the defence against haemorrhage, but if they are excessive or triggered in the absence of blood loss, they become damaging and ultimately lethal (Michell, 1989b; Michell *et al.*, 1989).

In the advanced stages of shock, vascular resistance falls (Schadt, 1994). This could be viewed as loss of arteriolar tone or as an adaptive

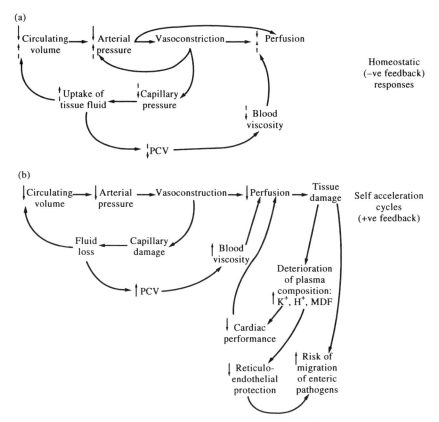

FIG. 8.5. Vasoconstriction: adverse and beneficial effects; (a) defence (dotted arrow pointing upwards or downwards) against loss (arrow pointing downwards) of circulating volume; (b) some of the vicious cycles of shock.

response to sustain regional blood flow or minimise further blood loss; the reasons are unclear, but the threat to survival is obvious. Severe hypovolaemia causes reflex inhibition of cardiac output and reduction of peripheral resistance; one of the benefits of hypertonic saline (see below) may be to reduce this response (Kirkman and Little, 1994; Secher and Friedman, 1994). Cardiac function is also depressed by a factor (or factors) released as a result of splanchnic, particularly pancreatic vasoconstriction (myocardial depressant factor, MDF; Michell, 1989b). The ability to avoid hypotension in response to hypovolaemia is severely compromised by anaesthesia, as is the protection of renal perfusion (Schadt, 1994).

Since **endotoxic shock**, in many ways the product of inflammatory mediators in circulation, involves maldistribution of blood volume and abnormal capillary permeability, it comes to share many of the features of hypovolaemic shock. Similarly, hypovolaemic shock, as a result of gut damage and impaired hepatic function associated with splanchnic vasoconstriction, often has an element of endotoxaemia. Shock usually occurs alongside other complications of disease or injury, for example, the metabolic response to trauma (Michell, 1985c; Barton and Cerra, 1991; Katz *et al.*, 1993). Some of these directly affect therapy, for example, chest injury or the inhalation of smoke or secretions predispose to pulmonary oedema (see above). Shock is, predominantly, a cardiovascular problem with widespread systemic effects but among these, the cellular aspects, whether caused by shock or by the metabolic response to trauma, pose some of the most intractable problems. These underlie many of the instances where shock remains 'irreversible', despite the highest standards of monitoring and therapy. The adverse effects of catecholamines are usually attributed to excessive vasoconstriction but they also result from changes in cell metabolism.

Four of the main controversies surrounding the treatment (prevention) of shock are:

1. the relative merits of colloid or crystalloid for repair of circulating volume;
2. the importance of blood for haemorrhagic shock;
3. the benefits of high-dose corticosteroids;
4. the role of small volume hypertonic saline.

Only the last of these relates directly to sodium, therefore the others are only briefly summarized; fuller reviews are widely available, including Michell (1989b) and Secher *et al.* (1994). The latter makes the important clinical point that, while textbook accounts of shock emphasise the associated tachycardia, this is true of mild and severe hypovolaemia, whereas intermediate stages commonly present with bradycardia. Metabolic acidosis is a frequent accompaniment of shock as a result of tissue hypoxia and impaired hepatic clearance of lactic acid.

Crystalloid vs Colloid?

It makes little difference provided both are used to full effect, i.e. assessed by improvement of monitored variables. Much of the controversy relates to fixed-dose experiments, especially where insufficient allowance was made for the fact that the entire dose of colloid remains intravascular until excreted, whereas the majority of a crystalloid distributes into interstitial fluid (ISF). Since pulmonary oedema results from excessive capillary pressure, especially where there is local damage, rather than haemodilution, the choice of colloid reflects its greater efficacy in remaining within the intravascular compartment, not its superior safety. Moreover, since there are constraints on the volume used, often due to fear of interference with clotting, there is good reason to use both. Not only does ISF require repletion sooner or later, but the circulatory system in shock demands large volumes of fluid because capillaries are leaky; such large volumes cannot be given as colloid alone.

Pulmonary oedema is not the only cause of impaired gas exchange in shock; loss of surfactant, collapse of alveoli and thus shunting of venous blood into the systemic circulation are even more important (Schumer, 1986; Seeger, 1987). The preoccupation with pulmonary oedema has, perhaps, diverted attention from the potential adverse effects of peripheral oedema on tissue oxygenation (Geheb *et al.*, 1994). In a pig model of hyperdynamic endotoxin shock, Kreimeir *et al.* (1993) found clear advantage of colloid (dextran 60) over crystalloid (Hartmann's solution), even when the latter was used at three times the volume. Both were used to maintain the same pulmonary capillary wedge pressure, but the colloid sustained a greater cardiac output and better renal and splanchnic perfusion. This probably reflected its superior ability to stay in circulation, its beneficial effects on capillary blood flow and its lesser tendency to cause peripheral oedema.

Recently, a question has been raised concerning the fundamental assumptions of the management of circulatory shock, namely the pre-eminent necessity for urgent replacement of plasma volume. It has been suggested that an exception may arise where there is hypotension due to haemorrhage and difficulty in controlling the bleeding (Bickell *et al.*, 1994). This concern is not new:

Injection of a fluid that will increase blood pressure has dangers in itself.
(Cannon *et al.*, 1918)

The fear is that, until the blood loss is controlled, restoration of normal arterial pressure may displace clots or impede clot formation, the latter problem being exacerbated by dilution of coagulation factors. The issue raised is important, but the evidence supporting the challenge is, as yet, insufficient; further studies are needed to resolve the dilemma (Jacobs, 1994).

Blood

The reason for supplying blood is because haemoglobin is already so low as to impair tissue oxygenation or liable to become so because of a rapid fall and an undetected or uncontrolled source of haemorrhage or haemolysis. Even with haemoglobin levels as low as 10 g/dl, not every patient needs blood (Faber *et al.*, 1994). Mild haemodilution in shock is beneficial because the packed cell volume (PCV) that is sufficient for full activity is well above the level needed for maintenance during recumbency. The adverse effect of an unnecessarily high PCV is to impede capillary blood flow because as PCV rises there is an accelerating increase of blood viscosity (Fig. 8.6). A high PCV in a shocked patient is therefore not simply an index of haemoconcentration due to loss of plasma, let alone an indication of generous oxygen carrying capacity, but a warning of declining peripheral perfusion. The increased viscosity is compounded by microthrombi and increased red cell adhesion to other red cells or to endothelium where circulation is sluggish, together with loss of red cell elasticity. The essence of effective perfusion is not the concentration of red cells but their ability to flow through capillaries. Other constituents of blood, notably plasma proteins, are extremely useful in the treatment of shock and there are now a number of ways of supplying them therapeutically. Eventually the dangers and difficulties of maintaining oxygen carrying capacity by blood transfusion will be solved by artificial bloods, of which some have already become available, e.g. haemoglobin-containing liposomes ('neo red cells', Usaba *et al.*, 1991).

Corticosteroids

The problem with corticosteroids is a fascinating clash between theory and practice, data and clinical impression. The theoretical

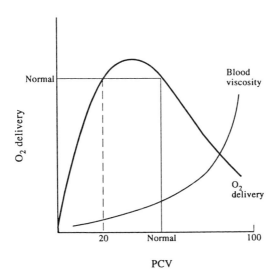

FIG. 8.6. Haematocrit, oxygen delivery and blood viscosity.

benefits of high-dose corticosteroids for the treatment of shock are overwhelmingly impressive in their variety (Michell, 1985c, 1989b); almost the only weakness is their inability to repair circulating volume. There is also a strong and frequently critically-based clinical impression of their beneficial effects, though there are also those who are sceptical. The problem is a lack of well-designed clinical trials demonstrating efficacy. Moreover, too many of the experimental data involve pre-treatment rather than reversal of established shock. There is also an element of risk, albeit in a small minority of patients with unsuspected loss of renal function. The result is that in human medicine, corti-costeroids are now contraindicated for the form of shock in which most of the evidence of benefit was obtained, i.e. endotoxic shock. The problems of proving their efficacy are substantially the ethical problems of double-blind trials in a potentially lethal condition with the hope of benefit from the trial drug; the decision to put patients in the control group has implications beyond good scientific design. Moreover, steroids are not intended as sole therapy and other aspects of treatment will inevitably vary considerably, according to individual need. Again, ethical considerations limit the scope to constrain legitimate clinical choice in an acute life-threatening condition.

The verdict on corticosteroids may yet prove premature, even unfair, because the problem may be one of proof rather than effectiveness (Sjolin, 1991). Shock disrupts the relationship between dose rate and drug concentration, affects receptor numbers and occupancy and hence both pharmacological effects and clinical responses. As a result, a fixed dose may have quite different effects and persistence in different patients (Neugebauer et al., 1992). If corticosteroids have been convicted in the therapy of shock, it may be on unsafe and unsatisfactory evidence.

Hypertonic Saline

Here we have an opposite controversy; substantial evidence of benefit but continuing debate as to its physiological basis. A number of potential benefits have been suggested, including reduction of endothelial swelling and hyperkalaemia, but the main effect seems to be a reflex improvement of cardiac output and renal and splanchnic perfusion; it also moderates hypotension caused by anaesthesia (Michell, 1989b; Dyson et al., 1990). Evidence that this depended on reflexes originating in the pulmonary circulation is weakened by the fact that it hinged on the results of vagotomy. This abolishes the effects of hypertonic saline; unfortunately, it also increases the impact of shock and weakens the response of dogs to isotonic saline where reflexes are not the issue (Schertel et al., 1991). Curiously, even intraosseous hypertonic saline is effective (Chavez-Negrete *et al.*, 1991) and some of the benefit may result from the reduced secretion of vasoconstrictors such as catecholamines and ADH (Wade *et al.*, 1991). There are also direct stimulatory effects on the myocardium in septic shock (Ing *et al.*, 1994).

Because it is used in small volume but high concentration (approximately 8%), hypertonic saline does not sufficiently increase ECF volume to explain its benefits, even when allowance is made for additional fluid drawn from ICF (Nakayama *et al.*, 1984). The impact on circulating volume is increased when hypertonic saline is combined with a colloid. Both solutions increase cardiac output, but the combination produces greater improvements in renal and splanchnic perfusion (Kreimeier *et al.*, 1993). There has been concern that by restoring arterial pressure, hypertonic saline may cause additional haemorrhage, but this seems to result from rat models involving transection of the caudal artery, rather than clinical use (Krausz *et al.*, 1992; Rabinovici *et al.*, 1991). In a canine haemorrhagic shock model the use of hypertonic saline (with dextran) or isotonic saline to produce the same restoration of cardiac output needed 8 or 47 ml/kg, respectively. Even with an experimental design which excluded differential effects on cardiac output, the hypertonic saline showed beneficial effects on tissue oxygenation, perhaps by reducing swelling of capillary endothelial cells and by causing less interstitial oedema (Tobias *et al.*, 1993). Improvement of enteric perfusion may be a particularly important benefit because it reduces bacterial translocation and the risk of endotoxaemia (Reed *et al.*, 1991). Not all studies confirm these beneficial effects on renal and enteric perfusion (Kreimeier *et al.*, 1990). There has also been concern, especially in horses, that the beneficial effects may be preceded by transient hypotension and a similar fall has been seen in humans (Prien *et al.*, 1993). Perhaps this reflects a direct depression of myocardial performance by the initial high concentration as seen *in vitro* in rat hearts (Brown *et al.*, 1990). Whatever its merits, hypertonic saline is an adjunct, not an alternative, to adequate restoration of circulating volume (Wade *et al.*, 1992), which remains the cornerstone of the treatment or prevention of shock. Moreover, since shock involves accelerating vicious cycles (positive feedback loops), delay or indecision undermine the effectiveness of all forms of therapy; early treatment is a vital element of success. Future progress is likely to hinge on the development of antibodies or receptor antagonists to the endogenous and exogenous mediators involved.

Adrenal Insufficiency (Addison's Disease; White, 1994)

Adrenalectomy deletes both mineralocorticoids and glucocorticoids, but clinical defects of adrenal function usually cause selective impairment of adrenal corticosteroid synthesis. The obvious reason for hyponatraemia in adrenal insufficiency is in response to the hypovolaemia caused by sodium depletion which results from defective mineralocorticoid synthesis or release (Gines *et al.*, 1994). Enhanced reabsorption in the proximal tubule reduces delivery to the diluting segments of the nephron and both thirst and ADH secretion are enhanced by hypovolaemia. It has long been suspected, however, that additional impairment of water regulation is involved either because glucocorticoids normally suppress ADH release or because they maintain the impermeability of the collecting duct, i.e. they oppose the effect of ADH. Loss of either or both of these interactions would allow excessive water conservation and further depress plasma sodium concentration. The factors responsible thus include ADH secretion and impaired urinary dilution, as well as glucocorticoid deficiency (Oh and Carroll, 1992; Bichet *et al.*, 1993; Robertson, 1994). The latter reinforces the increased secretion of ADH (Gines *et al.*, 1994; Sterns *et al.*, 1994). It is possible, but unproven, that glucocorticoid deficiency directly increases collecting duct permeability; it certainly impedes water excretion by mechanisms additional to any involving ADH or reduced GFR (Weiss and Robertson, 1984).

A variety of enzyme defects can cause selective aldosterone deficiency (De Fronzo and Smith, 1994). It can also arise from defective renin secretion; the reason for this condition is unknown, but it does not result from ECF volume expansion. Suggested causes have included juxtaglomerular damage, inhibition of prostaglandin synthesis and drugs such as non-steroid anti-inflammatories (NSAIs) (*Chapter 11*). Hyperkalaemia, *per se*, suppresses renin secretion. There are also three forms of pseudohypoaldosteronism, including Gordon's syndrome (Type II). Some cause renal sodium loss (Types I and III), all cause hyperkalaemia. Type II does not involve salt wasting but excessive reabsorption and volume expansion (Klemm *et al.*, 1991; De Fronzo and Smith, 1994). Unlike the other forms of hypoaldosteronism, pseudohypoaldosteronism is resistant to exogenous as well as endogenous mineralocorticoids (Daniel and Henrich, 1990).

Heat and Exertion

While the most important clinical disturbances are probably those associated with disease, there are fascinating disturbances and adaptations associated with heat and exertion. Here there are important species differences; sweating is a potent source of sodium loss in humans, whereas in most non-primates, the main route of thermoregulatory evaporative cooling is the upper respiratory tract (during visible or less obvious panting).

Dogs have a highly efficient panting mechanism, minimising effort (and therefore generation of unwanted heat) by utilising the resonant frequency of the chest, and they have no significant thermoregulatory sweating, hence heat or exertion cause pure water loss. Skin secretion in sheep contains potassium, rather than sodium, and sweating is trivial compared with respiratory water loss in evaporative cooling. Cattle, on the other hand, can produce substantial quantities of sweat, especially heat-adapted breeds at high temperatures (thus at modest U.K. summer temperatures they sweat less than British breeds).

The most familiar species of professional athletes are humans, dogs and horses. They provide an interesting contrast: humans lose sodium and water, whereas dogs lose only water as a result of heat and exertion. Human sweat, like urine, reaches low sodium concentration under the salt retaining influence of aldosterone.

Supplementary salt is therefore probably not necessary simply because of a move to a hot climate; it may merely postpone this adaptation. On the other hand, exertion at high temperature may induce sufficiently profuse sweating to defeat the effectiveness of the aldosterone-induced reduction in sweat sodium concentration, hence the justification for the use of salt tablets.

What are the potential fluid and electrolyte problems of exertion? Certainly not renal function, unless it is already compromised or the exertion is severe. The trade-off between salt and water loss seems crucial. A pure sprinter can afford water loss because it could not possibly amount to enough to cause hypernatraemia. But a marathon runner has a conflict; sodium loss contracts ECF volume and even if there is some mobilisation of ISF, plasma volume is likely to fall. Nevertheless, the serious threat is hypernatraemia resulting from continuing loss of hypotonic sweat. Moreover, if it becomes sufficiently severe it interferes with thermoregulation as well as its various other effects (Maughan, 1994).

There are a few instances of hyponatraemia at the end of prolonged exertion; these are usually associated with consumption of low or zero-sodium fluids (Noakes, 1992). The vast majority of athletes in difficulty at the end of severe protracted exertion are hypernatraemic. Thus, there is little reason for electrolyte replacement during exercise for most athletes (Maughan, 1994). There may, however, be a reason to include sodium in the oral rehydration fluid because, as in oral rehydration therapy (ORT) for diarrhoea (*Chapter 3*), it is needed to promote uptake of glucose and water (Rehrer, 1994). It will also prevent hyponatraemia if excess fluid is consumed. High glucose solutions deliver more energy despite slower gastric emptying, but enhance the risk of hypernatraemia by drawing water into the intestine. Moreover, their absorption could be restricted by lack of sodium for cotransport, although Hargreaves *et al.* (1994) found no evidence of this. The redistribution of blood flow during intense exercise could also restrict absorption; this effect is probably trivial unless exercise is extremely intense and protracted (Schedl *et al.*, 1993). Training, especially at high temperature, increases the sustainable effort and one of the benefits is expansion of plasma volume; plasma volume also expands, however, in response to a single session of strenuous exercise and the increase persists for several days. It allows a higher sweat rate as well as boosting cardiac output (Maughan, 1994).

In humans, water is not the best post-exertion rehydration fluid because it may cause hyponatraemia and hasten its own excretion while suppressing thirst (Maughan, 1994). It is clear that the inclusion of sodium accelerates the restoration of plasma volume and, as in diarrhoea, glucose–sodium mixtures seem to be most suitable. Where in the range 20–90 mmol/l, the optimum sodium concentration lies remains the subject of debate. Concern is often expressed about the palatability of sodium-containing oral rehydration solutions (ORSs) for athletes. The natural effect of exercise on the palatability of sodium solutions is interesting. In the first hour after exertion, the palatability of low sodium solutions was enhanced but after 3 h, by which time plasma sodium had fallen back to normal (as a result of water intake), the palatability of even hypertonic sodium solutions was strikingly increased (Takamata *et al.*, 1994).

The case of the horse is interesting. The sodium concentration in sweat is high and apparently unresponsive to aldosterone (whereas fear or anticipation, catecholamine secretion, apart from heat or exertion, can massively increase sweat production). As with any athlete, the exact consequences of exertion depend on its intensity and duration; the problems of sprinting are entirely different from those of marathon running. Moreover, in evolution, the incentive was not a prize, but survival. Why, then, do horses, typical herbivores with a low dietary sodium intake, 'waste' so much precious sodium in their sweat? In theory, even mild degrees of hypovolaemia could result in sufficient vasoconstriction to impede exercise tolerance; this occurs in chronic heart failure (Michell, 1988d). Perhaps during exertion it is outweighed by local and reflex effects on the muscle vasculature. For wild horses, like other herbivores, the main reason for a marathon (or a sprint) would have been to outrun predators. Human marathon runners in recent years have taken to topping up their water during the race. No such luxury would be permissible for herbivores escaping their predators; water might be available but not the time to drink it. Since the main impact of a pure water loss is hypernatraemia rather than hypovolaemia, avoidance of hypernatraemia may be a key determinant of survival and dumping of sodium into sweat may be one way of doing it (Michell, 1982b). This perhaps makes sense of the higher sodium content of equine sweat—the benefit of protection from hypernatraemia at the expense of some sacrifice of ECF volume. The sodium concentration of equine sweat may have been exaggerated by techniques which allowed evaporation to occur before sample collection, but it is undoubtedly high. Horses are able to sustain greater water losses than humans without sacrifice of athletic performance (Andrews *et al.*, 1994).

Short, fast races in hot weather cause slight dehydration but substantial lactic acidosis; losses of water and electrolytes are of the order of 2–10 l of water (up to 2% of bodyweight) with 200–500 mmol of sodium. Protracted endurance rides in a moderate climate cause less acidosis, but water losses of the order of 7% of bodyweight; loss of chloride in sweat may cause metabolic alkalosis (Michell *et al.*, 1989). Even before sweating begins, plasma osmolality increases and plasma volume undergoes isotonic contraction, as in humans (McKeever *et al.*, 1993). This may reflect an increased ISF in active muscle beds, not simply fluid loss (Fanestil, 1994). Whether plasma sodium increases or decreases depends on the nature of the exercise and whether or not water is taken (Michell *et al.*, 1989; McKeever *et al.*, 1993). Plasma potassium increases as a result of release from active muscles (Andrews *et al.*, 1994).

In humans, exertion in hot climates certainly requires additional salt during initial acclimatization, but excessive sodium is hazardous and the overriding need is for adequate water intake (Knochel and Read, 1994). Ingestion of excess salt and insufficient water is potentially a lethal cause of heatstroke in athletes. Apart from hyperthermia and hypernatraemia, there is usually rhabdomyolysis and this, combined with acute renal failure, readily causes fatal hyperkalaemia (Knochel and Read, 1994). The effect of exercise on renal function differs between species. In dogs, even intense exercise has little effect, whereas in humans and horses intense exercise encroaches on renal perfusion; humans are more successful at limiting the resulting fall in GFR by increasing their filtration fraction (Schott *et al.*, 1991). Our own observations sugest that even mild degrees of exercise may substantially reduce renal perfusion in horses (Gleadhill *et al.*, 1995); perhaps this is adaptive, since normal renal blood flow is exorbitantly high and a temporary reduction would allow greater perfusion of active muscles. Unlike humans, horses increase their sodium excretion after exercise (McKeever *et al.*, 1991), perhaps a further protection against hypernatraemia.

On top of other adverse effects, hypernatraemia may cause further elevation of body temperature, hence further water loss. Moreover, even if sweat contains more sodium than plasma, the danger of hypernatraemia still exists because of the massive increase in respiratory water losses. On the other hand, the magnitude of respiratory water loss and dermal water loss (not just sweat) both depend considerably on high humidity; the benefit of air conditioning is not so much that it cools the air as that it dries it and allows efficient evaporative cooling. In arid conditions, the skin may seem dry, but sweating can still be profuse. One of the benefits of rehydration during exercise is that it reduces hypernatraemia and permits more generous skin blood flow and more effective thermoregulatory sweating (Montain and Coyle, 1992). Since hypernatraemia, rather than hypovolaemia, is the main danger of excessive exertion, it would be wrong to assume that oral rehydration solutions formulated for diarrhoea *(Chapter 3)* are also suitable for athletes; unlike diarrhoea, the optimum formulation for absorption of water rather than sodium may be more important. The ideal sodium concentration for sports drinks remains uncertain (Burke and Read, 1993).

Another aspect of circulating volume in exercise which involves a trade-off is PCV. Additional RBC can be mobilised from spleen and a rise of PCV, also resulting from dehydration, might be argued to increase circulatory efficiency—more oxygen carrier per volume of blood pumped. The problem arises from the virtually exponential rise in viscosity which accompanies increased PCV. In shock, a high PCV not only warns of loss of plasma volume, but directly implies difficulty in maintaining peripheral perfusion in the face of high blood viscosity. In shock, where exertion is irrelevant, it is clear that the optimum PCV for oxygen delivery is actually subnormal. Further considerations enter into this optimum, notably factors affecting the deformability of RBC. Fit horses have a lower blood viscosity than sedentary horses with the same PCV; similar observations have been made in humans, but not without conflicting data (Sommardahl *et al.*, 1994). The optimum PCV for exertion will depend on the species and the type of exercise, but it is unknown, except perhaps for humans. In horses, PCV increases by up to 50% during exercise and blood viscosity (but not plasma viscosity) increases substantially (Sommardahl *et al.*, 1994). This is probably due to an increase in high molecular weight plasma proteins and their effect on red cell rouleaux formation, as well as the rise in PCV. There is, however, an important difference between the flow characteristics of blood in shock and in exertion. Blood, like non-drip paint, is thixotropic: movement reduces its viscosity. Thus, in shock, at low 'shear-rates' there is the maximum impact on viscosity. In exercise, the high shear rates offset the rise in viscosity which would otherwise result from the increased PCV (Fedde and Wood, 1993).

Less familiar but long-standing (and recently, equally professional) athletes include camels. In their thermoregulatory adaptations, as in most matters, they seem to be a law unto themselves. The camel has a lower PCV than other athletic species, including humans, horses and dogs (Knight *et al.*, 1994a). The red cells are unusual not only in being elliptical, but also highly resistant to hypotonic haemolysis, perhaps an adaptation to rapid rehydration in these animals (Swenson, 1993; Knight *et al.*, 1994a). PCV remains low despite exercise, rising by little more than 10% (Evans *et al.*, 1994).

A subtle form of water conservation in camels is their ability to permit a diurnal rise in temperature, dissipating heat to their cool nocturnal surroundings rather than wasting water on the evaporative cooling which would be needed during the high temperatures of the day (Schmidt-Nielsen *et al.*, 1957). Unlike horses or humans, camels maintain a stable or increased plasma volume during exertion; plasma sodium is also unchanged or slightly increased, but potassium increases, probably as a result of loss from active muscles; blood urea and creatinine remain unaffected by exertion (Snow *et al.*, 1988; Rose *et al.*, 1994). Alkalosis develops, but it is respiratory rather than metabolic. High intensity exercise (to fatigue) causes lactic acidosis and a slight rise in PCV (Knight *et al.*, 1994b). It also caused a substantial rise in plasma sodium concentration (8 mmol/l), mostly within 2 min; plasma potassium doubled. Extreme exertion makes even camels human!

Lest anyone imagines that camels are mere curiosities of the desert, not true athletes, they can go up to 10 km (6 miles) at up to 12 m/sec (27 mph) or up to 50 km (31 miles) at up to 3 m/sec (7 mph). Even carrying 200 kg (0.2 tonnes) their range is up to 32 km/day (20 miles). These figures (Rose *et al.*, 1994) would be remarkable for any species, let alone one able to sustain them carrying loads approaching half their body weight in some of the hottest, most arid environments on earth, with maximum exposure to solar radiation. In the characteristic understated prose of the British Military:

> *Properly selected and cared for, they make excellent transport, are not more difficult to deal with than other animals and on occasion are capable of a sustained effort which cannot be accomplished by any others.*
>
> (War Office, 1908)

It is time to return to more prosaic disturbances of fluid balance.

4. Renal Dysfunction

Chronic Renal Failure (Swartz, 1990; Tuso et al., 1994)

Chronic renal failure results from a relentless, progressive, irreversible loss of nephrons. GFR falls, although for some time, the associated rise in plasma creatinine or urea remains indistinguishable within the normal range. The measure of renal efficiency, however, is clearance—the rate of solute excretion at a given GFR:

$$C = \frac{U \times F}{P}$$

where U=urinary concentration, P=plasma concentration, F=urine output/time.

The clearance of solutes, such as creatinine, excreted virtually entirely by glomerular filtration, without secretion (or reabsorption), is a marker of GFR and therefore falls. The excretion of creatinine or other nitrogenous end-products remains virtually normal, even when the increase in plasma creatinine becomes obvious. The clearance formula emphasises that as chronic renal failure advances, solute output can be maintained, but by raising the concentration in plasma and, therefore, in unit volume of glomerular filtrate. Whether or not the condition becomes symptomatic depends on how well a rise in the concentration of the solute can be tolerated.

In fact, creatinine and urea are extremely well tolerated, as one would expect of a safe excretory product, and symptoms do not usually appear until two-thirds to three-quarters of renal function has been lost. There may thus be many years in which renal function is compromised but the patient appears and feels virtually normal. Renal function can, however, be more readily decompensated by factors which further reduce RBF, e.g. hypovolaemia, salt depletion, dehydration, hypotension. The combination of hypovolaemia and hypotension, e.g. due to anaesthesia in a dehydrated patient, is particularly potent in reducing GFR and the hazard is increased by NSAIs. These prevent the prostaglandin-induced vasodilation which normally protects renal perfusion against excessive vasoconstriction (see above).

The basis of the kidneys' compensatory response to CRF is the increased GFR in each of the surviving intact nephrons. This moderates but cannot prevent the overall reduction in GFR. Compensatory hyperfiltration distorts renal function because of the high flow rates within the nephrons. Equilibration times are reduced and extremes of concentration or dilution can no longer be attained. Patients become susceptible to fluid overload as well as dehydration. Moreover, the interference with renal acid–base regulation not only predisposes to metabolic acidosis, but also undermines the ability to respond to chronic respiratory acidosis or alkalosis with compensatory changes in plasma bicarbonate. Despite the reduced nephron cell population for the secretion of potassium and the trend towards acidosis, plasma potassium remains normal until CRF is well advanced. This is because of adaptive increases in Na–K ATPase activity *(Chapter 1)*. Indeed, in cats with CRF, the combination of inappetence and sustained excretion of potassium may cause sufficient potassium depletion to inflict muscle damage (Dow and Fettman, 1992).

In both cats and dogs, which normally drink to requirement rather than consuming flavoured fluids for pleasure, urine is usually concentrated and the first sign of CRF may be the loss of concentrating capacity, revealed by compensatory polydipsia (Michell, 1988c). Since the maintenance of concentrated ISF in the renal medulla depends on an intact population of loops, the onset of polyuria is not surprising and it is reinforced by the osmotic diuresis associated with increased flow rates and, once azotaemia is sufficiently severe, reduced responsiveness of the collecting duct to ADH. (Azotaemia is a measurable increase in the plasma concentration of nitrogenous end products, e.g. creatinine. Uraemia is the syndrome which results, once the rise is sufficiently severe.)

The endocrine functions of the kidney are also impaired. Activation of vitamin D by conversion of 1-OH-dihydroxycholecalciferol (DHCC) to 1,25-DHCC is depressed and contributes to the onset of hyperparathyroidism by depressing intestinal absorption of calcium. The increase in parathyroid hormone (PTH) mobilises bone mineral, minimising the fall in plasma calcium concentration. There is, however, a rise in plasma phosphate because, unlike normal kidneys which simply excrete the additional phosphate (because PTH decreases reabsorption in the proximal tubule), those with a low GFR can no longer do so. Secondary hyperparathyroidism, once it is sufficiently severe, causes bone damage and soft tissue calcification; because the latter also affects the kidney, it contributes to the progression of CRF. PTH may also be a 'uraemic toxin', contributing particularly to the CNS effects of uraemia. When renal failure in dogs was formerly allowed to become sufficiently advanced without appropriate dietary adjustments (reduced phosphate intake), secondary hyperparathyroidism could be so severe that the mandible became flexible ('rubber jaw').

Diminished secretion of erythropoietin is the main cause of the anaemia of CRF (though reduced production and survival of RBC also contribute). Therapeutic administration of the synthetic hormone (EPO) may cause PCV to rise too far and result in hypertension.

Anaemia can provide a stimulus to salt retention (*Chapter 2*), but since this is mediated by erythropoietin, it is unlikely to result from CRF where the hormone causes the anaemia, rather than responding to it.

Renal disease is frequently associated with hypertension, for a variety of reasons (Edmunds and Russell, 1994). Firstly, the kidney is the long-term regulator of arterial pressure; secondly, pressure natriuresis is unlikely to be normal in hyperfiltrating nephrons, and, in addition, the reduced renal mass may be less able to produce vasodilator lipids including prostaglandins. Nephrons with impaired perfusion generate renin and, as a result, concentrations of the potent vasoconstrictor, angiotensin II, are increased. Above all, sodium excretion is likely to be abnormal.

What, then, of sodium balance in CRF? The 'job specification' for the surviving intact nephrons is to maintain the excretion of the daily load of excess sodium. Since their GFR is increased, glomerulo-tubular balance demands that reabsorption is also increased, i.e. more energy is needed in each cell. Since overall GFR is reduced, however, the necessary excretion of sodium constitutes a greater proportion of the filtered load. Therefore, despite the increase in **absolute** reabsorption, **fractional** reabsorption is reduced. This increase in **proportional** sodium excretion per nephron as GFR declines is known as the 'magnification phenomenon' (Delmez *et al.*, 1990). Not surprisingly, it is all too easy for a slight mismatch to occur in this complex readjustment, hence CRF may result either in salt retention or salt wasting and dietary sodium will affect this. More important, however, is the fact that, given time, the kidney can adjust to various levels of salt intake, even with CRF (Eknoyan, 1990). It is, however, unable to adjust to **sudden** changes or to sudden depletion caused, for example, by diarrhoea. This again illustrates the ability of the kidneys in CRF still to cope with everyday life in a stable environment, but their restricted ability to respond to emergencies—the 'inflexible' kidney of CRF. Some of these, e.g. vomiting, may result from uraemia and create a vicious cycle if the associated dehydration further decompensates renal function.

It is not known how sodium excretion is adjusted to the reduced nephron population of CRF. An obvious candidate is the rise in ASTI/EDLI (*Chapter 6*) associated with uraemia and providing a strong potential link with hypertension; ANP also increases in CRF (Smith *et al.*, 1986; Eknoyan, 1990). The rise in ANP may reflect reduced breakdown as well as a response to volume expansion (e.g. in patients towards the end of the period between dialysis sessions). It occurs as a result of CRF, whether or not patients are dialysed

(Ardaillou and Dussaule, 1994). The increased secretion of natriuretic hormones is not simply a consequence of CRF but a response to it; the extent of the response varies with salt intake (Delmez *et al.*, 1990). The 'trade-off' hypothesis of CRF (Bricker and Licht, 1980) has emphasised the balance between beneficial effects of responses (e.g. maintenance of sodium excretion) and their adverse effects (e.g. predisposition to hypertension as a result of changes in electrolyte transport). Apart from arterial pressure, there is a further 'trade-off' between sodium intake and residual renal functio⅃ in CRF. A higher intake supports GFR, but at the cost—the energy cost—of greater reabsorptive work (Michell, 1995b). Moreover, the surviving nephrons not only increase their oxygen consumption, but do so disproportionately to the increase in reabsorptive load and may be at risk from an associated increase in free radical production (Schrier *et al.*, 1994b). The higher GFR may also sustain proteinuria, itself a factor in the progression of chronic renal disease (Brunskill and Klahr, 1993; Michell, 1995b). High salt intake sustained proteinuria, glomerular expansion and increased mesangial matrix in a rat subtotal nephrectomy model of CRF; these changes were avoided at lower intakes of sodium (Modi *et al.*, 1993).

As well as adaptive alterations of renal sodium excretion in CRF there are consequential changes. These include the effects of hypoproteinaemia, anaemia, osmotic diuresis, secondary hyperparathyroidism *(Chapter 7)* and metabolic acidosis (Eknoyan, 1990). Apart from functional factors, lesions can also affect sodium excretion. In the early stages of CRF the prevailing disturbance may be sodium retention or 'salt wasting', but once the reduction of GFR becomes extreme, there is usually evidence of ECF expansion and sodium retention (Delmez *et al.*, 1990). There is also hyponatraemia because, even if concentrating capacity is maintained, the reduction of GFR restricts the excretion of excess water (Meyer-Lehnert and Schrier, 1990). The hypertension seen in CRF is frequently a reflection of volume overload (Kaplan, 1990). It may also reflect an increased production of endothelins and their constrictor effects may further reduce GFR (Takahashi *et al.*, 1994b). As well as overall solute retention, endocrine responses may be triggered by an altered rhythm of excretion so that solutes are retained for a greater proportion of the day (Bourgoignie *et al.*, 1981).

Whether or not patients with CRF are retaining sodium, the inflexibility of renal function implies that ill-judged antihypertensive therapy or sudden reductions of sodium intake are likely to decompensate residual renal function; the same is true of excessively deep anaesthesia, especially if it is combined with dehydration.

Acute Renal Failure (ARF)

ARF involves a catastrophic decline in RBF and GFR in all nephrons, without time for compensation, and involves diversion of the remaining blood flow from the cortex towards the medulla (Stein *et al.*, 1978; Conger and Schrier, 1980; Kleinknecht, 1980; Brezin *et al.*, 1986). Tubular obstruction can also contribute and the condition can be functional, without lesions (e.g. hepatorenal syndrome; Gordon and Anderson, 1981), or involve tubular necrosis, depending on the underlying causes. These mainly involve severe hypotension (or hypovolaemia) or toxins, including many poisons and some drugs secreted by the nephron. ARF can be superimposed on CRF. Despite the availability of dialysis for over 40 years, the mortality of acute tubular necrosis (ATN) has scarcely changed and it remains above 50% (Lieberthal and Levinsky, 1990). While this partly reflects the rising number of elderly patients with complex underlying disturbances, 25% of young, healthy victims of ATN still die.

Typical ARF presents with azotaemia, uraemia, hyperkalaemia and minimal urine output. Unusually, however, the small volume of urine is poorly concentrated yet it contains substantial amounts of sodium (Blachley and Henrich, 1981). Reduced urine output (oliguria) in response to hypovolaemia (pre-renal failure) involves high concentration (in response to ADH) and virtual absence of sodium (in response to aldosterone). The unusual composition of ARF urine reflects the compromised cortical perfusion (interfering with sodium retention) and excessive medullary blood flow (impeding urine concentration). Less typically, ARF can present with increased urine output; GFR is still reduced but reabsorption is not reduced in proportion (Douglas, 1985). Recovery from ARF, assuming it occurs despite the high mortality, also involves a period of high urine output and electrolyte loss. This reflects the gradual nature of the recovery as well as excretion of solutes retained during the oliguric phase. Throughout ARF and the recovery period, the virtual paralysis of renal function demands avoidance of both dehydration and overhydration. Despite the high mortality, recovery can be complete without any subsequent loss of renal function.

The exact reasons for the catastrophic redistribution of residual blood flow in ARF remain uncertain, but suggestions include tubular obstruction where lesions occur, altered glomerular permeability, back leakage of glomerular filtrate (once the tubule is damaged) and intrarenal effects of angiotensin (Michell, 1988c; Lieberthal and Levinsky, 1990; Blantz, 1993). There may also be ischaemic injury involving superoxide injury in the reperfusion phase, cell swelling, and ATP depletion due to leakage of sodium into cells and the consequent increase in Na–K ATPase activity (Leaf *et al.*, 1986). Renal injury activates a range of vasoconstrictors, both intrarenally and systemically, including angiotensin, thromboxane, vasopressin, catecholamines and endothelins; it also suppresses production of nitric oxide. In addition, there may be afferent arteriolar constriction triggered from the macula densa. Many vasoconstrictors also cause mesangial contraction and reduce the surface area for filtration. Regardless of the cause, acute renal failure causes a dramatic fall in sodium excretion, despite the residual sodium in urine, because of the massive reduction of GFR (Delmez *et al.*, 1990). Whether this fall should cause clinical concern depends largely on whether the patient is hypertensive or not and the presence of signs of cardiac overload or oedema.

Anuria is not typical of acute real failure but rather of complete urinary obstruction ('post-renal failure'). It is followed by intense diuresis, formerly attributed to tubular damage (Wilson, 1980) but now believed to reflect appropriate excretion of excess sodium, water and nitrogenous end-products retained during the anuric period (Franklin and Klein, 1994). The immediate effect of relief of obstruction depends whether it was unilateral or bilateral. In unilateral obstruction the intense renal vasoconstriction present before release continues but recovers almost to normal in 24 h; the reduction of GFR is especially severe in deep nephrons. In bilateral obstruction, strangely, there is less vasoconstriction before release but afferent arteriolar resistance increases after release, inner medullary blood flow is greatly reduced by bilateral obstruction if it persists. The seemingly paradoxical increase in afferent resistance after release of bilateral obstruction has been attributed to feedback suppression of GFR once flow to the macula densa is restored (Wright, 1982). Unilateral obstruction lasting 24 h has little effect on the output of urine and sodium following release, but following a similar period of bilateral obstruction the increase is dramatic (Yarger and Buerkert, 1982). This post-obstructive diuresis varies in intensity with the duration of anuria and depends on haemodynamic changes rather than renal damage. It involves the effects of volume expansion, natriuretic factors and azotaemia and a temporary impairment of acid excretion. There may also be a temporary reduction of renal Na–K ATPase activity (Kurokowa *et al.*, 1982).

9

Interactions Between Sodium, Potassium, Calcium and Magnesium

Introduction

It is not surprising that there are important interactions between sodium, potassium, calcium and magnesium; these are the major univalent and divalent ions of extra- (ECF) and intracellular fluid (ICF). Sodium and calcium are both extruded from ICF by ATP consuming pumps, but intracellular calcium can also exchange for extracellular sodium. Active extrusion of sodium by Na–K ATPase brings potassium into ICF; this enzyme, like Ca-ATPase, is Mg^{2+} dependent and inhibition of sodium transport by cardiac glycosides leaves residual Mg-ATPase activity. Na–K ATPase utilises ATP as a magnesium complex. Magnesium is an essential cofactor for many enzymes, not only those involved in cation transport (Bara et al., 1993; Quamme and Dirks, 1994).

Sodium and potassium are freely dissolved in plasma, unlike calcium and magnesium, which are substantially bound to albumin. It is the freely dissolved calcium and magnesium, together with calcium complexed to small organic anions, which crosses capillary membranes and which enters glomerular filtrate. The biological effects, including those on membrane potentials, like the effects on ion-specific electrodes, depend on the activity of freely dissolved ions. Much of the clinical interpretation of disturbances in these ions is still based on their plasma concentration. This is fraught with pitfalls for the following reasons:

1. Sodium concentration most directly informs us about water balance rather than sodium balance, as already discussed (*Chapter 8*).
2. Potassium concentration in plasma informs us about potentially important clinical consequences of hypo- or hyperkalaemia, e.g.

for excitability of cardiac muscle. Severe hypokalaemia is unusual in the absence of ICF K depletion, unless caused by drugs promoting K^+ uptake (e.g. insulin, β-agonists) or by stress. The latter is particularly important because, for example, adrenaline concentrations comparable to those seen after acute myocardial infarction are sufficient to drop plasma potassium concentration by 0.7 mmol/l. Patients with underlying cardiac disease are particularly susceptible to dysrhythmias caused by even mild hypokalaemia (Schuster, 1989; Knochel, 1989).

Unfortunately, severe hyperkalaemia can co-exist with severe cellular deficits of potassium and frequently does so, for example when metabolic acidosis, hypovolaemia and reduced renal perfusion result from diarrhoea *(Chapter 3)*. Correction of acidosis and hypovolaemia readily causes a rapid transition from hyperkalaemia to hypokalaemia. In mild hypovolaemia aldosterone promotes both urinary and faecal loss of potassium, but in severe hypovolaemia, as discussed previously, potassium excretion is restricted by the reduced delivery of sodium from the proximal nephron. Thus, hypokalaemia is a hallmark of primary aldosteronism but secondary aldosteronism can be accompanied by hyperkalaemia (Linas and Berl, 1989). This further stimulates aldosterone secretion and may allow extrarenal effects of aldosterone, e.g. on cells, salivary glands, or colon, to alleviate the hyperkalaemia.

3. Plasma magnesium concentration, as a clinical guide, suffers from the same fundamental problem as potassium—unreliability as a guide to intracellular magnesium—with the additional problem of binding to albumin. It is perilous to interpret readings of either plasma magnesium or calcium in the absence of a measurement of plasma albumin unless the readings were obtained with ion-selective electrodes. Hypomagnesaemia reduces parathyroid responsiveness to hypocalcaemia; it also promotes renal potassium loss, for reasons which are not well understood (Linas and Berl, 1989; Schuster, 1989). Correction of both hypocalcaemia and hypokalaemia, therefore, may depend on diagnosis and correction of concurrent hypomagnesaemia.

Interactions between these ions clearly arise from interdependent aspects of transcellular transport and also renal excretion. The latter is particularly important because renal function is based upon glomerular filtration and tubular reabsorption of sodium; reabsorption is highly attuned to the maintenance of sodium balance. Changes in sodium balance, therefore, readily predispose to inappropriate secondary changes in the excretion of potassium, calcium and magnesium. In particular, excretion of excess sodium involves reduction of fluid reabsorption in the proximal tubule which readily enhances the concurrent losses of other ions. With calcium, a specific hormone (PTH) is available to promote conservation in the distal nephron, but no hormone has been linked to homeostatic retention of potassium or magnesium. Perhaps this reflects the fact that only a minority exists in ECF and is therefore readily available to glomerular filtrate. With potassium, however, the filtrate is largely reabsorbed and excretion depends on secretion in the distal nephron, therefore on the concentration of potassium in renal cells.

Interactions between ions also arise from shared hormonal mechanisms, e.g. the involvement of aldosterone in the regulation of both sodium and potassium excretion. Both the secretion of renin and the effects of angiotensin II on vascular tone and renal sodium excretion almost certainly involve changes in intracellular calcium (Seeley and Laragh, 1990; Laragh and Seeley, 1990; Koeppen, 1990). The stimulation of aldosterone by potassium is mediated by increases in intracellular calcium. PTH also stimulates aldosterone secretion by a direct effect on the adrenal cortex, probably by facilitating increases in ICF calcium caused by potassium or angiotensin (Hulter, 1994).

There is also scope for interaction between these ions in their enteric absorption or their reliance on bone as either a reservoir or a store for excess. The latter might seem particularly important for Ca^{2+} and Na^+, since intracellular fluid can only act to a very minor degree as a temporary haven for excess or as a store (Michell, 1976b). Nevertheless, bone also contains substantial quantities of potassium and magnesium. Enteric interactions are little better understood, although, for example, in ruminants, excess potassium can impede magnesium absorption. Granted that herbivorous diets are invariably rich in potassium, it has been suggested that low sodium intake, by causing replacement of salivary (and therefore ruminal) sodium by potassium, under the influence of aldosterone, predisposes to hypomagnesaemia (Care, 1988). Much less consideration has been given to the possibility that consumption of excessive supplementary salt may predispose to both hypocalcaemia and hypomagnesaemia as a result of increased urinary losses of these ions (Michell, 1989d). Further interactions can arise, in all species, as a result of the use of drugs and the effects, for example, of diuretics, are considered in *Chapter 11*.

Cellular Interactions

Both potassium and magnesium have important effects on cell metabolism, but calcium's main role in ICF is probably as a 'second messenger' (Ausiello and Bonventre, 1984). Thus, the intracellular calcium concentration is normally low but can rise rapidly as in excitation–contraction coupling of muscle; the source of the additional calcium is either ECF or intracellular stores, e.g. sarcoplasmic reticulum, depending on the type of muscle. The calcium is returned to stores or ECF by active transported mediated by Ca-ATPase. With sodium, the most important impact on cellular processes is not so much the result of its ICF concentration **per se** but of the transport mechanisms by which it enters or leaves (*Chapter 1*).

The second messenger role of calcium frequently involves binding to proteins such as calmodulin and troponin C, leading to changes in the activity of enzymes such as adenylate kinase or phosphodiesterase. These control the formation and breakdown of cAMP. Other actions of calcium involve protein kinases, the enzymes controlling phosphorylation. Calcium enters mitochondria on the potential gradient arising from the extrusion of protons, which results from the activity of the electron transport chain (Humes, 1984). It may leave mitochondria by exchange for sodium or by mechanisms independent of either sodium or ATP (Bourdeau and Attie, 1994).

The main impact of magnesium on potassium is probably via changes in Na–K ATPase activity, allowing cells to retain additional potassium in hypermagnesaemia (Solomon, 1987). Because of the efficiency of renal magnesium excretion, this is rare in the absence of renal dysfunction. In contrast, hypomagnesaemia is greatly under-diagnosed and may impede the correction of potassium depletion. This was formerly attributed to renal potassium 'wasting'. Most probably this represents the excretion of potassium, which cells are unable to retain when their Na–K ATPase activity is depressed by hypomagnesaemia (Solomon, 1987). This is clinically important because many of the problems attributed to hypokalaemia and depletion of cell potassium may be caused or exacerbated by hypomagnesaemia. These include cardiac dysrhythmias resulting from the side effects of diuretics (*Chapter 11*). It is important to remember that, while the problem is described as hypomagnesaemia, the effect on Na–K ATPase results from depletion of ICF magnesium, which is not necessarily reflected by changes in plasma magnesium. Patients on chronic diuretic therapy are vulnerable to a fall in intracellular magnesium, potassium and a reduction of Na–K ATPase pump sites, for example in skeletal muscle. Magnesium supplements can correct these abnormalities (Dorup *et al.*, 1993). Not surprisingly, magnesium deficiency predisposes to cardiac glycoside cardiotoxicity, since both depress Na–K ATPase (Kobrin and Goldfarb, 1990). Newman and Amarasingham (1993) have speculated that functional magnesium deficiency and the resulting impairment of Na–K ATPase and Ca-ATPase activity contributes to a number of clinical conditions including diabetes mellitus and pre-eclampsia. Acidosis is a predisposing factor.

Apart from its effects on Na–K ATPase, magnesium also affects both sodium and potassium channels and Na–K–2Cl cotransport (Ryan, 1993). Sodium–calcium exchange is inhibited by ECF magnesium (which competes with calcium), but effects on the Na–H antiporter are probably indirect (Bara *et al.*, 1993). The effects of magnesium depend on whether the membrane is 'tight' or 'leaky' and whether magnesium acts at the ICF or ECF site. Both transcellular and paracellular pathways can be affected. Generally, raised extracellular magnesium (Mg_e) favours paracellular movement of sodium and potassium in leaky membranes but blocks it in a tight membrane such as the myocyte. In **potassium channels**, internal magnesium (Mg_i) blocks the outward current, whereas Mg_e increases their permeability in leaky membranes. In **sodium channels**, Mg_i also blocks outward current. The Na–K–2Cl cotransporter is stimulated by Mg_e. The blocking effects on cation channels probably result, at least in part, from screening of their surface negative charges or by direct competition for occupancy of the channel.

The effect of calcium on Na–K ATPase is inhibitory in broken cell preparations, but in intact cells intracellular concentrations may inhibit, stimulate or show biphasic effects *(Chapter 1)*. Similarly, earlier ideas that the Na–H antiporter can be stimulated by a calcium–calmodulin mediated response now seem doubtful (Decker and Dieter, 1988). Such effects appear to result instead from cell shrinkage and do not affect protein kinase C, which is involved in stimulation of the Na–H antiporter. Instead, the antiporter is probably involved in the regulation of the free calcium concentration of ICF, via changes in intracellular pH. The effect of PTH on bone is inhibited by ouabain, suggesting that the release of calcium from bone cells is mediated by exchange with sodium (Humes, 1984). The role of such exchanges in the control of vascular tone has already been discussed *(Chapter 6)*.

The interactions between sodium, potassium, calcium and magnesium are also important in the function of excitable membranes, both in nerve and muscle, because they affect resting potentials, action potentials and the release of neurotransmitters. Voltage gated sodium channels are responsible for the regenerative electrical activity necessary for impulse generation. They are virtually impermeable to magnesium and slightly permeable to calcium and potassium (Pappone and Cahalan, 1986). The selective permeability for sodium rather than potassium does not result from ionic size, since the hydrated sodium ion is larger, but rather from the difference in the energy required for dehydration, since both ions are probably too large to pass through in fully hydrated form. Transmitter gated potassium channels and calcium channels are also essential to the function of excitable membranes and details of the relative ionic permeabilities vary with the associated receptor (cholinergic, nicotinic or muscarinic, adrenergic, adenosine, GABA, glycine, etc.); chloride channels are also involved (Clapham, 1989). There are a number of different types of potassium channels, including some activated also by calcium (Palmer, 1989). Calcium-sensitive potassium channels are also involved in epidermal barrier function (Lee *et al.*, 1994). The activities of these channels, and the effects upon them of drugs, are absolutely central to the physiology of nerve and muscle and its therapeutic manipulation. They are, however, beyond the scope of a book concerned primarily with sodium regulation.

Renal Interactions

Filtration of divalent cations involves only those which are not bound to albumin or larger molecules. Small organic complexes, however, are filtered but are reabsorbed less readily than the free ion. Thus, for example, with calcium the proportion which is complexed to citrate and other anions increases as the urine progresses through the nephron. This helps to maintain solution stability and prevent urolithiasis, even at calcium concentrations close to or sometimes somewhat above saturation (Michell, 1989e).

In the **distal nephron** there is differential control of the re-absorption of sodium and calcium, the latter stimulated by PTH and the former by aldosterone. PTH also favours the reabsorption of magnesium, but since hypomagnesaemia reduces the responsiveness of the parathyroid gland there is no appropriate feedback loop to govern magnesium excretion via PTH (or, within current knowledge, any other hormone). This contrasts with potassium, for which the direct sensitivity of the adrenal gland to hyperkalaemia provides a feedback loop independent of sodium. Although sodium shares aldosterone as a regulator of excretion, the renin–angiotensin system provides a feedback loop which is independent of potassium (and which operates using angiotensin as well as aldosterone as its effectors). Although high salt intake reduces aldosterone secretion, it facilitates potassium excretion by increasing the delivery of potassium to the distal nephron and by stimulating ANP secretion which also promotes urinary excretion of potassium (*Chapter 4*).

The independent control of magnesium excretion occurs in the **loop of Henlé**, but the regulatory mechanism is unknown (Quamme and Dirks, 1993). Acute changes in plasma magnesium superficially seem to produce the appropriate change in PTH (hypomagnesaemia favouring PTH secretion and magnesium retention) but, unlike chronic hypomagnesaemia or hypermagnesaemia, acute changes in plasma magnesium are clinically rare (Bourdeau and Attie, 1994). Whether or not the regulatory mechanism is independent of sodium, magnesium excretion is not necessarily so, thus loop diuretics increase the excretion of both ions *(Chapter 11)*.

The reabsorption of magnesium in the loop, like phosphate in the proximal tubule, is believed to depend on a T_m mechanism, i.e. a maximum rate of reabsorption. Thus, a fall in filtered load (hypomagnesaemia) reduces the proportion which escapes reabsorption, while an increased filtered load increases it, effectively by swamping the T_m. In the case of phosphate, however, the T_m is responsive to hormones; when PTH mobilises bone mineral in response to hypocalcaemia it also promotes the renal retention of the necessary calcium, but the excretion of the 'unwanted' phosphate. The latter is ensured by PTH's ability to reduce the T_m for phosphate reabsorption. In chronic renal failure (CRF), unlike primary or nutritional hyperparathyroidism, this mechanism fails because glomerular filtration rate (GFR) is so low that dietary phosphate cannot be excreted, whatever happens to T_m. Since hyperphosphataemia stimulates the parathyroid glands, the only way out of this vicious cycle is reduction of dietary phosphate.

The only clinically plausible cause of non-iatrogenic hypermagnesaemia is CRF. In the search for which hormones, if any, regulate magnesium excretion, perhaps the key question is which can depress or increase the T_m for magnesium excretion. Obvious responsiveness to changes in plasma magnesium is not necessarily important because it is so poorly related to ICF magnesium, i.e. a soundly-based endocrine feedback loop might well monitor some outcome of changes in intracellular magnesium. Apart from PTH, calcitonin, glucagon and ADH are known to affect magnesium reabsorption (Quamme and Dirks, 1994). Among other hormones with less persuasive evidence are thyroid hormone, catecholamines, vitamin D, growth hormone and angiotensin II. Mineralocorticoids are without effect (except via ECF volume), as are glucocorticoids. It is not surprising that ADH might enhance reabsorption of magnesium within the loop of Henlé, since it stimulates sodium transport there *(Chapter 2)*. PTH, however, promotes magnesium conservation within the loop without affecting sodium transport (Quamme and Dirks, 1994).

Renin secretion is affected by magnesium, but this is not surprising since it is affected by changes in intracellular calcium and, therefore, indirectly by Na–K ATPase activity (Laragh and Seeley, 1992). Chronic hypomagnesaemia stimulates renin release (Quamme and Dirks, 1994) but acute hypermagnesaemia also does so (Ichihara *et al.*, 1993); perhaps this conflict reflects differences in the impact on cell magnesium. The acute stimulation of renin secretion is prostaglandin-dependent, whereas concurrent suppression of aldosterone secretion is mediated by changes in intracellular calcium. Since aldosterone does not affect magnesium excretion and the evidence for effects of angiotensin II is flimsy, the influence of magnesium on renin and aldosterone secretion looks more like an incidental effect of its involvement in cellular cation transport, rather than a component of a regulatory system.

In the **proximal tubule**, the excretion of calcium and magnesium is subservient to the primary determinants of sodium excretion; circulating volume (or the signals which represent it), renal perfusion pressure and glomerulotubular balance. In particular, expansion of ECF volume suppresses reabsorption, appropriately for sodium, inappropriately for calcium and magnesium (Costanzo and Windhager, 1992; Quamme and Dirks, 1993). In any case, magnesium is less readily reabsorbed than calcium or sodium because, although it is a smaller ion, its hydration shell is larger (Quamme and Dirks, 1993). Curiously, despite the fact that the primary target of ANP is not the proximal tubule, it evidently has sufficient effect to increase the excretion of both calcium and magnesium (Atlas and Maack, 1993).

If natriuresis causes calciuresis it is reasonable to ask whether pressure natriuresis in hypertensives affects calcium excretion, and whether this has implications for urolithiasis. A relationship between arterial pressure and calcium excretion has been found (Vagelli *et al.*, 1994) and it may well explain the fact that patients with essential hypertension are prone to urinary stones (MacGregor and Capuccio, 1993). It also suggests that essential hypertensives would be more prone to the adverse effects of excess sodium on bone mineral (see below). High salt intake could also predispose to urolithiasis in normotensives (Sakhaee *et al.*, 1993). Any parallel increase in magnesium excretion might alleviate some of the effects of hypercalciuria by inhibiting the tendency for calcium salts to crystallise (Michell, 1989e).

It is also reasonable to consider whether calcium can affect sodium reabsorption; there is evidence that hypercalcaemia can cause natriuresis and that this is mediated by prostaglandins (Ruilope *et al.*, 1994). This relates to acute infusions, rather than clinical hypercalcaemia.

Physiological and Clinical Implications

Apart from cellular and renal interactions, there are also interactions affecting bone, enteric absorption and the behavioural regulation of salt intake. Thus salt appetite is responsive to changes in potassium and calcium as well as sodium *(Chapter 4)*. Loading of bone mineral with one cation during the period of skeletal growth may restrict the storage of other cations and, potentially, the endogenous load of these cations available later in life *(Chapter 1)*. In diarrhoeic calves there is hypomagnesaemia and also hypocalcaemia; while it is not sufficiently severe to cause symptoms, its clinical importance remains uncertain (Michell *et al.*, 1992).

In ruminants, it has been suggested (Grace, 1988) that the high potassium content of the diet impedes the absorption of magnesium (which mainly occurs in the 'third' stomach, the omasum; the 'fourth' stomach or abomasum corresponds to our own★. Low sodium intake reinforces this interaction because aldosterone replaces salivary, hence ruminal, sodium by potassium *(Chapter 5)*. On the other hand, excess salt, by suppressing proximal reabsorption of sodium, could also predispose to hypomagnesaemia, as is known to occur with calcium (see below). Interestingly, in another herbivore, the horse, high salt intake does not affect renal calcium excretion and, if anything, it promotes enteric absorption of calcium (Schryver *et al.*, 1987).

Parturient hypocalcaemia ('milk fever') remains a common disease of cattle. It is an enigmatic condition since it is precipitated by the demand for calcium which arises from lactation and which has been greatly increased by the selective breeding of the high yielding dairy cow. The enigma lies in the fact that despite these demands the bones still contain vast amounnts of calcium, but, somehow, PTH fails to achieve sufficient mobilisation. One factor that could undermine the renal effectiveness of PTH would be excessive delivery of calcium to its site of action in the distal nephron, as a result of expansion of ECF volume and suppression of proximal tubular reabsorption. This would be analogous to the way in which excess delivery allows the distal nephron to escape from the salt-retaining effects of mineralocorticoids. Since ruminants are routinely kept on exorbitantly high salt intakes *(Chapter 5)*, it is surprising that there has been little consideration of the possibility that sodium is a risk factor in milk fever (Michell, 1992c).

★The rumen, the large fermentation vat, is the 'second' stomach and the reticulum is the 'first'.

In human medicine, the possibility that excess salt causes sufficient negative calcium balance to be clinically significant has recently received considerable attention. The particular context is the rapid accumulation of evidence that it exacerbates the loss of bone mineral in post-menopausal women (Shortt and Flynn, 1990; Goulding *et al.*, 1993). Evidence arises from studies in rats, dogs and humans, involving not only calcium excretion but markers of bone de-mineralisation such as forearm bone density and urinary excretion of hydroxyproline. The risk of bone demineralisation is increased by the acid load associated with the metabolism of dietary protein, and oral potassium bicarbonate therefore reduces bone resorption (Sebastian *et al.*, 1994).

The rat data, as ever, are problematical, perhaps because the 'normal' diet is so high in sodium that, at least in some experiments, it is debatable what influence any further excess could have. Moreover, some of the rat experiments involved 8% salt diets; this dose rate scaled to human bodyweight is 6400 mmol/day—enough for 50 pregnancies per week! The comparison is no more absurd than the chosen intake. If we are prepared to look at the effects of intakes exceeding daily requirement by 100-fold, what would be the effect, on calcium or anything else, of the daily consumption of 275 l of water? Apart, that is, from death within hours from hyponatraemia or even haemolysis. If these are rat 'models' they are 'spitting images' of reality—recognisable but preposterous.

In contrast, the evidence in humans is almost unambiguous: there is a strong correlation between sodium and calcium excretion in both young and adult humans of either sex and a relationship between increments of sodium intake and increases in calcium excretion, despite the various factors beyond experimental control. The relationship is clearer in acute salt loading experiments because they allow stricter control of other variables. Indeed, "sodium intake appears to be a major determinant of urinary calcium excretion" (Shortt and Flynn, 1990). It is likely, though less clear, that secondary hyperparathyroidism results from the loss of calcium, hence the potential for mobilisation of bone mineral. This response seems to vary between individuals. The evidence that chronic salt loading mobilises bone mineral in humans is strong, particularly in females. It is imporant to remember, however, that bone remodelling involves both mobilisation and deposition of bone mineral; salt loading may also increase bone formation (McParland *et al.*, 1989), perhaps as a secondary response to skeletal weakening.

Despite the strength of the evidence even in 1990, Shortt and Flynn concluded that it was "not clear or strong enough at present to justify intervention at an individual or population level". We are fascinatingly resistant to any reasons for trying to moderate our addiction to salt (Michell, 1984a). In response to the recent COMA (1994) Report on nutritional risk factors in cardiovascular disease (*Chapter 6*), we read in *The Times* (Le Fanu, 1994):

> *The blood pressure is therefore the best defended of all physiological functions . . . with blood pressure so well protected, modest changes in salt consumption are unlikely to be very influential, not least because the concentration of salt itself in the body is rigorously controlled, so any excess is rapidly excreted . . .*

That is the response to an expert Report citing hundreds of references, compiled by specialists and taking evidence from others with specialist knowledge. Why quote it? Because even if it were read by only 1% of the readership that will far exceed the number likely to read the Report which, unlike the newspaper report, also discusses interactions with potassium. It would, perhaps, be nearer the mark to conclude that salt intake is the best defended of all nutritional prejudices.

Conclusion

This account has mainly concerned the interactions between monovalent cations (e.g. Na^+) and divalent cations (e.g. Mg^{2+}). It has not been concerned with interactions between divalent cations, important though these are (e.g. hypomagnesaemia as a factor predisposing to hypocalcaemia by desensitising the parathyroid secretory response to hypocalcaemia). These are beyond the periphery of a book primarily focused on sodium. So, too, are the interactions between electrolyte physiology and acid–base disturbances, though we encounter them, for example in their effects on renal function (*Chapter 2*), their importance in diarrhoea (*Chapter 3*) and the direct interactions between the transport of sodium and hydrogen ions both in health (*Chapter 1*) and disease (*Chapter 10*). Interactions between sodium and potassium, however, are extremely important and have received substantial comment in various chapters concerned, for example, with renal or enteric physiology, salt appetite, or hypertension.

Absolutely central to many of these interactions is the indissoluble link between sodium and potassium in the activity of the 'sodium' pump, Na–K ATPase, which is also sensitive to the effects of magnesium and calcium. Transport is the constant focus of the physiological importance of sodium and while Na–K ATPase is not the only route for the movement of sodium, or even for its interlinked movement with potassium, it is the one which makes major demands on the 'currency' of energy metabolism, ATP, and on which the others secondarily depend.

Sodium and potassium are also interlinked because they exploit, with partial but incomplete independence, aldosterone as a key link in their homeostatic pathways. Broadly, the relationship is reciprocal; cells lose sodium when they gain potassium, and kidneys, colons, sweat glands and salivary glands, according to species, shed potassium when they retain sodium. The renal interaction, however, is unique, because it relies on adequate delivery of sodium from the proximal nephron, which itself depends on ECF volume and therefore on sodium *(Chapter 2)*. Thus, diarrhoea causes potassium depletion but, once hypovolaemia and acidosis become sufficiently severe, hyperkalaemia is readily superimposed *(Chapter 3)*. That, perhaps more than anything else, creates the urgency of additional sodium intake *(Chapter 4)*. It also reinforces the importance of interactions between nephron segments, rather than mechanisms within nephron segments, as determinants of renal cation excretion.

A particular preoccupation in several chapters has been the importance of a gross excess of sodium in its effects on a variety of physiological mechanisms, including the excretion of divalent ions. Its ability to predispose to hypertension probably rests on interactions with divalent cations and may be offset by interactions with potassium *(Chapter 6)*. Further interactions will be encountered, not surprisingly, in considering the effects of drugs *(Chapter 11)*, notably diuretics.

10

Clinical Disturbances of Sodium Transport

Introduction

It is already clear, from previous chapters, that a broad range of clinical conditions can involve changes in sodium transport. These may be secondary effects, part of the underlying pathology or responses to therapy (*see Chapter 11*). Thus, abnormalities of Na–K ATPase activity may contribute to the problems of hypovolaemic shock or diabetes mellitus, but they would not be seen as part of the primary problem, whereas in hypertension they could well be key determinants of the increased pressure. This chapter is mainly concerned with some diseases not considered in earlier chapters where disturbances of electrolyte transport are believed to be among the main effects. In some cases, as with the mode of action of loop diuretics, there has historically been some ambiguity as to whether the primary changes concerned the transport of sodium or chloride. An obvious example is cystic fibrosis.

Cystic Fibrosis

Defective epithelial ion transport is the basic pathophysiological defect in cystic fibrosis (Noone *et al.*, 1994). This common hereditary disease of children and young adults was first recognised by Fanconi, who also discovered the syndrome for which he is better known, involving generalised failure of proximal tubular reabsorption (see below). Cystic fibrosis (CF) causes serious respiratory impairment together with intestinal obstruction, salivary gland enlargement and a number of other systemic effects. Steatorrhoea (fatty diarrhoea) results from impaired pancreatic function. The common factors in the various changes are abnormal composition of mucus and abnormal electrolyte concentrations in exocrine secretions. Yet electrolyte transport in brain, kidney, muscle and heart remains normal unless affected by complications of the disease (Mangos, 1986).

It has been known for over 40 years that sweat is abnormally rich in salt in cases of CF; recently, it has become clear that this results from defective chloride absorption in the ducts of the sweat glands (Mangos, 1986). This results from a factor, present within sweat, which is not yet identified. The abnormalities in sweat composition make CF patients vulnerable to the effects of thermal stress (Knochel and Reed, 1994). In salivary glands, it appears to be sodium transport which is inhibited, independent of Na–K ATPase, which is unaffected. The glycoprotein content of saliva is also increased and its physico-chemical characteristics are abnormal. Some attribute this to increased salivary calcium, which they view as a key defect. Pancreatic secretion is rich in chloride, low in bicarbonate and poorly responsive to secretin (Mangos, 1986).

The respiratory problems of CF could potentially arise from increased mucus secretion, defective mucociliary clearance or abnormal mucus composition. Both secretion and clearance could show secondary changes, e.g. in response to infection. The upper airway secretions are normally present as a superficial gel, propelled by the cilia, and an underlying sol. In CF, sodium content is reduced, calcium increased; this may reflect enhanced reabsorption of both sodium and water, thickening the secretions and predisposing to alveolar collapse (Mangos, 1986). There may also be a circulating factor which depresses ciliary activity in CF, thus impeding mucociliary clearance.

The airway surface liquid is an integral part of the pulmonary defence mechanisms and its volume and composition are regulated by ion transport. Thus defective ion transport could potentially contribute to a number of diseases involving respiratory dysfunction, including cystic fibrosis (Noone *et al.*, 1994). Like milk, the sodium concentration of this fluid is unaffected by aldosterone. At birth, the airway changes from a secretory tissue, driven by chloride secretion, to an absorptive tissue driven by active transport of sodium. This transition is slower where birth is by Caesarean section. Studies in a genetic mouse model suggest that the basic secretory defect in CF is excessive absorption of sodium from airway epithelia with a compensatory increase in chloride secretion (Grubb *et al.*, 1994). The increase in sodium permeability could result from increased conductance per channel, an increased number of channels or an increase in the probability that channels are open; the latter appears to be the cause (Chinet *et al.*, 1994). Similar changes occur in human airway epithelial cells (Jiang *et al.*, 1993). A logical therapeutic response to this sodium channel abnormality is the use of an amiloride aerosol, but with conflicting results (Tomkiewicz *et al.*, 1993; Middleton *et al.*, 1993; Graham *et al.*, 1993). The intended effect is to improve the visco-elastic properties of mucus in the airway and restore the normal mucociliary defence mechanisms (Noone *et al.*, 1994).

As well as changes in sodium transport, cystic fibrosis involves changes in chloride channels. Loss or dysfunction of a phosphory-lation-regulated chloride channel impedes chloride secretion (Jiang *et al.*, 1993). The defect may be an inability of the chloride channel to respond to activation by cAMP (Field, 1993). There is also retention of chloride channels within the endoplasmic reticulum; this leads to subnormal acidification of the Golgi apparatus and therefore, perhaps, to defective synthesis of glycoproteins (Barasch and Awqati, 1993). While cystic fibrosis affects a range of epithelial cells, chloride channels are present in all cells; the gene involved in cystic fibrosis controls the synthesis of a large protein (cystic fibrosis transmembrane regulator) present particularly in epithelial cells; it may be the link between cAMP and activation of chloride channels (Field, 1993), or it may function as a chloride channel, or both (Sferra and Collins, 1993). Its role in the control of sodium transport remains unclear (Noone *et al.*, 1994).

Disorders of Muscle Excitation

A number of abnormalities of skeletal muscle contraction have been linked to dysfunction of membrane ion channels (Barchi, 1993). Thus **hyperkalaemic periodic paralysis** and **paramyotonia congenita** (cold-induced muscle contraction) have been described as 'sodium channelopathies' (Zhou and Hoffman, 1994). These conditions occur in animals as well as humans. The hyperkalaemia is a result, not a cause, of paralytic attacks (Ruff, 1989). The defect involves the gene controlling the adult isoform of the voltage-sensitive sodium channels of adult skeletal muscle; a different gene controls these channels in cardiac muscle and foetal skeletal muscle (Zhou and Hoffman, 1994). The condition would be lethal if the diaphragm were involved, but the reason why it is spared remains unknown; it may relate to cellular and physiological differences rather than sodium channel genes. The potassium efflux which causes the hyperkalaemia is not caused by changes in insulin, catecholamines, glucagon, glucocorticoids or mineralocorticoids, but it affects red cells as well. Basal Na–K ATPase activity in muscle is normal (Ruff, 1989).

There are other myotonias (delayed relaxation resulting from persistent sarcolemmal electrical activity) due to abnormal sodium channel function, e.g. atypical painful myotonia and Schwartz–Jampel syndrome, as well as those caused by abnormal function of chloride channels (Barchi, 1993). The latter include myotonia congenita, recessive generalised myotonia, hereditary goat myotonia, toxic myotonia and a genetic mouse model.

In **hypokalaemic periodic paralysis** the cause of the hypokalaemia accompanying the attacks is unknown. Aldosterone may be increased, but inconsistently (Ruff and Gordon, 1986). In fact, the loss of circulating potassium is into muscle, not into urine, but the cause of the influx is not known. Basal Na–K ATPase activity is normal and there is no documented link with enhancement by insulin or adrenaline, although if this were episodic, it would be hard to detect. Thyrotoxicosis causes a similar condition, but despite the stimulatory effect of thyroid hormone on Na–K ATPase *(Chapter 1)*, there is no change in muscle potassium, indicating parallel increases in passive efflux. Thyrotoxic patients with periodic paralysis, unlike other thyrotoxic patients, are also hyperinsulinaemic (Lee *et al.*, 1991). Both in hyperkalaemic and hypokalaemic paralysis, the muscle weakness results from loss of sarcolemmal excitability rather than subsequent stages of excitation—contraction coupling (Ruff, 1989). The loss of excitability is probably due to abnormal voltage inactivation of sodium channels or spontaneous activation. This may be associated with a fall in the potassium conductance of the depolarised membrane, an effect exacerbated by insulin.

It may seem paradoxical that both hypo- and hyperkalaemic periodic paralysis should involve such a similar fundamental defect, i.e. abnormal sodium permeability of the muscle membrane. The changes in potassium are results rather than causes of the condition for which they are named. In hyperkalaemic paralysis, membrane potential can be increased and hyperkalaemia reduced by stimulation of Na–K ATPase with β-agonists such as salbutamol (Sterns and Spital, 1987).

Hypokalaemic and hyperkalaemic periodic paralysis have been seen in dogs and cats (Leon *et al.*, 1992; di Bartola and de Morais, 1992) and hyperkalaemic periodic paralysis in horses; in the latter, the defect in electrolyte transport has been linked to a gene controlling ion transport through sodium channels (Rudolph *et al.*, 1992; Zeilmann, 1993). Neither the number nor the affinity of sodium pump sites is affected (Pickar *et al.*, 1993).

There has also been evidence for the involvement of Na–K ATPase in both Duchenne and myotonic muscular dystrophy, with a fall in ouabain binding sites in the latter (Plishker and Appel, 1986). Evidence concerning muscle Na–K ATPase in both conditions is conflicting and open to doubt over the experimental conditions used (Plishker and Appel, 1986). Nevertheless, there is substantial evidence for abnormal red cell Na–K ATPase activity in these conditions although, again, it is not consistent. Muscle from a mouse model of Duchenne muscular dystrophy shows an increase in intracellular sodium (Dunn *et al.*, 1993).

Down's Syndrome: Alzheimer's Disease

Down's syndrome is the result of an abnormality of human chromosome 21; patients are susceptible to Alzheimer's disease as early as 35 years of age (Galdzicki *et al.*, 1993). Both in humans and a genetic mouse model of Down's syndrome, action potentials of dorsal root ganglia are shortened, with the speed of both depolarisation and repolarisation increased. These differences are very specific and do not occur in a number of other human chromosomal disorders. In hippocampal neurons, depolarisation is slower (in the mouse model) and this probably reflects a smaller inward sodium current. The difference between dorsal root ganglia and hippocampal neurons may be due to cell-specific regulation of ion channels or to the influence of nerve growth factor (Galdzicki *et al.*, 1994). A reduced density of sodium channels in hippocampal neurons would not only affect their electrophysiological characteristics, but also differentiation, morphogenesis and the pattern of interconnections. These changes would potentially affect learning and memory.

In Alzheimer's disease it is likely that a polypeptide (amyloid beta protein) plays a key role in the pathogenesis. It has recently been suggested that this effectively functions as an ionophore, forming cation selective channels and that this may be the basis of its adverse neural effects (Pollard *et al.*, 1993). Patients with Alzheimer's disease also have a higher fractional excretion of sodium and their plasma contains a factor which is natriuretic in rats (Maesaka *et al.*, 1993). In addition, the sodium concentration in their sweat is 50% higher than in normal controls (Elmstahl and Winge, 1993). This probably reflects autonomic dysfunction rather than a specific transport defect. Thus, perhaps for different reasons, cystic fibrosis, Alzheimer's disease and Down's syndrome all increase sweat sodium concentration but diminish the ability to accelerate sweat production in response to increased temperature (Geeth and Skelty, 1987).

Papillary Necrosis

Necrosis of the renal papilla is the typical outcome of excessive analgesic consumption, but it also results from other conditions such as obstructive uropathy or diabetes mellitus. It particularly affects juxtamedullary nephrons and a number of mechanisms have been suggested, including inhibition of Na–K ATPase (Sabatini, 1989). A prior step is postulated to be cell damage, allowing increased sodium entry, a compensatory increase in Na–K ATPase but, as a result, excessive consumption of ATP.

Polycystic Kidney Disease

This is one of the most common single-gene disorders of humans, with widespread effects outside the kidneys, although it can remain asymptomatic. Although the genetics are well understood, the link between the gene defect and its effects remains to be explained (Reeders and Germino, 1989). Ultimately, as many as 10% of patients on dialysis may be those with this disease (Carone, 1988). Extra-renal manifestations can include brain aneurysms and hepatic cysts; 50% of patients are hypertensive (Bennett, 1990).

The cysts are not truly cysts initially but local expansions of unobstructed tubules (Carone, 1988); eventually, however, they become sealed cavities. Loss of renal concentrating capacity is an early result of the condition. Cyst formation may involve abnormalities in both the activity and distribution of Na–K ATPase (Avner, 1993; Nelson, 1993; Carone *et al.*, 1994). Because it becomes abnormally located on the apical, rather than the basolateral membrane, and is present at unusually high concentrations, it transports sodium into the lumen of the cyst (Wilson, 1991). Clearly, the question remains not only why does the genetic defect have these effects on Na–K ATPase, but also why only in certain cells; if it were generalised it would be lethal. There may also be disturbances of Na–Li countertransport, especially in patients who are hypertensive, even though their renal function is normal (Guarena *et al.*, 1993; Edmunds and Russell, 1994). The hypertension has been attributed to renin secretion within the cysts or in ischaemic areas caused by the cysts (Chapman and Schrier, 1991; Wilkinson, 1994). The increased arterial pressure is sensitive to sodium loading and there may be an underlying expansion of ECF volume; ANP levels are increased (Wilkinson, 1994). Cyst formation or other structural abnormalities are also seen in the liver, colon and heart valves (Kaehny and Everson, 1991). An acquired cystic disease of the kidneys is quite common in patients with end-stage renal disease. It has much in common with the genetic form and speculations as to its cause include electrolyte abnormalities (Ishikawa, 1991).

Gastric Cancer

There seems little doubt that high salt intake increases the risk of gastric cancer (Nazario *et al.*, 1993). Since intracellular sodium seems to be a key factor in mitogenesis (*Chapter 1*), this might appear to offer a plausible link. Nevertheless, high salt diets also increase the prevalence of *Helicobacter pylori* infection. This might simply be a marker of high sodium intake but it has also been suggested as a cause of gastric cancer (Tsugane *et al.*, 1994). Some have considered that nitrates, rather than sodium, might be the factor predisposing to gastric cancer. This seems unlikely on epidemiological grounds because the incidence has been falling for 45 years whereas the reduction of nitrates in processed food is much more recent and their concentration in water has probably increased as a result of agricultural run-off. Moreover, the potential for salt to act as a tumour promoter or co-carcinogen has been demonstrated in rat stomachs (Nazario *et al.*, 1993). Thus, when tumours were initiated with an organic carcinogen ('MNNG'), there was dose-dependent enhancement by high salt intake of their ability to cause gastric adenomas (benign but pre-malignant) and adenocarcinomas (malignant). More perturbingly, even 2–5% dietary salt supplements had a similar tendency to cause tumours independent of MNNG, though in this case the result was not statistically significant (Takahashi *et al.*, 1994a).

It is implied, but not stated, that the basal diet was sodium-free and the experiments lasted for nearly two years; for most of this time the rats weighed 0.4 kg or more. Food intakes were *ad lib.* and unmeasured, but it is reasonable to assume that these results arose from a daily salt consumption of approximately 32 mmol/kg. In human terms this is an intake of 2 lb (0.9 kg) per week and cynics might conclude that if you feed enough of anything to rats it may be carcinogenic, particularly since the intakes producing statistically significant results were equivalent to 0.6–1.2 lb/day. Thus, the higher of the intakes producing significant results would give a 70 kg human a daily salt intake of 8960 mmol. Such a prodigious intake would produce symptomatic hypernatraemia (an increase of 15 mmol/l) in 100 min, but for a spectacular increase in water intake and sodium excretion. Since the rats were on tap water, it is reasonable to wonder what impurities it may have contained and what effects they may have had when water intake, although unmeasured, also increased in parallel with salt intake.

The basis of the promoting effect of salt on gastric carcinogenesis may be direct irritation leading to replicative DNA synthesis in the proliferative zone of the gastric mucosa. It may also decrease the viscosity of hyaluronic acid in gastric secretions, thereby reducing their ability to protect the mucosa from damage. There is also evidence of increased lipid peroxidation, which causes increased production of free radicals which damage membrane phospholipids (Takahashi *et al.*, 1994a).

While there are grounds for scepticism regarding the salt intakes in these rat experiments, their relevance to human gastric cancer cannot be dismissed. The prevalence of gastric cancer is high in Japan, where intakes up to 800 mmol/day have been linked to hypertension *(Chapter 6)*. While this is only a tenth of the dose rate received by the rats, the Japanese diet traditionally includes items such as salted pickles or salted fish guts with a salt content exceeding 30% (Takahashi *et al.*, 1994a). Thus, the stomach would be repeatedly exposed to high concentrations of salt, regardless of total daily intake.

Whether or not the Na–H antiporter is involved in gastric cancer as a mediator of mitogenic effects, it is almost certainly involved in renal tubular acidosis.

Renal Tubular Acidosis (and Fanconi Syndrome)

There are two classical forms of renal tubular acidosis, though further subdivisions are possible. In proximal tubular acidosis, the major impact of defective proton secretion is to impede the reabsorption of bicarbonate. This condition stabilises once the acidosis reduces the plasma bicarbonate far enough for the resulting fall in filtered load to offset the reduction in reabsorption. In contrast, the main impact of distal tubular acidosis is to prevent the final reduction of urine pH to a maximally acidic range.

Most hydrogen ion excretion results not so much from the low pH (which still represents an infinitesimally small concentration), but the parallel loading of urinary buffers such as phosphate and ammonia. In fact, the physiological importance of buffers, in contrast to their chemical importance, is not so much that they stabilise pH, but that they enable hydrogen ions to be **transported** at an acceptable pH, whether in plasma or urine. The magnitude of this power is seen from the fact that excretion of the proton load in a wine glass of gastric acid (100 mmol/l), in the absence of urinary buffers, would require 1.23 tons of urine at a pH of 5.0. The main expression of distal renal acidosis is therefore the reduction in the excretion of ammonium ions. It may therefore arise from a reduced availability of ammonia buffer or an inability to load them with secreted protons. An important influence on ammonia synthesis is aldosterone (Arruda and Kurtzman, 1978), thus low sodium intakes potentially enhance and high sodium intakes potentially restrict the renal capacity to excrete acid loads. Na–H antiporter activity is also increased by low salt diets, further favouring excretion of acid (Seldin *et al.*, 1991). On the other hand, a primary deficiency of aldosterone, by causing hypovolaemia, would restrict the delivery of fluid to the distal nephron. This might be reinforced by restricted renal perfusion, resulting in reduced delivery of glutamine, the substrate for ammonia synthesis, to the nephrons (Sabatini and Kurtzman, 1994). In chronic metabolic acidosis, there is enhanced ammonia secretion and this may involve, as well as increased synthesis, carriage of ammonium ions by the Na–H antiporter. Aldosterone may also stimulate proton secretion by H-ATPase or H–K ATPase (Hulter, 1994).

In proximal tubular acidosis, defective proton secretion, and hence inadequate reabsorption of bicarbonate, may occur in isolation or as part of the Fanconi syndrome. The most obvious explanation for a generalised failure of proximal reabsorption would be a permeability change, allowing increased backleakage from tubular interstitial fluid into primitive urine; on kinetic grounds, this is implausible (Halperin *et al.*, 1994). A generalised interference with sodium reabsorption either requires separate impairment of a number of antiporters or cotransporters, e.g. those linking the movement of sodium to protons, glucose or amino acids, or alternatively a defect which affects them all. In the latter category, the obvious possibilities are impairment of Na–K ATPase, which provides the sodium gradients on which the cotransport and antiport mechanisms depend, or an inadequate supply of ATP to sustain its activity. An argument against the latter is that it would leave Fanconi patients extremely vulnerable to sodium depletion resulting from even a modest rise in glomerular filtration rate (GFR); they would have insufficient ATP to permit the normal compensatory increase in reabsorption. Essentially the same argument applies to impaired Na–K ATPase activity—with the additional implication that a minor reduction in GFR should eliminate the problem by reducing sodium delivery to the proximal tubule (Halperin *et al.*, 1994).

Faced with these shortcomings in proposed mechanisms for the Fanconi syndrome, Halperin *et al.* (1994) examine two further possibilities:

1. A normal amount of Na–K ATPase but reduced affinity for sodium. This would allow intracellular sodium to rise and reduce the gradient for reabsorption from proximal tubular fluid without also implying that a rise in GFR would cause salt wasting. The Basenji dog model would allow this hypothesis to be tested.

2. Decreased activity in the proximal tubular cells of H-ATPase, the Na–HCO$_3$ symporter or the Na–H antiporter (or a defect in carbonic anhydrase activity).

Haemolytic Anaemia

A rare form of human haemolytic anaemia resulting from excess intracellular sodium is not invariably associated with overhydration of the red cells. In some cases, the cell water is reduced (Parker and Berkowitz, 1986). The basic defect is increased red cell permeability to sodium and, as a result, Na–K ATPase activity is increased. This contrasts with the low potassium (LK), high sodium red cells which are the norm in many mammals (e.g. dogs, cats, cattle and a genetically determined majority of sheep). Here the mature cells lose their sodium pumps, hence becoming LK; in regenerative anaemia, immature red cells come into circulation which still have their pumps, thus high sodium cells become less usual than in normal animals. Even in predominantly high potassium species, such as cattle or dogs, there are individuals or breeds with red cells which are normally LK (di Bartola and de Morais, 1992).

Bartter's Syndrome

While it has been suggested that this syndrome results from decreased reabsorption of sodium and chloride in the ascending thick limb of the loop of Henlé, its pathophysiology remains uncertain (Schuster, 1989; Krishna *et al.*, 1994). It involves hypokalaemia, metabolic alkalosis, raised levels of renin, angiotensin and aldosterone, insensitivity to exogenous angiotensin II and hyperplasia of the juxtaglomerular apparatus. Arterial pressure is usually normal; hypomagnesaemia may also occur. As a result of the potassium deficits there is growth failure, polyuria and weakness. Atrial natriuretic peptide (ANP) levels are increased (Kenyon and Jardine, 1989). The juxtaglomerular hyperplasia may be a secondary response to a primary renal defect which impairs sodium conservation, acc rding to Laragh and Seeley (1992) but, if so, it is hard to understand why ANP levels should be high.

The condition closely resembles the effects of excessive use of loop diuretics and the impaired chloride reabsorption is not a result of the potassium deficit. Both conditions increase urinary prostaglandin excretion, but inhibitors of prostaglandin (PG) synthesis reduce the elevated renin and aldosterone levels without stopping renal potassium loss. Angiotensin-converting enzyme (ACE)-inhibitors, however, correct the potassium deficits and their effects (Schuster, 1989).

There are a number of variants of Bartter's syndrome, e.g. without raised aldosterone, and Liddle's syndrome is characteristically hypertensive; it resembles, but does not involve, excessive secretion of aldosterone (Krishna *et al.*, 1994). It is a specific defect of distal nephron function, since sodium and potassium in saliva and sweat are normal and the condition is cured by renal transplantation (Botero-Velez *et al.*, 1994). Pseudo-Bartter's syndrome mimics the appearance of the genuine condition, but results from non-renal chloride loss, e.g. due to vomiting or other gastrointestinal disturbances, or sweating (Koshida *et al.*, 1994).

11

Drugs Affecting Sodium Transport and Sodium Balance

Many drugs potentially affect sodium balance without necessarily doing so to a degree causing clinical concern. Any drug, including anaesthetics, affecting renal function either directly or via renal perfusion, is liable to cause changes in sodium excretion; the same is potentially true of drugs altering neural input to the kidney, e.g. by altering sympathetic tone. Drugs affecting the renin–angiotensin–aldosterone axis will readily affect tubular reabsorption of sodium. Drugs affecting water excretion (or thirst) will predispose to changes in plasma sodium concentration, without necessarily affecting sodium balance. This chapter is primarily concerned with drugs intended to affect the balance or distribution of sodium or those with unintended effects which merit clinical concern.

The archetypal sodium transport inhibitors, the cardiac glycosides, first came to attention through their diuretic effects. This seems logical in view of what we now know about tubular reabsorption of sodium, but at clinical dosage the effect is probably via improved renal perfusion, secondary to increased cardiac output. At high dosage or by close arterial injection, however, there is direct inhibition of renal Na–K ATPase. It remains true, even of modern diuretics, that their main use is in cardiovascular rather than renal disease; the main exception arises in acute renal failure where restoration of urine output becomes a primary objective.

Diuretics

The most widely used drugs which alter sodium transport and sodium balance are diuretics which remain cornerstones of the treatment of hypertension and, except for tranquillisers, are the most

259

widely prescribed drugs (Hymans, 1986; Gifford, 1986; Laski, 1986; Ram, 1988; Alderman et al., 1993). In veterinary medicine their main use is to treat oedema, particularly in patients with cardiac failure, and this is also their main indication in the treatment of human oedema (Brater, 1994). The term 'diuresis' simply means an increased urine output,but the hallmark of a clinically effective diuretic is that it increases sodium excretion. Recently, two appalling terms were coined to distinguish between drugs promoting water excretion (by opposing antidiuretic hormone, ADH) and those promoting sodium excretion (Mavichak and Sutton, 1986; Imbs et al., 1987). Mercifully, the life of these terms, which sounded more like a new sport ('aquaretics') and a health food ('salidiuretics') has been short. The basis of diuretic therapy is natriuresis. The differences between diuretics lie in their potency, duration of effect, side effects, mode of action and site of main effect. In this context, there has been a tendency to identify nephron segments with Roman numerals (Site I, proximal tubule, Site II, loop, Sites III and IV, early and late distal nephron) but it adds nothing to our understanding and seems best avoided (Michell, 1988d).

The early history of diuretics is a fascinating testimony to how, often, science actually works as opposed to the popular mythology of scientific advance (Beyer, 1976, 1977; Michell, 1988d). It also explains why there are so many diuretics already and such a continuing flow of new ones. The key to the discovery of most of the major types of diuretic was a mixture of keen clinical observation or fortuitous outcomes of incorrect hypotheses; despite detailed knowledge of renal physiology, it was seldom accurate deductive logic. The clinical observations were frequently of side effects of antimicrobial drugs (mercurials, sulphonamides), and even when mercurials and carbonic anhydrase inhibitors were superseded, the associated concepts remained the basis for research underlying thiazides, ethacrynic acid and furosemide (frusemide). Side effects remain the motivation for further development of new diuretics; usually the existence of a vast range of drugs for a common purpose implies that none is both dependable and safe. With diuretics, however, effective drugs are available, either mild or potent, and side effects are mainly (though not exclusively) inconvenient rather than life-threatening. Their importance, however, is increased by the fact that patients may need to take these drugs for decades and, while clinicians are often dismissive of minor side effects, their impact on quality of life can be distressing (Anon, 1977).

The effects of diuretics on renal sodium excretion strengthened belief in the involvement of sodium in the pathogenesis of hypertension. Diuretics also have direct vascular effects, however, both on arteries and veins (Young and Roberts, 1986), and anti-hypertensive drugs have been based on diuretics but with minimal remaining effect on sodium excretion, e.g. the non-diuretic thiazide, diazoxide (Fried and Kunau, 1986; Friedman, 1988).

The frequency of side effects is not, perhaps, surprising. Firstly, diuretics are often used at unnecessarily high dosage, when their diuretic effects are already maximal but the likelihood of side effects is increased (Lant, 1986; Birkenhager, 1990). The key point is the sigmoid dose–response curve; no response occurs until sufficient delivery to the tubule is achieved. Further increases in dose sharply accelerate the natriuresis, but a plateau is then reached (Voelker and Brater, 1990). Side effects are more often seen in women (Davidson, 1985; Nader *et al.*, 1988), and while this may reflect greater use, one wonders whether it also reflects a smaller bodyweight—it is second nature to veterinarians to dose accurately to bodyweight whereas human clinicians are more inclined to think in terms of 'adult' dosage, whether for a heavyweight boxer or a ballerina. Since excessive dosage, beyond the level at which natriuresis has reached its plateau, is the main reason for side effects, it seems irrational to use diuretics without facilitating their task by prudent reduction of sodium intake (Hura *et al.*, 1988).

The more fundamental reason for the importance of side effects is that, not only do diuretics act primarily on sodium transport mechanisms which are not confined to the kidney but that within the kidney, tubular reabsorption of sodium underlies numerous other functions. Moreover, because it occurs in several segments, and responds sensitively to changes in sodium balance, diuretics with a primary site of action in one segment will be opposed by compensatory increases in sodium conservation in other segments (Bock and Stein, 1988). These may then produce further side effects, e.g. kaliuresis, due to heightened aldosterone secretion and increased distal delivery of sodium. Increased production of angiotensin not only promotes distal sodium conservation by stimulating aldosterone secretion but directly increases proximal reabsorption, thus angio-tensin-converting enzyme (ACE)-inhibitors may potentiate the effect-iveness of diuretics (Abraham and Schrier, 1994). In the case of oedema, the less the remaining surplus of interstitial fluid, the more difficult it will be to mobilise it into plasma, hence, if renal protection of plasma volume remains effective, the effectiveness of the diuresis becomes blunted. This is the basis of the familiar clinical observation that a fixed dosage of diuretic has a declining effect ('braking phenomenon') as patients reach their 'dry weight' (Quamme, 1986). Excessive dosage of a potent diuretic carries the risk that the rate of natriuresis exceeds the speed at which sodium can be mobilised from interstitial fluid with the result that hypovolaemia, even circulatory shock, becomes a risk. Thus, humans can only be expected to mobilise approximately 0.8 l of ascitic fluid per day if hypovolaemia is to be avoided (Hura *et al.*, 1988).

There is no simple answer to the search for an 'ideal' diuretic. Except for osmotic diuretics which are active in all nephron segments, no modern diuretic has a primary site of action in the main site of sodium reabsorption, the proximal tubule. Even if it did, there is considerable scope for 'compensatory' increases of sodium reabsorption in the loop and distal nephron to oppose its action. Carbonic anhydrase inhibitors, no longer used primarily as diuretics (Laski, 1986), block sodium reabsorption in the proximal tubule, but only the minority which is associated with bicarbonate as opposed to the majority which is accompanied by chloride. Not surprisingly, loop diuretics are more potent than those affecting the distal nephron, where least sodium is absorbed.

Segmental balance is not the only issue, however; ever since Withering observed the diuretic effects of cardiac glycosides in patients with 'the dropsy' (ascites due to cardiac failure), the ability to increase urine output, like that to lower a fever, has been a convincing demonstration of the power of the physician, whether or not it was a rational objective. A brisk, intense diuresis may combine hazard with futility. It is likely to be short-lived, if only because hypovolaemia will occur if it continues for too long. During the hours when the diuretic is not acting, however, there is ample time for renal sodium conservation in defence of plasma volume to nullify its effects (Kramer, 1987; Rose, 1987). It may well be that a less intense diuresis, sustained for longer, is more effective and more compatible with the speed of mobilisation of oedema fluid. Thus, loop diuretics, used as a single dose, are not particularly potent when 24-h sodium excretion is considered (Reyes and Leary, 1993). In fact, thiazides may be more potent because they produce a less intense but more sustained diuresis with less rebound (Reyes, 1993). The antihypertensive effects of diuretics involve different considerations since in the long term they do not depend on reduction of extracellular fluid (ECF) volume.

If we believe that intracellular electrolyte disturbances within the arterioles are a major contributory cause of hypertension (*Chapter 6*), then direct or indirect effects of diuretics on these would legitimately appear to attack the underlying problem. With oedema, however, renal disease, whether through glomerular protein leakage or inappropriate glomerotubular balance (e.g. in chronic renal failure, CRF) is not usually the primary cause: cardiac or hepatic disease provide the usual clinical context for severe oedema. Recently, if we prefer the 'overfill' to the 'underfill' concepts of oedema formation, renal dysfunction, arising from hepatic disease or other causes of oedema, is regarded as the main drive, i.e. expansion of interstitial fluid is a natural result of excessive sodium retention (*Chapter 8*). The more traditional view is that renal sodium retention is a homeostatic response protecting plasma volume when factors in the capillary beds favour an increased outflow into interstitial fluid. Whichever is correct, it remains true that a large compartment cannot be expanded at the expense of a small compartment unless the latter is continuously replenished. **The kidney, therefore, is always the 'enabling' cause of oedema** and, in that sense, the use of diuretics is not simply symptomatic but directed towards an essential component of the problem.

There are less obvious benefits of diuretics. Firstly, if there is a tendency towards hypovolaemia, it will, to some extent, be self-correcting, not only through renal effects, perhaps through increased vasoconstrictor tone (counterproductive in hypertension), but through the resulting increase in oncotic pressure if plasma volume contracts. More important, substantial benefit accrues from increased venous capacitance, which may precede diuresis (Lant, 1986; Ramires and Pileggi, 1986; Michell, 1988d); this may be particularly helpful with pulmonary oedema. It also emphasises that the adverse effect of venous congestion is not the increased volume of blood on the venous side of the circulation which, after all, is the 'storage' or 'capacitance' side. The harm lies in a raised venous pressure and its effect on capillary pressure (Michell, 1988d). Similarly, the problem created by cardiac failure, in the early stages, is not the inability to maintain a reasonable cardiac output, but to do so without an increased filling pressure (and hence, elevated venous pressure). At that stage, if there is no other support for cardiac function, sodium retention and increased venous pressure is sustaining cardiac output (Young and Roberts, 1986). Direct cardiovascular effects probably underlie other beneficial effects of diuretics:

1. decreased breathlessness associated with improved cardio-pulmonary function (Taylor, 1987);
2. better exercise tolerance with a reduction of arteriolar sodium removing an important constraint on muscle perfusion (Hampton, 1987).

In addition, adequate cardiac output is achieved at a lower ventricular volume. As a result of La Place's Law, broadly that in an elastic container, the wall tension (T) to sustain the contained pressure (P) increases with radius (R):

$$P \propto \frac{T}{R}$$

Hence, a reduction of the increased size of the dilated heart enables the same output to be sustained at a lower workload, i.e. myocardial oxygen demand is reduced. Since the myocardium is especially sensitive to ischaemic damage, this is an important benefit (Ramires and Pileggi, 1986; Reifart and Hampton, 1987).

The initial antihypertensive effects of diuretics depend on a reduction of ECF volume, but this may secondarily increase peripheral resistance. The chronic response depends less on ECF volume, which tends to recover, though not completely; its importance is superseded by vasodilator effects which also depend on a maintained reduction of body sodium (Birkenhager, 1990). It remains unclear why these vasodilator effects should not prevent the initial vasoconstriction. Vasodilator prostaglandins increase soon after the initiation of therapy (Birkenhager, 1990); whether there are delayed effects on the release of either nitric oxide or endothelins remains to be shown.

Both receptor-active and osmotic diuretics have been shown to have **prophylactic** effects in various models of acute renal failure (ARF) and advocated for its treatment. Despite obvious potential benefits, e.g. increased nephron flow rates to minimise tubular blockage, improved renal blood flow, reduced renal energy consumption and, in the case of osmotic diuretics, reduced cell swelling (Lieberthal and Levinsky, 1990), the effectiveness of diuretics in reversing established ARF remains controversial (Daniels and Ferris, 1988). A key factor in established ARF is the unpredictable delivery of receptor-active diuretics to their site of action, since this requires proximal secretion and adequate delivery to more distal sites. Dopamine may alleviate this restriction (Lieberthal and Levinsky, 1990). As in shock, there is also the ethical problem of properly designed studies in a clinical condition with a high mortality; to consign patients to the control group, whatever other treatment this may allow, raises grave questions of conscience.

It is convenient to begin with osmotic diuretics since their indications and mode of action are so different from receptor-active diuretics. Not least, they are contraindicated in heart failure or hypertension because their primary effect is to expand ECF volume, including plasma (Wells, 1990; Brater, 1994). They can, however, reduce endothelial swelling and other forms of cellular rather than interstitial oedema, e.g. in brain. Hypertonic solutions are also used in the treatment of circulatory shock but this does not relate to any diuretic effects and is considered separately (*Chapter 8*).

Osmotic Diuretics (Michell, 1979d; Wilcox, 1989; Martinez-Maldonado and Benabe, 1990)

Among solutes normally present in plasma, glucose, urea and saline can all be used to induce diuresis by intravenous infusion and may do so spontaneously, for example in diabetes mellitus or renal failure. The most commonly used osmotic diuretic is mannitol; unlike natural solutes it remains exclusively in ECF, whereas urea, in particular, equilibrates into intracellular fluid (ICF) unless it is rapidly infused. Osmotic diuretics affect all segments of the nephron but have their main effect in the proximal tubule; they also increase renal perfusion (Mavichak and Sutton, 1986; Raymond *et al.*, 1986).

Mannitol expands ECF volume (which promotes sodium excretion by reducing proximal absorption) and is freely filtered at the glomerulus. By holding water in the proximal tubule it dilutes the sodium in tubular fluid and reduces the gradient for movement of sodium into ICF. It also increases flow rates throughout the nephron and reduces time for osmotic equilibration, thus favouring increased loss of water. This effect is reinforced by increased medullary blood flow which reduces the concentration of medullary interstitial fluid ('medullary washout'). The effects on water excretion therefore outweigh those on sodium excretion and, unlike other diuretics, osmotic diuretics tend to raise rather than reduce plasma sodium concentration. Potassium loss is also increased in parallel with sodium, although plasma potassium may rise as a result of increased ECF osmolality (Morrison and Singer, 1994).

Thiazides

Carbonic anhydrase inhibitors, from which thiazides were originally derived, caused metabolic acidosis by promoting loss of bicarbonate. Apart from its undesirable effects, the metabolic acidosis limited the effectiveness of carbonic anhydrase inhibitors by diminishing the filtered load of bicarbonate. Thiazides, however, like loop diuretics, tend to cause metabolic alkalosis as a result of contraction of ECF volume without a proportionate increase in bicarbonate excretion. In addition, should there be significant potassium depletion, there is the usual risk of a vicious cycle involving mutual reinforcement of metabolic alkalosis and potassium depletion (Michell *et al.*, 1989). The increased drive to reabsorb sodium in the distal nephron (in response to aldosterone, as a result of hypovolaemia) enhances loss of potassium or hydrogen ions at a time when both are depleted. Hypovolaemia undermines the normal effectiveness of the renal defence against metabolic alkalosis.

Thiazides are believed to have their primary effect in the early distal nephron (Costanzo, 1988), since they enhance potassium loss whereas diuretics acting later in the distal nephron specifically reduce potassium excretion; in fact that is their main purpose ('potassium-sparing diuretics', see below). Thiazides impair the final stage of urinary dilution (sodium removal in the distal nephron) without impeding concentration, i.e. they affect neither the loop nor the collecting duct. Their site of action appears to lie beyond the macula densa, since they have no direct effect on renin secretion. They do not directly affect Na–K ATPase, but they inhibit electroneutral-coupled reabsorption of sodium with chloride. The gene controlling the thiazide NaCl cotransporter has now been isolated and distinguished from mechanisms for $Na^+:H^+$ or $Cl^-:HCO_3^-$ exchange. The latter are also affected by thiazides, perhaps because of residual anti-carbonic anhydrase activity (Ausiello, 1993); this probably explains their minor effect on the proximal tubule (Costanzo, 1988). The thiazide receptor gene is not restricted to the kidney and this is consistent with the extra-renal basis of their long-term anti-hypertensive effects.

Unlike most other diuretics, thiazides reduce renal perfusion and glomerular filtration rate (GFR); the mechanism seems to involve changes in intratubular pressure and tubuglomerular feedback mediated by angiotensin II (Fried and Kunau, 1986; Eknoyan and Suki, 1991). Triamterene (see below) also reduces GFR, but this is mediated by reduced production of prostaglandin E; the effects of thiazide and triamterene on renal haemodynamics are additive. Like other diuretics, except osmotic diuretics and spironolactone, thiazides enter the nephron by secretion into the proximal tubule; they reach concentrations well above plasma but are liable to competition from other secreted organic anions. This, rather than reductions in renal perfusion or GFR, underlies the reduced access of diuretics to the nephron in chronic renal failure (Chonko and Grantham, 1994). Side effects of diuretics are discussed below. Some diuretics, such as metolazone, chlorthalidone, which are not truly thiazides, have very similar properties (Wells, 1990).

Potassium-Sparing Diuretics

Like thiazides, these act in the distal nephron but at a more distal site. Their natriuretic effects are modest, but they significantly reduce the potassium depletion caused by other diuretics. They also reduce magnesium losses despite the fact that the main site of reabsorption is the loop (Martin and Milligan, 1987; Hollifield, 1989). Spironolactone is a competitive antagonist to aldosterone (developed from progesterone which has similar effects). Its effects are not confined to the kidney since it also reduces the effects of aldosterone on the colon and sweat glands and the direct effects on the myocardium (Krishna *et al.*, 1988). The active form of spironolactone is its metabolite, canrenone (Friedman, 1988) and perhaps other thio-metabolites (Fanestil, 1988). Amiloride and triamterene block sodium channels on the luminal side of cells in the distal convoluted tubule and collecting duct. They reach the nephron by secretion in the proximal tubule as organic cations. Inhibition of luminal entry of sodium restricts potassium excretion by reducing the luminal electronegativity and secondarily reducing Na–K ATPase activity as ICF sodium falls (Birkenhager, 1990). Magnesium excretion is also reduced, and to a lesser extent, calcium excretion (Kleyman and Cragoe, 1988). All three impede hydrogen ion secretion; this may be reinforced by increased intracellular potassium which reduces the secretion of ammonia and thus reduces the amount of urinary buffer available to carry hydrogen ions (Wilcox, 1989; Martinez-Maldonado and Benabe, 1990; Chonko and Grantham, 1994). Amiloride-sensitive sodium channels are not restricted to the kidney but are widely distributed, for example, in the colon (Ausiello, 1993) and, as mentioned previously, in taste cells and leukocytes *(Chapters 4 and 6)*. Amiloride has been used to block sodium channels in the respiratory tract in cystic fibrosis *(Chapter 10)*, but with mixed results.

While the *raison d'être* of potassium-sparing diuretics is to counter the potassium depletion caused by other diuretics, it is possible for them to predispose to hyperkalaemia. This is least likely with spironolactone, since it is a competitive antagonist and any increase in plasma potassium will tend to be self-limiting as a result of the consequent rise in aldosterone secretion. In fact, none of the potassium-sparing diuretics is likely to cause hyperkalaemia except as part of a combination of predisposing factors such as:

1. renal failure;
2. metabolic acidosis (e.g. caused by severe diarrhoea);
3. diabetes mellitus (reduced insulin stimulation of cellular uptake of K^+; Bargman and Jamison, 1986);
4. β-blockers (reducing β-adrenergic promotion of potassium uptake);
5. K^+ supplements (or, in veterinary medicine, K^+-supplemented prescription diets);
6. ACE-inhibitors.

It might be thought that β-blockers would also act by suppressing renin secretion and hence the angiotensin-driven secretion of aldo-sterone. This is less likely to be important, however, because hyperkalaemia is an independent stimulus to aldosterone secretion. Nevertheless, ACE-inhibitors, for which the same argument could be made, can predispose to hyperkalaemia, in combination with other factors, notably K-sparing diuretics (Anon, 1987; Stokes, 1989). It might be safest to regard ACE-inhibitors as K-sparing diuretics, albeit this is an unintended effect. It is also important to appreciate that while patients with stage II heart failure show signs of renal impairment by ACE-inhibitors in only 5% of cases, this rises to 40% in severe congestive heart failure (Francis, 1994) and, inevitably, the risk is heightened by hypovolaemia caused by excessive use of diuretics. It will also be increased in patients on NSAIs (see below).

Loop Diuretics

Whereas thiazides represent a chemically homogeneous group (with a few exceptions such as chlorthalidone) loop diuretics include a surprising variety of compounds with remarkably similar modes of action (Quamme, 1986; Michell, 1988d). These are not only the most potent diuretics, but perhaps the best understood. They include furosemide, ethacrynic acid, bumetanide, muzolimine and (historically) the mercurials (Friedman, 1988). Cotransport of so-dium, potassium and chloride is inhibited in the ascending thick limb of the loop of Henlé (Bankir *et al.*, 1987; Reeves and Molony, 1988) and, as a result, not only is sodium excretion increased but the ability to form either a fully dilute or a fully concentrated urine is impeded (i.e. the extremes of both positive and negative free water clearance are unattainable). While this is their main site of action, the extent of the natriuresis suggests additional effects, elsewhere in the nephron (Friedman, 1988). Not all loop diuretics have identical effects on the thick ascending limb, thus indacrinone affects chloride transport on the peritubular side whereas torasemide affects the luminal side as well (Reeves and Molony, 1988).

Furosemide and ethacrynic acid also improve renal perfusion and have other beneficial effects on the vasculature (Imbs *et al.*, 1987). The effectiveness of loop diuretics is blunted by proteinuria, probably because these drugs act from the luminal side of the nephron and may be bound to the protein abnormally present in tubular fluid (Quamme, 1986; Humphreys, 1994). Despite their potency, loop diuretics are short-acting, thus their effectiveness is undermined if they are given too infrequently and if excessive sodium intake is ignored, permitting extensive salt retention during periods when the diuretic is not active (Voelker and Brater, 1990; Brater, 1993). In addition, congestive heart failure, chronic renal failure and cirrhosis also reduce the renal response to diuretics. Factors which reduce renal perfusion undermine the effectiveness of diuretics since it is the concentration in tubular fluid, rather than plasma, which matters (Brater, 1993). In chronic treatment, the increased delivery of sodium to the distal nephron causes hypertrophy and increased sodium reabsorption, thus resisting the effectiveness of loop diuretics. This can be rectified by adding a thiazide (Brater, 1993). In CRF, delivery of diuretics is not only restricted by reduced perfusion but by increasing concentrations of organic acids which compete for secretion in the proximal tubule. Diuretic resistance in congestive heart failure, previously attributed to impaired intestinal absorption, is more likely to reflect inadequate delivery to the loop since it affects intravenous as well as oral administration. In patients with normal renal function despite their heart failure the response to intravenous diuretics is unimpaired, hence their diuretic resistance probably does reflect factors such as delayed gastric emptying or abnormal intestinal motility. An important aspect of diuretic resistance in heart failure patients is that where it occurs it reduces the **maximum** diuresis in response to the drug, i.e. higher doses of the same diuretic cannot help (Friedman, 1988; Brater, 1994).

The beneficial haemodynamic effects of loop diuretics, both renal and elsewhere, are substantially prostaglandin-dependent and are blunted by non-steroidal anti-inflammatory drugs (NSAIs) (Lant, 1986; Imbs *et al.*, 1987; Eknoyan and Suki, 1991). Pulmonary vasodilation is a valuable feature in patients with congestive heart failure (Martinez-Maldonado and Benabe, 1990), but does not occur otherwise. More prosaically, it is important to emphasise the potency of these diuretics; excessive use will readily cause hypovolaemia and therefore precipitate or decompensate renal failure. A natriuresis of 20% of filtered load is equivalent to approximately 50% of plasma volume per hour. Unlike ethacrynic acid, furosemide retains some anti-carbonic anhydrase activity, but the losses of sodium and chloride are so much more marked than any loss of bicarbonate that both cause contraction alkalosis (Wells, 1990).

The loop is also the main site of magnesium reabsorption (Quamme, 1986), hence these diuretics readily predispose to magnesium depletion (Ryan, 1987), though other diuretics also increase magnesium loss in parallel with sodium and potassium. Indeed, magnesium depletion frequently accompanies potassium depletion, may underlie many effects formerly attributed to it and, by depressing Na–K ATPase activity, exacerbates cellular potassium depletion (Dawson and Sutton, 1986; Dyckner and Wester, 1987). Hypomagnesaemia is grossly underdiagnosed in humans and animals treated with diuretics and its ability, like hypokalaemia, to cause ventricular dysrhythmias is a cause for serious concern (Wills, 1986; Hollenberg, 1987; Hollifield, 1987; Whang, 1987; Michell, 1988d; Cobb and Michell, 1992). It also exacerbates cardiac glycoside toxicity (Nader et al., 1988), probably by additive depression of Na–K ATPase activity.

A significant proportion of diuretic-treated patients experience hypokalaemia and this may be exacerbated by extremes of sodium intake, both low and high (Nader et al., 1988). Large changes are not the only problem; marginal hypokalaemia may be aggravated by stress (Kaplan, 1986) and adrenergic stimulation and oedematous states often increase the risk because of associated rises in aldosterone secretion (Seldin and Giebisch, 1989; Cobb and Michell, 1992). Depression of either myocardial or plasma potassium predisposes to ventricular dysrhythmias and the particular worry is the speed with which plasma potassium can change (Bargman and Jamison, 1986; Rose, 1987; Davidson, 1987). It is important to note that hypokalaemia seems to be more of a problem in patients treated for cardiac disease, rather than hypertension (Quamme, 1986; Freis, 1987; Nader et al., 1988) and with thiazides in particular (Young and Roberts, 1986; Poole-Wilson, 1987). Duration of action, rather than potency, governs the extent of the potassium loss (Bargman and Jamison, 1986). Cardiac muscle is especially vulnerable to the effect of acute hypokalaemia associated with chronic potassium depletion since the latter does not affect its ICF K so the effect on membrane potentials is not offset by a parallel reduction in ICF (Solomon, 1987).

It may be that diuretics exacerbate losses of intracellular potassium and magnesium associated with cardiac failure, hence the lack of evidence for the importance of such effects in patients treated for hypertension (Freis, 1990). Nevertheless, a recent study of the use of thiazides for the treatment of hypertension shows that although they generally reduce the mortality from coronary heart disease, increasing dosage carried a risk of death from cardiac arrest; this risk could be reduced by concurrent use of a potassium-sparing diuretic (Bigger, 1994; Siscovick et al., 1994).

The effects of diuretics on calcium excretion are more ambivalent; although calcium excretion is often dictated by sodium excretion, this usually results from changes in proximal tubular reabsorption. Thus, even though loop diuretics increase sodium and calcium excretion, any opposing increase in proximal tubular sodium reabsorption will tend to conserve calcium; more importantly, increased intestinal absorption prevents calcium depletion. Nevertheless, loop diuretics can be used to treat acute hypercalcaemia (Chonko and Grantham, 1994). If calcium depletion did occur, diuretics do not oppose the calcium-conserving effects of parathyroid hormone (PTH) in the distal tubule, indeed, thiazides restrict calcium excretion in humans (Fried and Kunau, 1986; Sutton, 1986), and dogs (Nader *et al.*, 1988; Daniels and Ferris, 1988). In contrast to their effect on calcium, thiazides promote magnesium loss (Wells, 1990).

Supplementation of loop diuretics with thiazides may help to counter diuretic resistance, one aspect of which is compensatory hypertrophy of the distal nephron in response to the increased delivery of sodium from the loop. This unwanted increase in the reabsorptive capacity of the distal tubule can be countered by the use of thiazides (Brater, 1994).

Other Side Effects

Apart from subjectively unpleasant effects, the main side effects beyond those already discussed are increased plasma lipids (Gifford, 1986; Dawson and Sutton, 1986), elevated uric acid concentration in plasma (a result of increased proximal fluid reabsorption in response to hypovolaemia; Kahn, 1988) and glucose intolerance, probably predisposed by potassium depletion. This not only causes insulin resistance, but also reduces insulin secretion (Prichard *et al.*, 1992). In addition, the importance of hyponatraemia has probably been underestimated, especially with thiazides (Sonnenblick *et al.*, 1993); it is common in humans being treated for severe heart failure (Young and Roberts, 1986) and it can impair renal function in dogs (Tyler *et al.*, 1987). Loop diuretics enhance renin secretion independent of hypovolaemia (Martinez-Maldonado and Benabe, 1990). While side effects are important to individual patients, their prevalence is probably less than many doctors believe; the effects of diuretics on blood lipids have mainly been shown in relatively short term studies and although diuretics raise blood levels of uric acid, gout is not a common result of their use (Moser, 1990).

An interesting application of a side effect is the paradoxical use of diuretics to treat a condition of exorbitant urine output, diabetes insipidus (Laski, 1986). The fundamental problem of diabetes in-sipidus is the production of excessively dilute urine, either because ADH secretion is inadequate or the renal response to it is defective. Hence the urine remains dilute instead of being concentrated in the collecting duct (*Chapter 2*). The volumes produced are potentially far greater than, for example, in CRF where urine may be little more concentrated than plasma. That is not simply because CRF reduces GFR, but because maximally dilute urine is five times as dilute as plasma, hence the enormous increase in volume. If, therefore, the ability to concentrate urine cannot be restored, an important objective is to prevent urine becoming maximally dilute. This can be done by restricting the supply of solute to the loop, i.e. by increasing proximal reabsorption in response to mild hypovolaemia (Michell, 1979d; Martinez-Maldonado and Benabe, 1990; Morrison and Singer, 1994). This approach is dangerous if the problem is actually primary polydipsia rather than diabetes insipidus because prevention of the formation of fully dilute urine will impede the excretion of excess water (Sutton and Drance, 1986). Inhibition of urinary dilution is also exploited in the use of loop diuretics, combined with saline, to correct symptomatic hyponatraemia (Daniels and Ferris, 1988).

Other Diuretics

The xanthines, which closely followed mercurials (Lant, 1986) have long outlived them and are now the 'oldest' diuretics, with over 60 years of clinical experience, although few would use them for this purpose any longer. They are still used as anti-asthmatics. They blockade adenosine receptors and, at higher dosage, they increase cyclic nucleotides and calcium within cells (Rall, 1990). Diuretic effects are most likely to be encountered as an adverse result of intravenous use in asthmatics; this substantially increases urinary loss of sodium, calcium and magnesium, thereby causing reductions in the plasma concentration of divalent cations (Knutsen et al., 1994). Like corticosteroids, xanthines increase GFR via renovascular effects, but this would not increase sodium excretion unless they also prevented a compensatory increase in tubular reabsorption, probably by inhibitory effects in the proximal tubule and perhaps beyond. Similarly, dopamine, which enhances GFR and urine output, also blocks sodium reabsorption in the proximal tubule (Chonko and Grantham, 1994). It does so via a cAMP-mediated depression of $Na^+:H^+$ exchange, but it may also depress Na–K ATPase activity indirectly (Felder et al., 1989). Thus, while a number of drugs may enhance GFR, particularly where it has been compromised by inadequate cardiac output, the key to an effective clinical diuretic, whether intended to reduce blood pressure or ECF volume, or simply to increase urine output, remains its ability to block tubular reabsorption of sodium. Lastly, it is not widely realised that heparin can act as a distal tubular diuretic, i.e. it causes natriuresis and, potentially, sufficient potassium retention to result in hyperkalaemia. These effects result from suppression of aldosterone synthesis (Oster et al., 1994a).

Cardiac Glycosides

Cardiac glycosides are specific inhibitors of Na–K ATPase. This does not directly underlie their natriuretic effect, except at high dose, but it underlies their cardiac effects, which lead to diuresis through improved renal perfusion and mobilisation of oedema fluid. The accepted mode of action closely parallels the explanation of the hypertensive effects of sodium transport inhibitors *(Chapter 6)*. Uncertainties remain, however, because of evidence that inotropic effects may be obtained in the absence of inhibition and, conversely, inhibition can occur without inotropic effects (Rosenberg *et al.*, 1986). Indeed, it can be argued that the effects of cardiac glycosides on cardiac rhythm are more readily explained on the basis of Na–K ATPase inhibition than the inotropic effects (Rosenberg *et al.*, 1986); clinically, the chronotropic effects (slowing of ventricular rate) have escaped the controversy surrounding the inotropic benefits of chronic cardiac glycoside therapy (Malcolm, 1986; Smith, 1989). Doubts about the benefits of cardiac glycosides in patients with normal sinus rhythm remain unresolved (de Bono, 1994) despite recent large scale clinical trials (Young, 1994). One problem is that Na–K ATPase is a dynamic system and it responds to disturbances. Thus, the immediate consequence of inhibition is to increase intracellular sodium and produce, in the short term, an opposing increase in Na–K ATPase activity and, in the longer term, an increased synthesis of pump sites (McDonough *et al.*, 1990). Heart failure, *per se*, depresses Na–K ATPase activity independently of any cardiac hypertrophy (Fan *et al.*, 1993). This effect is mediated by high levels of β-adrenergic activity but, unexpectedly, the cardiac response to a cardiac glycoside was unimpaired. Interpretation of older studies of the chronic effects of cardiac glycosides is complicated by the likelihood that they have differential effects on the various isoforms of Na–K ATPase (Rose and Valdes, 1994). Whatever the uncertainties, the fact remains that cardiac glycosides remain the only orally active positive inotropes approved by the American FDA which are widely used for patients with congestive heart failure (Francis, 1994).

The cardiac glycosides bind to the external surface of the Na–K ATPase sodium–potassium pump. The resulting rise of intracellular sodium leads to augmentation of intracellular calcium by Na^+–Ca^{2+} exchange. The increased store of intracellular calcium enhances contractility of both normal and failing myocardium (Smith, 1989). Taken to excess, the increased calcium underlies the cardiac manifestations of digitalis toxicity (whereas the favourable electrophysiological effects are mainly mediated by changes in vagal tone). Evidence from guinea-pigs suggests that the atria are more sensitive effects to the effect of cardiac glycosides than the ventricles because they have a lower density of Na–K ATPase pump sites (Wang *et al.*, 1993). Cardiac glycosides also increase the release of noradrenaline at cardiac sympathetic nerve terminals and reduce its re-uptake, but this does not seem to relate to their clinical effects (Smith, 1989). The rise of intracellular sodium caused by Na–K ATPase inhibition is reinforced by exchange of extracellular Na^+ for intracellular H^+ (which rises in response to the increased intracellular Ca^{2+}). Since Na–K ATPase is essential to cell uptake of potassium, it might be anticipated that cardiac glycosides would predispose to hyperkalaemia. This is a minor effect at therapeutic dosage because of very limited binding to receptors in skeletal, as opposed to cardiac muscle; at toxic concentrations, however, it becomes important (Williams and Epstein, 1989).

Cardiac glycosides inhibit the sodium pump in its phosphorylated form (Blaustein, 1993). The binding site for cardiac glycosides (and endogenous inhibitors of sodium transport; *Chapters 2 and 6*) is on the external surface of the α subunit of Na–K ATPase, as is that for potassium (Lingrel *et al.*, 1994). Whether the latter is the same site which binds sodium on the intracellular surface is uncertain, indeed, both the binding sites for cations and their pathway through the pump, central to its mode of action, constitute a "conspicuous gap in our knowledge" (Glynn, 1993). Nevertheless, potassium is a non-competitive antagonist of cardiac glycoside inhibition (Doris, 1994). Both calcium and vanadate inhibit the pump from the internal side (Hoffman, 1986). The inhibitory effect of calcium is increased by an intracellular protein (calnaktin) which is distinct from calmodulin (Hoffman, 1986). The inhibitory effect of vanadate depends, at least in part, on alterations in the optimum pH and potassium sensitivity of Na–K ATPase, as do the effects of catecholamines *in vitro* (Michell and Taylor, 1982). Vanadium also inhibits cardiac Ca ATPase, gastric HK ATPase, but not renal H ATPase. It is a potent diuretic and may have a physiological role in potassium regulation (Dafnis and Sabatini, 1994).

Whether or not there is direct competition between cardiac glycosides and potassium, hypokalaemia reduces the number of Na–K ATPase sites in skeletal muscle, making it particularly susceptible to intracellular potassium deficits but helping to limit any further decline in plasma potassium (McDonough *et al.*, 1992). By limiting the fall in extracellular potassium and causing a parallel fall in intracellular potassium, the impact of potassium depletion on excitability may be reduced. This response depends on the α_2 isoform of Na–K ATPase and similar, though less marked effects would be expected in cardiac muscle which has less of this isoform. The ventricular myocardium predominantly contains the α_1 isoform, whereas the conduction pathways have mainly α_2 and α_3. It is likely that differential binding of cardiac glycosides to these isoforms and genetic or hormonal influences on the balance between them underlies some of the variability of clinical response. This may explain some of the differences, perhaps genetically determined, in susceptibility to toxic effects and also changes in response during the development of cardiac hypertrophy (Medford, 1993).

While doubts remain to be resolved concerning both the mode of action and the effectiveness of these veteran drugs, one fact stands out in relation to sodium transport; they have not been shown to have any primary biochemical effect other than inhibition of Na–K ATPase (Rosenberg *et al.*, 1986). Indeed, this may be the basis of their physiological function, as hormones *(Chapter 1)*, as well as their therapeutic effects.

NSAIs (Hall and Granger, 1994)

It has long been argued that prostaglandins, notably PGE_2, contribute to renal sodium regulation. Infusions of PGs increase sodium excretion, perhaps through direct effects on Na transport as well as perfusion. Moreover, antagonists restrict the natriuretic response to sodium infusions and there is a strong belief that clinically they cause salt retention. Since the effects of prostaglandins are often to modulate the impact of other factors, e.g. ADH, responses to NSAIs may vary according to a variety of background conditions (Zambraski and Dunn, 1993). Whether or not NSAIs cause salt retention in particular types of patient, they seem to have little impact on the ability to maintain sodium balance in healthy subjects (Zambraski and Dunn, 1993). They do inhibit the natriuretic effects of diuretics, though the reasons are unclear (Friedman, 1988). Perhaps the most important clinical effect on fluid balance is that they reduce GFR and predispose to acute renal failure by undermining the prostaglandin-dependent defence against excessive renal vasoconstriction. They can also cause hyperkalaemia because renin release is prostaglandin-dependent (Laragh and Seeley, 1992; Hackenthal and Nobiling, 1994), hence NSAIs cause hyporeninaemic hypoaldosteronism *(Chapter 8)*. Chronic use at high dose is well known to cause nephrotoxic lesions, as in 'analgesic nephropathy' (Levin, 1988); acute nephrotoxicity can also occur (Kleinknecht, 1993; Zambraski and Dunn, 1993). In extreme cases there may be papillary necrosis (Sabatini, 1988). This probably reflects increased susceptibility to ischaemic damage when prostaglandin (PG) synthesis is impaired (Levin, 1988). Some NSAIs, e.g. sulindac and salicylates, have less impact on renal function.

Other Drugs

A variety of other drugs can affect sodium balance, often via effects on Na–K ATPase and, therefore, on potassium too. Foremost among these is insulin, which may play a role in potassium homeostasis since its secretion rate is increased by hyperkalaemia and it stimulates cellular potassium uptake by enhancing Na–K ATPase activity (Williams and Epstein, 1989). This underlies its use to treat hyperkalaemia (Schuster, 1989). The antinatriuretic effect may also contribute to the hypertensive effects of insulin in insulin-resistant states, including obesity *(Chapter 6)*. This is, however, only one of a number of actions which may affect blood pressure (Ferrannini and Natali, 1994). Insulin deficiency may reduce the ability of the adrenal cortex to secrete aldosterone, perhaps by reducing its intracellular potassium (De Fronzo and Smith, 1994). A number of other hormones (e.g. thyroxine) discussed in *Chapter 7* are also used as drugs and may affect sodium balance, thus high dose corticosteroids can cause salt retention, as can oestrogens in some species. β-agonists such as **salbutamol** can cause hypokalaemia by stimulating Na–K ATPase and driving potassium into cells *(Chapter 1)*. Conversely, **β-blockers** may exacerbate other factors (such as ACE-inhibitors) which predispose to hyperkalaemia. As well as drugs which specifically affect Na–K ATPase, a number, including some antibiotics, are general inhibitors of membrane-bound ATPases; inhibition may be direct or indirect (Reynolds *et al.*, 1981).

Among other drugs affecting sodium transport are those such as **oligomycin** and **ethylmaleimide**, used to study the process rather than therapeutically (Rosenberg *et al.*, 1986). There are also numerous drugs which affect substances cotransported with sodium, e.g. anions (**DIDS**), glucose (**phlorizin** and **phloretin**) or **cocaine**, which inhibits sodium coupled uptake of noradrenaline at nerve terminals (Rosenberg *et al.*, 1986). Drugs such as **tetrodotoxin** and **tetraethylammonium** compounds have been used extensively in neurophysiology because of their specific effects on ion-conductance pathways. Moreover, transmitters such as **acetylcholine, catecholamines** and **GABA** exert their effects through receptor-mediated changes in ion permeability. A variety of drugs act by blocking voltage-sensitive sodium channels, including local anaesthetics, class I antidysrhythmics and some anticonvulsants (Rang and Dale, 1991). Other drugs affect the gating of sodium channels; these include DDT, veratridine and pyrethrins. **Ionophores**, which facilitate ionic transfer across lipid membranes, have also been used as drugs, usually to alter the distribution of calcium, but potentially of sodium, potassium and magnesium as well. They create static or moving channels through the membrane (Bowman and Rand, 1980); in the latter case their action frequently involves a ring structure which encloses the transported cation, e.g. **valinomycin**, which carries potassium, rubidium, caesium or protons but not sodium or lithium. **Nigericin** slightly favours potassium over sodium and **monensin** is primarily a sodium ionophore. The differences in specificity cannot completely be explained by differences in the size of the artificial pore created by the drug, but also depends on the formation of complexes (Reynolds *et al.*, 1981). Rather than creating pores, some ionophores form a double ring with a divalent cation between the layers. As well as being used to investigate membrane transport, ionophores have been used to alter the rumen microflora, as anticoccidials and as antibiotics (Brander *et al.*, 1991). **Amphotericin B**, an antifungal drug, forms transmembrane pores which prevent the maintenance of gradients of sodium, potassium or pH (Stokes, 1994).

Lithium is sufficiently similar to sodium to be carried initially by the sodium pump, but its toxic effects then cause inhibition. It is the smallest alkali metal and penetrates sodium channels as well as sodium but neither passes potassium channels (Bunney and Bunney, 1987). It interacts with the intracellular site of Na–K ATPase in a similar manner to sodium and is transported outwards; this is likely to be a minor effect at therapeutic concentrations, at least in red cells (Rodland and Dunham, 1980). The exchange of intracellular lithium for extracellular sodium is used as a marker for hypertension *(Chapter 6)* and may reflect the activity of the normal proton–sodium exchange mechanism. As well as inhibiting Na–K ATPase, lithium can also participate in some of its artificial reactions, e.g. K^+-K^+ exchange, allowing active uptake of lithium (Duhm, 1982). Toxic doses cause hyperkalaemia, possibly a reflection of Na–K ATPase inhibition (De Fronzo and Smith, 1994). Since its introduction as a treatment for affective disorders, lithium has become the most usual cause of nephrogenic diabetes insipidus in humans; it also stimulates thirst and inhibits ADH release, though this is less certain (Teitelbaum *et al.*, 1994; Morrison and Singer, 1994). Thiazides and other diuretics reduce the excretion of lithium and enhance its toxicity (Elliott, 1994). Lithium and sodium are handled similarly by the glomeruli and proximal tubule but, unlike sodium, there is little reabsorption of lithium in the distal nephron (Raskind and Barnes, 1984). Lithium is natriuretic because it competes with sodium for cotransport mechanisms and blunts the renal response to aldosterone, probably by inhibiting Na–K ATPase. At high dosage, lithium may cause renal damage (Raskind and Barnes, 1984). While the interaction between lithium and other cations may contribute to its therapeutic effects, these are more probably due to interactions with neurotransmitters such as dopamine, γ-amino butyric acid (GABA) and 5-hydroxytryptamine (5-HT) (Price and Heninger, 1994). As well as being involved in therapy of affective disorders, it has been suggested that abnormal lithium–sodium exchange may be a marker for the underlying condition in some patients, as suggested also for hypertension (Plishker and Appel, 1986). Until recently it was believed that this represented a mechanism which was only invoked in response to therapeutic or pharmacological administration of lithium; it now appears that, in **trace** amounts, lithium may be an essential nutrient (Szilagyi *et al.*, 1989; Pickett and O'Dell, 1992).

Diphenylhydantoin (dilantin, phenytoin, DPH), the anticonvulsant drug, was formerly thought to stimulate sodium transport (Woodbury and Kemp, 1971; Hertz, 1977). This hypothesis became the subject of much conflicting evidence (Michell, 1979c) and it now seems more likely that the mode of action depends on membrane stabilisation and a reduction of both resting fluxes of sodium and sodium currents during action potentials (Rall and Schleifer, 1990). Nevertheless, phenytoin is still cited as a drug which stimulates Na–K ATPase (Laragh and Seeley, 1992). Its ability to reduce ischaemic brain damage in experimental rats was associated with increased Na–K ATPase activity (Qi and Dong, 1992) and it blocked the effect of ouabain on the neuromuscular junction in cats (Riker *et al.*, 1990). In addition, when vanadate was used to stimulate insulin secretion and test the hypothesis that this was mediated by depression of Na–K ATPase, its effect was abolished by DPH (Fagin *et al.*, 1987).

Carbenoxolone, an anti-ulcer drug, is a drug with a steroid-like structure and significant mineralocorticoid activity. Interestingly, while other mineralocorticoids have no anti-inflammatory effect, spironolactone blocks the therapeutic action of carbenoxolone. As might be expected, the side effects of this drug include potassium depletion and retention of salt and water (Brunton, 1990).

A number of **drugs sometimes cause concern not because of their effects but their actual sodium content**; these concerns are perhaps greatest where excretion is limited by renal failure. The concern is not always proportional to the actual risk: antibiotics such as ampicillin, cephalothin, oxacillin and nafcillin contain sodium at 2.4–3.0 mmol/g (Reed and Sabatini, 1986). The latter, for example, even at 6 g/day, provides only 18 mmol to a 70 kg patient, i.e. 0.26 mmol/kg/day, perhaps 7% of the usual dietary intake, equivalent to about 120 ml of 'normal' saline. Carbenicillin contains 4.7 mmol/g and might be used at 30 g/day, so here the load is significant, 141 mmol/day; equivalent to slightly less than a litre of saline (910 ml) and roughly half the usual dietary sodium intake. The greatest likely load comes from ion-exchange resins used, for example, to treat hyperkalaemia; thus a dose of sodium polystyrene sulphonate (50 g) contains 200 mmol of sodium, of which 50 are likely to exchange, equivalent to 320 ml of normal saline.

Considering the lack of clinical concern over dietary salt and the gay abandon with which parenteral fluids are often used, concern over iatrogenic salt loading is not misplaced but, usually, exaggerated.

Finally, any drugs which affect renal function either directly, or by altering renal perfusion or interfering with autoregulation, potentially affect sodium and water balance. The same is true of drugs which alter the secretion of hormones such as ADH or the activity of the renin–angiotensin system. Again, such effects may reflect a primary action on the cardiovascular system or (e.g. β-blockers) a direct effect on secretion. These effects contribute to the changes in fluid and electrolyte balance during anaesthesia and surgery (*Chapter 8*).

12

Concluding Synthesis

The central questions addressed by this book have been: why do we need sodium? How much do we need? How do we regulate its balance and distribution? What are the adverse effects of depletion and excess? How do they arise? What are the natural defences and the clinical strategies which counter these problems? It seems strange that the biological 'economy' of sodium is seldom addressed in this way; rather, the focus is on mechanisms—salt appetite, renal sodium regulation, pathophysiology of diarrhoea, etc. We would not study the economy of oil in this way. Certainly, we would have access to specialist information concerning its extraction, refining, transportation, its various uses for energy and synthesis, etc. But if oil were our industry, and such a vital one, we would need to understand the ramifications and to seek a global perspective. That perspective would be false unless it rested on sound detail in a variety of contexts.

With the biology of sodium, however, we have a wealth of elements but little synthesis. Perhaps that reflects not only a trend towards specialisation in science but the tendency for it to be shaped, increasingly, by techniques rather than ideas. It is possible to make a distinguished scientific career by soundly and imaginatively building on a technique and its developments, whether micropuncture, micro-assay, *in vitro* perfusion, or single unit recording, without needing to consider what happens, for example, once the brain has perceived that the electrical activity triggered by certain ions interacting with structured sequences in taste receptors is attributable to sodium. How sodium is absorbed, distributed, excreted in appropriate amounts is someone else's problem, another specialist in another field. Each field needs proper husbandry but there also need to be overall concepts of agriculture.

The productivity of such questions, the yield and speed of their answers, is always slower than processing further refinements through powerful specialist techniques. But the value of asking them is not simply philosophical—the satisfaction of theoretical order—there is the benefit of cross-fertilisation, of problems or solutions which resonate in different areas of the subject. Does the prenatal environment have an important influence on adult blood pressure? If so, via taste sensitivity or preference for salt; or via baroreceptor sensitivity, or renal sodium excretion? Does early experience alter our sodium economy?

The idea of the 'economics' of body sodium is a salutary one. For in few disciplines other than economics do we so readily confuse an ability to alter outcomes with an understanding of underlying mechanisms. As a result, much importance is attached to 'signals', while much less is attached to their real significance. We have seen the extent to which rigid stereotyped assumptions about the meaning of low Na:K ratios, like indicators of money supply, can readily generate interpretations which are both confident and disastrously wrong. The same is true of widespread assumptions about the meaning of low sodium excretion in urine—or, in some circumstances, higher sodium excretion in faeces. In late pregnancy, when the accelerating demands of the conceptus and the impending demands of lactation are increasing the need for sodium, a paradoxical rise in sodium excretion implies that dietary sodium is too high, not that it should be increased (Michell *et al.*, 1988). The problem is that, while we can imagine, reason, or predict why the body needs to adjust its sodium balance, we still have no idea how the body's regulatory system assesses its sodium status, or where the assessment is made.

That may seem an extreme view but throughout time we have been satisfied by what our concepts—whether religious, superstitious or scientific—do explain rather than disturbed by what they don't: in ordinary life, 'in that way lies madness'. Perceived control, rather than actual control, determines our peace of mind and there is no doubt that it is not simply the intensity of a stressor but the victim's belief that it is, or is not, controllable which determines the actual degree of stress (Michell, 1988e).

We know that excess sodium triggers natriuretic mechanisms and that sodium depletion triggers adrenal, renal, sympathetic, enteric and behavioural responses which defend circulating volume. We can describe sensory mechanisms on the arterial and venous side of the circulation that respond to changes in circulating volume. But we do not know, independent of sodium intake, what establishes that an individual's circulating volume is normal—hour by hour or month by month. Worse, the quantity of sodium in plasma is small compared with that in the major compartment of extracellular fluid (ECF), interstitial fluid (ISF). It is interesting that, while the release of renin in response to haemorrhage is widely appreciated, renin secretion is rather insensitive to blood volume. Recent experiments in which, for example, repletion of ECF volume, but not blood volume is allowed, following a period of sodium depletion, show that repletion still reduces renin secretion. This strongly suggests the existence of extravascular receptors for ECF volume, whether within the kidney or elsewhere (Laragh and Seeley, 1992). Both bone and interstitial fluid could function as reservoirs or overflows, but as yet we have no knowledge of how their sodium stores might be monitored. Lymph flow, though normally small, is greatly increased by a rise in interstitial pressure. Perhaps we should be seeking neural or chemical signals from the lymphatic system, including lymph nodes, which might contribute to the regulation of extracellular volume.

> *There is no known mechanism for detecting either total body fluid volume or extracellular volume. Mechanisms do exist, however, for detecting changes in blood volume, but these do not detect blood volume per se. Rather, changes in blood volume are detected by changes in pressures in the cardiovascular system. . . .*

(Cowley and Roman, 1989)

Cowley and Roman go on to differentiate between three mechanisms capable of detecting changes in cardiovascular filling:

1. neural mechanoreceptors (e.g. atrial or arterial baroreceptors, including those in the afferent arteriole);
2. non-neural mechanoreceptors (cardiac myocytes, vascular endothelium or smooth muscle);
3. physical changes affecting renal function (colloid osmotic pressure, arterial blood pressure).

Atrial receptors provide information reflecting venous fill in relation to cardiac performance, whereas arterial receptors relate cardiac output to vascular compliance and peripheral resistance. We have the pieces but we do not know the picture which they form, nor how many pieces are missing.

When the volume of plasma and ECF expand in pregnancy we do not truly know the extent to which this is 'normal' and the extent to which it results from salt-loaded diets. Does 'the system' monitor the 'internal' demand for sodium during pregnancy or simply furnish a safe excess? And if we no longer believe that cardiac, hepatic or nephrotic oedema simply involve the defence of plasma volume against leakage into interstitial fluid or body cavities, but that there is a primary expansion of plasma, with overflow into interstitial spaces, what fools 'the system'? If the body can respond to salt loading by suppressing at least two salt-conserving hormones (angiotensin and aldosterone) and mobilising at least two types of natriuretic hormones (atrial natriuretic peptide, ANP, and active sodium transport inhibitors, ASTI/ endogenous digitalis-like inhibitors, EDLI), why is 'the system' so comprehensively defeated. In Peters' words, "The most crucial problem related to the formation of edema is the discovery of the criteria by which the kidneys determine whether to accelerate or retard the reabsorption of sodium". A vintage comment—when it was made Truman was President, Bradman had just retired, the milk ration in the U.K. was three pints and the Deutschmark was 11 weeks old. An important contrast with the situation nearly 50 years ago is that, while fluid space measurements in disease are still problematical, the measurement of sodium itself is simple, whereas then "The measurement of sodium has been a protracted, meticulous procedure for which few laboratories are equipped" (Peters, 1948).

As in economics, as in ancient civilisations, we respond to what we fear or do not really understand by naming it and making it seem more familiar. 'Effective circulating volume', like the reservoir hypothesis of salt appetite, creates a mythical site where 'the system' behaves as it should when, elsewhere, the measurements show that it does not. It is a close cousin of the 'underlying trend' in the economy or the 'real terms' in which output, inflation, wages and prices support the desired conclusion. The danger is the belief that we understand and that we do not need to know more or to re-examine our assumptions.

The fact that we do not know something does not mean that we need to know it or that it is worth the price of finding out. That is anathema to the traditional spirit of academic inquiry but, since expensive research is publicly funded or needs to show some pay-off for industry, there needs to be a justification for pursuing a question, not simply the fact that it is there. The problem is that with the vast amount which we seem to know about sodium regulation, our ability to describe what the kidney achieves rather than explaining how or why it does it, let alone how it interacts with other regulatory mechanisms, even those in the field may feel that the essential questions have already been answered, that the rest is a matter of detail.

Yet in the most recent *Handbook of Physiology*, the two volumes on renal physiology (updated in 1992) include a chapter summarising current knowledge of the control of sodium excretion. Drawn together, its conclusions on a number of the key factors include the following. The exact mechanisms whereby changes in peritubular Starling forces influence tubular reabsorption of sodium is unknown. The importance of pressure natriuresis is beyond dispute but its site is uncertain and its mechanism unknown. The importance of the renin–angiotensin–aldosterone system is the best understood factor (but even here, the rapid growth in evidence for the importance of angiotensin II makes the role of aldosterone less obvious than it was 20 years ago). The importance of prostaglandins in the acute or chronic regulation of sodium balance is uncertain. The importance of kinins has not been elucidated. The physiological role of ouabain-like sodium transport inhibitors, such as that originating from the hypothalamus, remains a matter of speculation. It is not clear whether ANP plays an important role in the physiological regulation of sodium excretion. It is likely that the nervous system is involved in renal sodium regulation, but perhaps it is only important in pathological states and subject to variations between different strains of rat, let alone species; it may or may not affect the influence of other factors such as angiotensin II.

> *Nervous control of renal function . . . which has been neglected in this discussion, might prove to be an important factor in the phenomena of circulatory disorders.*
>
> (Peters, 1948)

Perhaps the problem is precisely our scientific tradition of thinking in terms of clear experimental designs which isolate single factors. Perhaps the key lies in the interactions between factors, in the neural or endocrine environment, rather than its components. Computers allow the analysis of such complexity—the difficulty lies in designing the experiments and defining the relevant **combinations** of changes.

There is an analogy in taste physiology. The original expectation was that specificity concerning 'salt', 'sweet', 'bitter' and 'sour', the 'primary colours' of taste *(Chapter 4)* would arise from unique responses of particular receptors for each. Now it is realised that specificity, i.e. the unique character of the stimulus, is conveyed not by unique receptors but by the **pattern** of effects across a number of **different** receptors (Doetsch *et al.*, 1969; Scott, 1987). So it may be with sodium balance; not the separate signals which matter most but the **pattern** which they form. In a sense, we already accept this; we expect a different response when high angiotensin levels coexist with high or low plasma potassium, high or low ECF volume. But we see it, perhaps, as a complication of single feedback-loop concepts, an interaction between conflicting signals. We less readily consider that the **pattern** of the signals, ultimately, may be the stimulus on which responses are based.

For the moment, when question 3 in the physiology exam is 'How is sodium balance regulated?', we are really setting quite a challenge: the most unchallengeable answer is 'We remain uncertain'.

'The entrance or loss of sodium into the ECF appears to be monitored by a biologic control system which detects the translocations of some function of volume from the steady state, rather than a detector element sensitive to changes in the concentration of sodium per se. The design of this 'volume control system' is understood only in outline form and the specific elements and events involved in its operation largely remain to be delineated.

(Bricker and Licht, 1980)

Despite the fundamental importance and clinical relevance of the issues involved, the view remains true 15 years later.

Of the seven questions raised at the start of this chapter, the two most important are also those to which we most urgently need answers.

'How much?' or rather, how little 'do we need?' We might also give more serious attention to what drives us to take so much more and whether there are simple and effective ways of preventing or reducing this craving. 'How do we regulate its balance and distribution?' It is the paramount question, at the centre of all the others, but the truth is that it is also the furthest from convincing answers. We might almost profit from someone who could start fresh—an excellent scientist, perhaps from the field of guidance and control systems. We should specify where sodium is, how it gets there, why it matters and let them specify how best to monitor and control such a system. They might then ask us for the information which we most seriously lack. In a sense, there are echoes of the modelling approach best exemplified by Guyton's group. The difference would be the lighter initial emphasis on accommodating what we already know and the greater emphasis on trying to specify how an effective regulatory system ought to function. We might then, as a hypothetical example, be forced to accept that for all the daunting technical difficulty, the most urgent areas of inquiry turned out to be the sensory mechanisms monitoring bone sodium or interstitial volume, rather than those fine-tuning renal sodium excretion.

There was a time when the key to understanding endocrinology seemed to be the changes in the hormone content of endocrine glands. Then it seemed to be plasma levels, then rates and patterns of secretion and elimination, then interactions with receptors, their consequences, the regulation of receptor populations. There has been a constant redefining of what **matters** about hormones and their effects and, therefore, how they are capable of regulation. The question we have evaded is what **matters** about sodium. We are content that the answer is 'ECF and plasma volume'. There are even those who persist in believing that it is plasma sodium concentration. If we are enlightened enough to consider its role as a prime mover of water and solutes, as a key determinant of the activity of one of the main consumers of ATP, as a generator of potential differences, does it also matter that we are still not much further forward in explaining the various aspects of sodium regulation which cannot satisfactorily be explained in terms of ECF or plasma volume.

The mainstream view of sodium regulation is that it involves no more than an analysis of renal mechanisms and the neural or endocrine factors which control them.

The familiarity of renal function belies its eccentricity. The renal medulla and cortex have perfusion rates, respectively, among the most sparse and the most generous in the body. A single kidney, 0.25% of body mass, receives almost as great a share of cardiac output as the brain, yet it contains perhaps 50% more nephrons than are necessary to avoid the signs of renal failure. A gram of kidney receives seven times as much blood as a gram of brain. A fifth of renal plasma flow is converted to glomerular filtrate and since well over 95% is usually reabsorbed, depending on sodium intake, this 'unnecessary' filtration, some 120 ml/min, is roughly the same as the rate of plasma flow to the myocardium. The task of reabsorbing the entire plasma volume nearly three times an hour accounts for some 10% of the body's resting oxygen consumption and most of it is associated with the activity of the sodium–potassium pump, Na–K ATPase. Indeed, sodium transport may account for as much as 80% of renal energy consumption and 40% of resting energy consumption throughout the body (Katz, 1988). It should come as no surprise that such a profligate demand on resources is subject to regulation, but the nature of the regulatory process is unexpected, intriguing and an insight into a new dimension of cell physiology.

The 1950s brought Skou's discovery that the underlying mechanism of the 'sodium pump' was Na–K ATPase and, throughout the 1960s, De Wardener championed the concept that glomerular filtration rate and aldosterone were insufficient to explain the regulation of sodium excretion and that a third factor, a natriuretic hormone, must be involved. Such a hormone was likely to be an inhibitor of sodium transport. The precise origin of the experiments in search of such an inhibitor (Astrup, 1993) was a bet, in a coffee break, that aldosterone probably had little influence on the regulation of sodium excretion! The fact that science should involve fun as well as discovery could seldom have had stronger vindication than the origins of the concept of natriuretic hormones.

The resistance to this concept was spectacular, with emphasis on inhibition of sodium retaining mechanisms as an adequate explanation for excretion of excess sodium and peritubular physical factors as the 'third factor'. In part, this reflected the view that the key disturbance of body sodium, the focus of homeostatic defence mechanisms, was sodium depletion. It has become increasingly clear, however, in a growing range of species, that the everyday problem of sodium regulation is the excretion of a dietary salt load well above requirement.

The 1980s brought a proliferating wealth of knowledge of two families of natriuretic hormones. ANP (and related peptides from other tissues) allowed the heart to 'offload' sodium into urine and into interstitial fluid in response to a rise in atrial pressure; surprisingly, however, it did not inhibit sodium transport. While the detailed structure of natriuretic peptides was rapidly defined, uncertainty remained regarding the structure and origin of the second family of natriuretic hormones, those which inhibit active sodium transport (ASTI/EDLI). Initially, attention focused on areas of the brain such as AV3V, already implicated in various aspects of the regulation of sodium, water and arterial pressure.

Cardiac glycosides followed morphine as fascinating examples of drugs used for centuries whose action turned out to depend on their ability to mimic hormones whose existence was unknown until the last 25 years. The new dimension is that ouabain, originally the tool of the hunter, a literal toxin (arrow-poison), latterly the tool of the pharmacologist as the definitive inhibitor of Na–K ATPase, is itself a hormone. It is surprising that we could be so surprised that it is, in fact, yet another adrenal steroid, opposite in effect to aldosterone. Thus one group of natriuretic hormones, typified by ANP, opposes the effects of aldosterone while the other (ASTI/EDLI), typified by ouabain, opposes its fundamental action. The challenge now is to explain:

1. the factors which control the release of this type of hormone and their role in the regulation, not only of sodium but of cellular metabolism and the availability of ATP to other competing processes;
2. the distinction between roles played by ANP and hormones which inhibit sodium transport.

If the future accumulation of experimental evidence sustains the central role of endogenous cardiac glycosides in the regulation of sodium transport and the importance of the adrenal cortex as a source, the importance of this gland in electrolyte physiology will need reappraisal. Since the earliest ablation experiments, its role in the defence against sodium depletion has been obvious. The discovery of aldosterone led to an exaggeration of its undoubted importance in sodium regulation. Firstly, the importance of other aspects of sodium regulation, especially the factors affecting the proximal tubule, became very evident and, secondly, there is growing reason to perceive aldosterone as being perhaps more important in potassium regulation. This realisation was perhaps postponed by the relatively low potassium levels in Western human diets compared with those of herbivores. Yet once perceived, this enhances rather than diminishes the role of the adrenal cortex. By producing aldosterone or endogenous cardiac glycosides, it can regulate sodium transport in either direction. Moreover, it is directly responsive, where aldosterone is concerned, to small changes in plasma potassium. It seems highly probable that, where endogenous cardiac glycosides are concerned, it is also directly responsive to ECF volume expansion. If so, perhaps it is able to monitor ECF volume contraction by signals other than angiotensin II alone. Is it, perhaps, not just the key effector site, able to regulate sodium transport, the excretion of sodium and potassium, and possibly salt appetite, but also a key site for the receptors which monitor ECF volume and composition?

It is perhaps unavoidable to comment on the impact of molecular biology on recent progress. This has in some ways been similar to the impact of a new toy on a young child; the potential is so exciting that play is usually well in progress ahead of any detailed consideration of how it should be used.

The positive benefits are obvious and scarcely need repetition: the ability to define and compare biological molecules and their subtle variants, whether in the context of hormones, other mediators or their receptors. From this flows the ability to discuss distributions and their differences, whether between tissues, species or physiological or clinical states. We also potentiate our ability to look at biological activity in terms of active sites and their constituent sequences, rather than the larger molecules which carry them. The ability to identify receptor proteins, or those responsible for ion transport with great specificity, has revealed unexpected tissue distributions and unexpected kinships between diseases previously thought to be quite unrelated *(Chapter 10)*. Transgenic animals, while in the same lineage as ablation, lesion and implantation studies, provide a unique opportunity for predictive testing of hypotheses. And yet . . .

Whether we look at Na–K ATPase, aldosterone, ANP or other key molecules involved in the cell physiology of sodium, the impression of the last decade is of a proliferating wealth of detail but a declining ratio of detail to physiological context. It is as if we had believed that ever-increasing knowledge of the geology of the rocks would increase our understanding of the causes of shipwreck.

These views will reflexly be regarded as perverse, reactionary or Luddite by the most evangelistic of the new biologists—many of whom, as in any religion, are among the most recent converts. They are not, however, entirely idiosyncratic views.

Although basic research has been voluminous, the critical issues important for clinical medicine have scarcely been defined, so that little of the basic research has had much relevance to the practice of medicine.

(Katz, 1980, regarding ISF)

We all understand the sacred role of serendipity in research and the unpredictability of its benefits [as documented by Comroe and Dripps for the National Institute of Health (DHEW Publications 78-1521, 1977)], although our political leaders have never accepted this and too many of our research managers have found it expedient to echo their views. Nevertheless, we have choices over the questions which we ask and, though far less, influence over the questions which it is fashionable to answer. There is currently a dangerous tendency to believe that a measure of the importance of a question is the amount of molecular biology supposedly needed to answer it. It is a delicate balance; we cannot, for example, even speculate on the importance of the isoforms of Na–K ATPase until we have the means to identify them. But there is a danger that without an anchoring viewpoint, a physiological context, pursuit of isoforms can become, at a new level of sophistication, the oldest form of scientific 'butterfly collecting'.

Medical research has depended most heavily on a genetic view of life that has yielded remarkable progress in our understanding of the fundamental mechanism of biology, including evolution. However, increased understanding of cellular, hereditary and evolutionary mechanisms has progressively revealed important deficiencies in the genetic paradigm of biology. While cellular mechanisms now are amply understood, and further progress is expected, a mechanistic view alone will severely limit advances in our understanding of the cell or organism . . . viewed whole.

. . . The paradigm of biomedical research has been summarised as follows:

Genes cause disease; genes cause aging and death.

Genetic research will produce the means for eliminating disease and extending life. . . .

*. . . **Applied** biomedical efforts centred on molecular genetics have diverged from fundamental research issues. . . .*

. . . This cleavage between applied medical sciences and basic research biology is potentially dangerous to the public health.

. . . A strategy dedicated to single-gene causality and gene or gene product replacement is important but incomplete . . . a new biomedical research outlook would shift to a focus on the organism–environment interface. . . .

. . . The reality (of disease) is the inability of one person's homeostasis, conditioned by his genotype and a lifetime of special experiences to maintain equilibrium; neither genes nor environment 'cause' disease. . . .

(Strohman, 1993)

This extensive quotation comes from a Professor of Molecular and Cell Biology, at the University of the Berkeley. Cowley (1991) has commented wisely on these problems and his views also deserve exact quotation (the emphasis is added).

An understanding of both the cellular and integrative aspects of the organism is essential. Exciting developments in cell and molecular biology can provide important new insights about regulatory function. **But the development and testing of comprehensive hypotheses for blood pressure regulation is becoming more difficult as greater numbers of investigators work on smaller pieces of the system.** . . . *It is necessary that* **equally imaginative whole animal and cell/molecular research** *is carried out . . . Increasing efforts must be made to merge and integrate the data obtained at the cellular level with that obtained in intact systems.*

Patients still suffer from hypertension, oedema, and other basic disturbances of sodium regulation which elude our best attempts at a convincing explanation. To seek such answers is not a narrow utilitarian outlook but a recognition that in the adaptation and maladaptation inherent in disease we find challenges to our current concepts as powerful as those arising from experimental intervention. Indeed, it may require clinical insight to understand the significance of some of the feedback loops. For example, increased sodium excretion in response to hypernatraemia seems 'appropriate' until we recall that plasma sodium is actually a determinant of water excretion, not sodium excretion. This response occurs in dehydration so, physiologically, it appears to superimpose an inappropriate loss of ECF volume on the reduction already caused by loss of water. Some have even anticipated that this sodium 'depletion' should be 'corrected' by enhanced salt appetite (see Michell and Moss, 1994). What makes it an appropriate physiological response is the clinical problem which it solves, namely that the main threat from water loss arises from hypernatraemia rather than hypovolaemia.

Basic scientists are right to hope that new insights into basic mechanisms will one day help to understand or correct disease— nowhere is this more obvious than our new ability, conferred by molecular biology, to correct an increasing range of genetic disorders and to identify genetic determinants of multifactorial disease. But clinical scientists are also right to emphasise that until we can explain the mechanisms of organic disorders, we do not understand the underlying regulatory mechanisms.

The disorders encountered in disease may be regarded as normal physiologic responses to unusual conditions . . .

(Peters, 1948)

Aldosterone provides a touchstone of progress in our understanding of the physiology of sodium. It looked like the universal regulator, the key defence against depletion, yet the most interesting detail among its effects, one obvious from clinical disturbances, was the ability of the kidney to escape from the effects of excess. Not only were other factors involved but the revelation beyond 1980 was the recognition that these mechanisms, at least as sophisticated as those defending against sodium depletion, protected animals and people from the more familiar daily problem of excessive sodium loads.

Molecular biology has given us the means to dissect these mechanisms. But their context, the motivation to seek their existence (in the face of considerable scepticism) stemmed from the penetrating insight arising from the combination of curiosity about basic regulatory mechanisms and about their ability to precisely explain what is unusual in clinical patients. The development of fields of clinical science such as nephrology has rested on the efforts not just of basic scientists working alongside clinical scientists but on the fact that clinical scientists are inescapably committed to the understanding of basic biological phenomena: it is self-evident in the work of De Wardener, Guyton, Kleeman, Edelman, Epstein, Katz and many others, as it was in previous decades with Pitts and Homer Smith.

Perhaps this book should end where my interest began—with salt appetite—because ingestion of sodium is obviously the initial prerequisite for all its other effects. Much sodium intake is clearly unregulated—we are not necessarily aware of either the taste or the content of much of the sodium contained in the various ingredients of our food. Yet much is discernible and some is deliberately added or sought by the consumer, whether human or animal. The recent history of our understanding of water balance has seen increasing acknowledgement of the importance of thirst, rather than renal responses to ADH, provided that fluid intake is not dominated by hedonic factors, i.e. tasty fluids. But with sodium, overconsumption seems to be more fundamental than simply hedonic: it remains a powerful addiction even among those convinced of its adverse effects. Moreover it persists, whether in humans or animals, in the absence of direct evidence of need. Perhaps that is because our concept of need is too simplistic, too focused on individual mechanisms rather than the 'society' of mechanisms which they form. There may indeed be a need for sodium, even when sodium balance *per se* seems adequately defended, because the importance of sodium reaches far beyond the volume and composition of ECF. It involves a host of interactions with other solutes, transport pathways and physiological mechanisms, some not immediately or self-evidently linked to sodium.

The classical physiology of sodium has been too pure, too isolated from the more general hurly-burly of survival. Whatever else it may have done, the purpose of this book has been to correct this isolated, 'renocentric' view of sodium regulation. I believe that for all its massive achievements, this imbalanced view has misled us in some areas and held us back in many others. The quest for detail, however essential, can prejudice the growth of knowledge.

It is as true of ions and physiological mechanisms as it is of man himself that none ". . . is an island, entire of itself . . . each . . . is a part of the main". Loss or unawareness of any one diminishes our awareness of the whole.

References

ABER, G.M. 1985. The kidney in pregnancy. In: *Postgraduate Nephrology*, Chap. 20, pp. 499–525. MARSH F. (Ed.). Heinemann, London.

ABERDEEN, G.W., CHA, S.C., MUKADDAM-DAHER, S., NUWAYHID, B.S. and QUILLEN, E.W. 1992. Renal nerve effects on renal adaptation to changes in sodium intake during ovine pregnancy. *Am. J. Physiol.* **263**: F823–F829.

ABRAHAM, W.T. and SCHRIER, R.W. 1994. Body fluid volume regulation in health and disease. *Adv. Intern. Med.* **39**: 23–47.

ABRAMS, M., DE FRIEZ, A.I.C., TOSTESON, D.C. and LANDIS, E.M. 1949. Self selection of salt solutions and water by normal and hypertensive rats. *Am. J. Physiol.* **156**: 233–238.

AITKEN, F.C. 1976. Sodium and potassium nutrition of mammals. Commonwealth Agricultural Bureaux, Farnham Royal, Bucks.

AKSENTSEV, S.L., MONGIN, A.A., ORLOV, S., RAKOVICH, A.A., KALER, G.V. and KONEV, S.V. 1994. Osmotic regulation of sodium pump in rat brain synaptosomes; the role of cytoplasmic sodium. *Brain Res.* **644**: 1–6.

ALAM, N.H., AHMED, T., KHATUN, M. and MOLLA, A.M. 1992. Effects of food with two oral rehydration therapies: a randomised controlled clinical trial. *Gut* **33**: 560–562.

ALCANTARA, P.F., HANSON, L.E. and SMITH, J.D. 1980. Sodium requirements, balance and tissue composition of growing pigs. *J. Anim. Sci.* **50**: 1092–1101.

ALDERMAN, M.H., CUSHMAN, W.C., HILL, M.N. and KRAKOFF, L.R. 1993. International roundtable discussion of national guidelines for the detection, evaluation and treatment of hypertension. *Am. J. Hyperten.* **6**: 974–961.

ALPERN, R.J. and PREISIG, P.A. 1988. Bicarbonate transport mechanisms across peritubular cell membrane of proximal convoluted tubule. In: *Nephrology*, Vol. 1, pp. 202–213. DAVISON, A. M. (Ed.). Bailliere Tindall, London.

ALTURA, B.M. and ALTURA, B.M. 1987. Cardiovascular actions of magnesium: importance in etiology and treatment of high blood pressure. *Magnesium Bull.* **9**: 6–21.

AMEND, W.J.C. 1993. Pathogenesis of hepatorenal syndrome. *Transpl. Proc.* **25**: 1730–1733.

ANAND, I.S., FERRARI, R., KALRA, G.S., WAHI, P.L., POOLE-WILSON, P.A. and HARRIS, P.C. 1989. Edema of cardiac origin. *Circulation* **80**: 299–305.

ANAND, I.S., CHANDRASHEKHAR, Y., FERRARI, R., SARMA, R., GULERIA, R., JINDAL, S.K., WAHI, P.L., POOLE-WILSON, P.A. and HARRIS, P.C. 1992. Pathogenesis of chronic obstructive pulmonary disease: studies of body water and sodium, renal function, hemodynamics and plasma hormones during edema and after recovery. *Circulation* **86**: 12–21.

ANAND, I.S., CHANDRASHEKHAR, Y., FERRARI, R., POOLE-WILSON, P.A. and

HARRIS, P.C. 1993. Pathogenesis of oedema in chronic severe anaemia: studies of body water and sodium, renal function, haemodynamic variables and plasma hormones. *Br. Heart J.* **70**: 357–362.

ANDERSON, B. 1977. Regulation of body fluids. *Ann. Rev. Physiol.* **39**: 185–200.

ANDERSON, D.E. 1989. The role of sodium in behavioural hypertension in animals. In: *Perspectives in Emotional Stress*, Vol. 3, pp. 367–382. GANTEN, D. and NIKOLOV, N. A. (Ed.). Gordon and Breach, New York.

ANDERSON, R.J. 1994. Electrolyte, water, mineral and acid–base disorders in liver disease. In: *Maxwell and Kleeman's Clinical Disorders of Fluid and Electrolyte Metabolism*, 5th Edn, Chap. 37, pp. 1153–1174. NARINS, R. G. (Ed.). McGraw-Hill, New York.

ANDERSON, S. (Ed.) 1993. *Hypertension: Nonpharmacologic Management.* National Kidney Foundation, New York.

ANDREWS, F.M., RALSTON, S.L., SOMMARDAHL, C.S., MAYKUTH, P.L., GREEN, E.M., WHITE, S.L. *et al.* 1994. Weight, water and cation losses in horses competing in a three-day event. *J. Am. Vet. Med. Assoc.* **205**: 721–724.

ANON. 1977. Trials and tribulations of a symptom-free hypertensive physician receiving the best of care. *Lancet* **1**: 1358–1360.

ANON. 1978. Hypertension: salt poisoning? *Lancet* **i**: 1136–1137.

ANON. 1987. 'CONCENSUS' Trial Study Report. *New Engl. J. Med.* **316**: 1429–1435.

ANON. 1988. Intersalt: an international study of electrolyte excretion and blood pressure. Results for 24-hour urinary sodium and potassium excretion. *Br. Med. J.* **297**: 319–328.

ANTUNES-RODRIGUES, J. and COVIAN, M.R. 1963. Hypothalamic control of sodium chloride and water intake. *Acta Physiol. Latinoam.* **13**: 94–100.

APERIA, A., HOLTBACK, U., SYREN, M.-L., SVENSSON, L.-B., FRYCKSTEDT, J. and GREENGARD, P. 1994. Activation/deactivation of renal Na^+ K^+-ATPase: a final common pathway for regulation of natriuresis. *FASEB J.* **8**: 436–439.

APPEL, L.J., MILLER, E.R., SEIDLER, A.J. and WHELTON, P.K. 1993. Does supplementation of diet with 'fish oil' reduce blood pressure? A meta-analysis of controlled clinical trials. *Arch. Int. Med.* **153**: 1429–1438.

ARC 1980. *Agricultural Research Council: The Nutrient Requirements of Ruminant Livestock* (2nd Edn). Commonwealth Agricultural Bureaux, Farnham Royal.

ARC 1981. *Agricultural Research Council: The Nutrient Requirements of Pigs* (2nd Edn). Commonwealth Agricultural Bureaux, Farnham Royal.

ARDAILLOU, R. and DUSSAULE, J.C. 1994. Role of atrial natriuretic peptide in the control of sodium balance in chronic renal failure. *Nephron* **66**: 249–257.

ARDUINO, R.C. and DUPONT, H.L. 1993. Travellers' diarrhoea. *Bailliere's Clin. Gastroenterol.* **7.2**: 365–385.

ARGENZIO, R.A. 1985. Pathophysiology of neonatal calf diarrhoea. *Vet. Clin. N. Am. Food Anim. Prac.* **1**: 461–469.

ARGENZIO, R.A. 1991. Comparative physiology of colonic electrolyte transport. In: *Handbook of Physiology*, Section 6, *Gastrointestinal Tract*, Vol. IV, Chap. 9, Intestinal secretion and absorption, pp. 275–288. FIELD, M. and FRIZELL, R. A. (Eds). American Physiological Society, Bethesda, Maryland.

ARGENZIO, R.A. 1993. Intestinal transport of electrolytes and water. In: *Dukes' Physiology of Domestic Animals*, 11th Edn, pp. 376–386. SWENSON, M. J. and REECE, W. O. (Eds.) Comstock, Cornell University Press, Ithaca.

ARGENZIO, R.A. and ARMSTRONG, M. 1993. ANP inhibits NaCl absorption and elicits Cl secretion in porcine colon: evidence for cGMP and Ca mediation. *Am. J. Physiol.* **265**: R57–R65.

ARGENZIO, R.A. and CLARKE, L.L. 1989. Electrolyte and water absorption in the hind gut of herbivores. *Acta Vet. Scand. Suppl.* **86:** 159–167.

ARGENZIO, R.A., RHOADS, J.M., ARMSTRONG, M. and GOMEZ, G. 1994. Glutamine stimulates prostaglandin-sensitive Na/H exchange in experimental porcine cryptosporidiosis. *Gastroenterology* **106:** 1418–1428.

ARONSON, P.S., SOLEIMANI, M. and GRASSI, S.M. 1991. Properties of the renal Na^+–HCO_3^- cotransporter. *Sem. Nephrol.* **11:** 28–36.

ARROYO, V., GINES, P., NAVASA, M. and RIMOLA, A. 1993. Renal failure in cirrhosis and liver transplantation. *Transpl. Proc.* **25:** 1734–1739.

ARROYO, V., CLARIA, J., SALO, J. and JIMENEZ, W. 1994. Antidiuretic hormone and the pathogenesis of water retention in cirrhosis with ascites. *Sem. Liver Dis.* **14:** 44–58.

ARRUDA, J.A.L. and KURTZMAN, N.A. 1978. Relationship of renal sodium and water transport to hydrogen ion secretion. *A. Rev. Physiol.* **40:** 43–87.

ASCHERIO, A., RIMM, E.B., GIOVANNUCCI, E.L., COLDITZ, G.A., ROSNER, B., WILLETT, W.C. and STAMPFER, M.J. 1992. A prospective study of nutritional factors and hypertension among US men. *Circulation* **86:** 1475–1484.

ASHTON, W.M. and YOUSEF, I.M. 1966. A study of the composition of the Clun Forest ewe's milk. II. Mineral constituents. *J. Agric. Sci.* **67:** 77–80.

ASTRUP, P. 1993. *Salt and Water in Culture and Medicine*, pp. 137–138. Munksgaard: Radiometer, Copenhagen.

ATHERTON, J.C., DARK, J.M., GARLAND, H.O., MORGAN, M.R.A., PIDGEON, J. and SONI, S. 1982. Changes in water and electrolyte balance, plasma volume and composition during pregnancy in the rat. *J. Physiol.* **330:** 81–93.

ATLAS, S.A. and LARAGH, J.H. 1986. Atrial natriuretic peptide. *A. Rev. Med.* **37:** 397–414.

ATLAS, S.A. and MAACK, T. 1993. Atrial natriuretic factor. In: *Handbook of Physiology*, Section 8: *Renal Physiology*, Vol. 2, Chap. 33, pp. 1576–1673. WINDHAGER, E. E. (Ed.). American Physiological Society/Oxford University Press, New York City.

AUSIELLO, D.A. 1993. Renal tubular transport mechanisms. *Sem. Nephrol.* **13:** 472–478.

AUSIELLO, D.A. and BONVENTRE, J.V. 1984. Calcium and calmodulin as mediators of hormone action and transport events. *Sem. Nephrol.* **4:** 134–143.

AVERY, M.E. and SNYDER, J.D. 1990. Oral therapy for acute diarrhoea. *New Engl. J. Med.* **323:** 891–894.

AVIV, A. 1994. Cytosolic Ca^{2+} Na^+–H^+ antiport, protein kinase C trio in essential hypertension. *Am. J. Hyperten.* **7:** 205–212.

AVNER, E.D. 1993. Epithelial polarity and differentiation in polycystic kidney disease. *J. Cell Sci. Suppl.* **17:** 217–222.

BADALAMENTI, S., GRAZIANI, G., SALERNO, F. and PONTICELLI, C. 1993. Hepatorenal syndrome: new perspectives in pathogenesis and treatment. *Arch. Int. Med.* **153:** 1957–1967.

BADR, K.F. and BRENNER, B.M. 1993. Renal prostaglandins and kinins. In: *Maxwell and Kleeman's Clinical Disorders of Fluid and Electrolyte Metabolism*, 5th Edn, Chap. 5, pp. 443–476. NARINS, R. G. (Ed). McGraw-Hill, New York.

BAILIE, M.P. 1983. Nephrotic edema. *Sem. Nephrol.* **3:** 249–255.

BALMONT, R.J., BRIMBLE, M.J. and FORSLING, M.L. 1980. Release of oxytocin induced by salt loading and its influence on renal excretion in the male rat. *J. Physiol.* **208:** 439–449.

BANKIR, L., BOUBY, N., TAN, M.T. and KAISSLING, B. 1987. The thick ascending limb. *Adv. Nephrol.* **16:** 69–102.

BANKIR, L., BOUBY, N. and TRINH-TRANG-TAN, M.-M. 1989. The role of the kidney in the maintenance of water balance. *Bailliere's Clin. Endocrinol. Metab.* **3**: 249–311.

BARA M., GUIET-BARA, A. and DURLACH, J. 1993. Regulation of sodium and potassium pathways by magnesium in cell membranes. *Magnesium Res.* **6**: 167–177.

BARASCH, J. and AWQATI, Q. 1993. Defective acidification of the biosynthetic pathway in cystic fibrosis. *J. Cell Sci.* Suppl. **17**: 229–233.

BARBAGALLO, M. and RESNICK, L.M. 1994. The role of glucose in diabetic hypertension effects on intracellular cation metabolism. *Am. J. Med. Sci.* **307**: Suppl. 1., S60–S65.

BARBER, B.J. and NEARING, B.D. 1990. Spatial distribution of protein in the interstitial matrix of mesenteric tissue. *Am. J. Physiol.* **258**: H556–H564.

BARCHI, R.L. 1993. Ion channels and disorders of excitation in skeletal muscle. *Curr. Opin. Neurol. Neurosurg.* **6**: 40–47.

BARELARE, B. and RICHTER, C.P. 1938. Increased salt appetite in pregnant rats. *Am. J. Physiol.* **121**: 185–188.

BARGMAN, J.M. and JAMISON, R.L. 1986. Disorders of potassium homeostasis. In: *Diuretics: Physiology, Pharmacology and Clinical Use*, Chap. 16, pp. 297–319. DIRKS, J. H. and SUTTON, R. A. L (Eds). Saunders, Philadelphia.

BARRETT, K.E. and DHARMSATHAPHORN, K. 1994. Transport of water and electrolytes in the gastrointestinal tract: physiological mechanisms, regulation and methods of study. In: *Maxwell and Kleeman's Clinical Disorders of Fluid and Electrolyte Metabolism*, 5th Edn, Chap. 17, pp. 493–520. NARINS, R. G. (Ed). McGraw-Hill, New York.

BARRON, W.M. 1987. Volume homeostasis during pregnancy in the rat. *Am. J. Kidney. Dis.* **9**: 296–302.

BARRON, W.M. and LINDHEIMER, M.D. 1984. Renal sodium and water handling in pregnancy. *Obstet. Gynecol. A.* **13**: 35–69.

BARRON, W.M., STAMOUTSOS, B.A. and LINDHEIMER, M.D. 1989. Role of volume in the regulation of vasopressin secretion during pregnancy in the rat. *J. Clin. Invest.* **73**: 923–932.

BARTON, R.G. and CERRA, F.B. 1991. Metabolic and nutritional support. In: *Trauma*, 2nd Edn, Chap. 62, pp. 965–994. MOORE, E. E., MATTOX, K. L. and FELICIANO, D. V. (Eds). Appleton and Lange, New York.

BAUMAN, H.E. 1982. Industry views the role of sodium in the diet. In: *Sodium Intake—Dietary Concerns*, pp. 73–82. FREEMAN, T. M. and GREGG, O. W. (Eds). American Association of Cereal Chemists, St Paul, MN.

BAYLIS, C. 1982. Glomerular ultrafiltration in the pseudopregnant rat. *Am. J. Physiol.* **234**: F300–F305.

BEARD, T.C., COOKE, H.M., GRAY, W.R. and BARGE, R. 1982. Randomised controlled trial of a no-added-sodium diet for mild hypertension. *Lancet* **2**: 455–458.

BEATON, G.H. 1988. Nutritional requirements and population data. *Proc. Nutr. Soc.* **47**: 63–78.

BEILHARZ, S. and KAY, R.N.B. 1963. The effects of ruminal and plasma sodium concentrations on the sodium appetite of sheep. *J. Physiol.* **165**: 468–483.

BELL, F.R. and WILLIAMS, H.Ll. 1960. The effect of sodium depletion on the taste threshold of calves. *J. Physiol.* **151**: 42–43P.

BELL, P.D. and NAVAR, L.G. 1982. Intrarenal feedback control of GFR. *Sem. Nephrol.* **2**: 289–301.

BELL, R.J., LAURENCE, B.M., MEEHAN, P.J., CONGIU, M., SCOGGINS, B.A. and WINTOUR, E.M. 1986. Regulation and function of arginine vasopressin in pregnant sheep. *Am. J. Physiol.* **250**: F777–F780.

BENABE, J.E. and MARTINEZ-MALDONADO, M. 1993. Dietary modification of the renin–angiotensin system. *Sem. Nephrol.* **13:** 567–572.

BENNETT, W.M. 1990. Diagnostic considerations in autosomal dominant polycystic kidney disease. *Sem. Nephrol.* **10:** 552–555.

BERETTA-PICCOLI, C. 1990. Body sodium in normal subjects predisposed to hypertension. *J. Cardiovasc. Pharmacol.* **16;** Suppl. 7., S52–S55.

BERETTA-PICCOLI, C., DAVIES, D.L., BODDY, K., BROWN, J.J., CUMMING, A.M.M., EAST, B.W., FRASER, R., LEVER, A.F., PADFIELD, P.L., SEMPLE, P.F., ROBERTSON, J.I.S., WEIDMANN, P. and WILLIAMS, E.D. 1982. Relation of arterial pressure with body sodium, body potassium and plasma potassium in essential hypertension. *Clin. Sci.* **63:** 257–276.

BERL, T. and BETTER, O.S. 1979. In: *Contemporary Issues in Nephrology*, Vol. 4, *Hormonal Function and the Kidney*, pp. 200–214. BRENNER, B.M. and STEIN, J.H. (Eds). Churchill Livingstone, Edinburgh.

BERNARD, R.A., DOTY, R.L., ENGELMAN, K. and WEISS, R.A. 1980. Taste and salt intake in human hypertension. In: *Biological and Behavioural Aspects of Salt Intake*, pp. 397–409. KARE, M.R., FREGLY, M.J. and BERNARD, R.A. (Eds). Academic Press, New York.

BERNARDI, M., TREVISANI, F., GASBARRINI, A. and GASBARRINI, G. 1994. Hepatorenal disorders: role of the renin–angiotensin–aldosterone system. *Sem. Liver Dis.* **14:** 23–34.

BERNSTEIN, I.L. 1993. Amiloride-sensitive sodium channels and salt preference in rats. *Proc. 11th Int. Conf. Physiology of Food and Fluid Intake*, p. 64. Oxford.

BERON, J. and VERREY, F. 1994. Aldosterone induces early activation and late accumulation of Na–K ATPase at surface of A6 cells. *Am. J. Physiol.* **266:** C1278–C1290.

BERRY, Y.C.A. and RECTOR, F.C. 1991. Mechanism of proximal NaCl reabsorption in the proximal tubule of the mammalian kidney. *Sem. Nephrol.* **11:** 86–97.

BERTORELLO, A.M. and KATZ, A.I. 1993. Short-term regulation of Na–K ATPase activity: physiological relevance and cellular mechanisms. *Am. J. Physiol.* **265:** F743–F755.

BEVAN, J.A. 1993. Flow regulation of vascular tone. Its sensitivity to changes in sodium and calcium. *Hypertension* **22:** 273–281.

BEYER, K.H. 1976. Lessons from the discovery of modern diuretic therapy. *Perspec. Biol. Med.* **19:** 500–508.

BEYER, K.H. 1977. Discovery of the thiazides. *Perspec. Biol. Med.* **20:** 410–420.

BIANCHI, G., CUSI, D. and VEZZOLI, G. 1988. Role of cellular sodium and calcium metabolism in the pathogenesis of essential hypertension. *Sem. Nephrol.* **8:** 110–119.

BICHET, D.G. 1989. Water disturbances in cardiac failure. *Bailliere's Clin. Endocrin. Metab.* **3:** 559–574.

BICHET, D. and SCHRIET, R.W. 1984. Water metabolism in edematous disorders. *Sem. Nephrol.* **4:** 325–333.

BICHET, D., KLUGE, R., HOWARD, R.L. and SCHRIER, R.W. 1993. Pathogenesis of hyponatraemic states. In: *Clinical Disturbances of Water Metabolism*, Chap. 9, pp. 169–187. SELDIN, D.W. and GIEBISCH, G. (Eds). Raven Press, New York.

BICKELL, W.H., WALL, M.J., PEPE, P.E., MARTIN, R.R., GINGER, V.F., ALLEN, M.K. and MATTOX, K.L. 1994. Immediate versus delayed fluid resuscitation for hypotensive patients with penetrating torso injuries. *New Engl. J. Med.* **331:** 1105–1109.

BIGGER, J.T. 1994. Diuretic therapy, hypertension and cardiac arrest. *New Engl. J. Med.* **330:** 1899–1900.

BINDER, H.J. and TURNAMIAN, S.G. 1989. Differential effects of corticosteroids on active electrolyte transport in the mammalian distal colon. *Acta Vet. Scand. Supp.* **86**: 174–180.

BIRD, E. and CONTRERAS, R.J. 1986. Dietary salt affects fluid intake and output patterns of pregnant rats. *Physiol. Behav.* **37**: 365–369.

BIRKENHAGER, W.H. 1990. Diuretics and blood pressure reduction: physiologic aspects. *J. Hyperten.* **8**: Suppl. 2, S3–S7.

BLACHLEY, J.D. and HENRICH, W.L. 1981. The diagnosis and management of acute renal failure. *Sem. Nephrol.* **1**: 11–20.

BLACKBURN, R.E., DEMKO, A.D., HOFFMAN, G.E., STRICKER, E.M. and VERBAKS, J.G. 1992. Central oxytocin inhibition of angiotensin-induced salt appetite in rats. *Am. J. Physiol.* **263**: R1347–R1353.

BLAIR-WEST, J.R., COGHLIN, J.P., DENTON, D.A., GODING, J.R., WINTOUR, M. and WRIGHT, R.D. 1963. The control of aldosterone secretion. *Rec. Prog. Horm. Res.* **19**: 311–383.

BLAIR-WEST, J.R., DENTON, D.A., MCKINLEY, M.J. and WEISINGER, R.S. 1989. Sodium appetite and thirst in cattle subjected to dehydration. *Am. J. Physiol.* **257**: R1212–R1218.

BLANTZ, R.C. 1993. Glomerular blood flow. *Sem. Nephrol.* **13**: 436–446.

BLAUSTEIN, M.P. 1977. Sodium ions, calcium ions, blood pressure and hypertension; a reassessment and a hypothesis. *Am. J. Physiol.* **232**: C165–C173.

BLAUSTEIN, M.P. 1993. Physiological effects of endogenous ouabain: control of intracellular Ca^{2+} stores and cell responsiveness. *Am. J. Physiol.* **262**: C1367–1387.

BLOOD, D.C., RADOSTITS, O.M. and GAY, C.C. 1994. In: *Veterinary Medicine*, 8th Edn, pp. 1499–1502. Bailliere, London.

BOBIK, A., NEYLON, C.B. and LITTLE, P.J. 1994. Disturbances of vascular smooth muscle cation transport and the pathogenesis of hypertension. In: *Textbook of Hypertension*, pp. 175–187. SWALES, J.D. (Ed.). Blackwell, Oxford.

BOCK, H.A. and STEIN, J.H. 1988. Diuretics and the control of extracellular volume: role of counterregulation. *Sem. Nephrol.* **8**: 264–272.

BODEY, A.R. and MICHELL, A.R. 1995. An epidemiological study of canine blood pressure. *J. Sm. Anim. Prac.* (submitted).

BODEY, A.R., YOUNG, L.E., BARTRAM, D.H., DIAMOND, M.J. and MICHELL, A.R. 1994. A comparison between direct and indirect oscillometric. measurement of arterial pressure in anaesthetised dogs, using both tail and limb cuffs. *Res. Vet. Sci.* **57**: 265–269.

BOEGEHOLD, M.A. and KOTCHER, T. 1991. Importance of dietary chloride for salt sensitivity of blood pressure. *Hypertension* **17** Suppl. 1, I158–I161.

BOHR, D.F. and DOMINICZAK, A.F. 1991. Experimental hypertension. *Hypertension* **17** Suppl. 1, I39–I44.

BONVENTRE, J.V. and LEAF, A. 1982. Sodium homeostasis: steady states without a set point. *Kidney Int.* **21**: 880–885.

BOOTH, I.W. and MCNEISH, A.S. 1993. Mechanisms of diarrhoea. *Bailliere's Clin. Gastroenterol.* **7**: 215–242.

BORER, K.T. and EPSTEIN, A.N. 1965. Disappearance of salt and sweet preference in rats drinking without taste and smell. *Physiologist* **8**: 118–126.

BORING, E.G. 1942. *The History of Experimental Physiology*. Appleton Century, New York.

BORON, W.F. 1982. Cell pH regulation and acidification mechanisms in the proximal nephron. *Sem. Nephrol.* **2**: 336–342.

BOTERO-VELEZ, M., CURTIS, J. and WARNOCK, D.G. 1994. Liddle's syndrome revisited—a disorder of sodium reabsorption in the distal tubule. *New Engl. J. Med.* **330**: 178–181.

BOULANGER, B.R., LILLY, M.P., HAMLYN, J.M., LAREDO, J., SHURTLEFF, D. and GANN, D.S. 1993. Ouabain is secreted by the adrenal gland in awake dogs. *Am. J. Physiol.* **264**: E413–419.

BOURDEAU, J.E. and ATTIE, M.F. 1994. Calcium metabolism. In: *Maxwell and Kleeman's Clinical Disorders of Fluid and Electrolyte Metabolism*, 5th Edn, Chap. 11. NARINS, R.J. (Ed). McGraw-Hill, New York.

BOURGOIGNIE, J.J., JACOB, A.A., SALLMAN, A.L. and PENNEL, J.P. 1981. Water, electrolyte and acid–base abnormalities in chronic renal failure. *Sem. Nephrol.* **1**: 91–111.

BOVEE, K.C. 1984. *Canine Nephrology*, pp. 175–217, 555–642. Harwal, U.S.A.

BOVEE, K.C. 1993. Genetic essential hypertension in dogs: a new animal model. In: *The Advancement of Veterinary Science*, Vol. 4: *Veterinary Science, Growth Points and Comparative Medicine*, pp. 185–194. MICHELL, A.R. (Ed.). C.A.B. International, Wallingford.

BOWMAN, W.C. and RAND, R.J. 1980. *Textbook of Pharmacology*, 2nd Edn, pp. 2.6–2.7. Blackwell Scientific, Oxford.

BRADLEY, R.M. and MISTRETTA, C.M. 1975. The developing sense of taste. In: *Olfaction and Taste V*, pp. 91–98. DENTON, D.A. and COGHLAN, J.P. (Eds). Academic Press, New York.

BRANDER, G.C., PUGH, D.M., BYWATER, R.J. and JENKINS, W.C. 1991. *Veterinary Applied Pharmacology and Therapeutics*, 5th Edn, pp. 290, 555–557. Bailliere Tindall, London.

BRANDS, M.W. and HALL, J.E. 1992. Insulin resistance, hyperinsulinemia and obesity-associated hypertension. *J. Am. Soc. Nephrol.* **3**: 1064–1077.

BRANDS, M.W., HILDEBRANDT, D.A., MIZELLE, H.L. and HALL, J.E. 1992. Hypertension during chronic hyperinsulinemia in rats is not salt-sensitive. *Hypertension* **19**: Suppl. 1, I83–I89.

BRATER, D.C. 1993. Resistance to diuretics: mechanisms and clinical implications. *Adv. Nephrol.* **22**: 349–369.

BRATER, D.C. 1994. The use of diuretics in congestive heart failure. *Sem. Nephrol.* **14**: 479–484.

BRENNER, B.M. and ANDERSON, S. 1989. Filtration surface area, salt intake and hypertension. *Contrib. Nephrol.* **75**: 45–49.

BRENNER, B.M. and RECTOR, F.C. 1986. *The Kidney*, 3rd Edn, pp. 343–432, 1553–1586. Saunders, Philadelphia.

BRENSILVER, J.M., DANIELS, F.H., LEFAVOUR, G.S., MALSEAK, R.M., LORCH, J.A., PONTE, M.L. and CORTELL, S. 1985. Effect of variations in dietary sodium intake on sodium excretion in mature rats. *Kidney Int.* **27**: 497–502.

BREWER, N.R. 1983. Nutrition of the cat. *J. Am. Vet. Med. Assoc.* **180**: 1179–1182.

BREZIN, M., ROSEN, S. and EPSTEIN, F.H. 1986. Acute renal failure. In: *The Kidney*, 3rd Edn, Chap. 19. BRENNER, B.M. and RECTOR, F.C. (Eds). Saunders, Philadelphia.

BRICKER, N.S. and LICHT, A. 1980. Natriuretic hormone: biological effects and progress in identification and isolation. In: *Hormonal Regulation of Sodium Excretion*, pp. 399–408. LICHARDUS, B., SCHRIER, R. and PONEC, J. (Eds). Elsevier, Amsterdam.

BRICKER, N.S., KLAHR, S., PUEKERSON, M., SCHULTZE, R.G., AVIOLI, L.V. and BIRGE, S.J. 1968. *In vitro* assay for a humoral substance present during volume expansion and uraemia. *Nature* **219**: 1058–1059.

BRIER, M.E. and LUFT, F.C. 1994. Sodium kinetics in white and black normotensive subjects: possible relevance to salt-sensitive hypertension. *Am. J. Med. Sci.* **307**: Suppl. 1, S38–S42.

BRIER, M.E., BUNKE, C.M., LATHON, P.V. and AARDNOFF, G.R. 1993. Erythropoietin-induced antinatriuresis mediated by angiotensin II in perfused kidneys. *J. Am. Soc. Nephrol.* **3**: 1583–1590.

BRIGGS, J.P., SAWAYA, B.E. and SCHNERMAN, N. 1990. Disorders of salt balance. In: *Fluids and Electrolytes*, 2nd Edn, pp. 70–138. KOKKO, J.P. and TANNEN, R.L. (Eds). Saunders, Philadelphia.

BROD, J. 1978. Hypertension and renal parenchymal disease: mechanisms and management. In: *Hypertension; Mechanisms, Diagnosis and Treatment*, pp. 137–164. ONESTI, G. and BREST, A.N. (Eds). Davis, Philadelphia.

BRODY, M.J., FINK, G.D., BUGGY, J., HAYWOOD, J.R., GORDON, F. and JOHNSON, A.K. 1978. The role of the anteroventral third ventricle AV3V region in experimental hypertension. *Circ. Res.* **43**: Suppl. I., 2–13.

BROOKS, H.W., MICHELL, A.R., WAGSTAFF, A.J. and WHITE, D.G. 1995. Fallibility of faecal consistency as a criterion of success in the evaluation of oral fluid therapy for calf diarrhoea. *Br. Vet. J.*, in press.

BROWN, D. 1991. Structural–functional features of antidiuretic hormone-induced water transport in the collecting duct. *Sem. Nephrol.* **11**: 478–506.

BROWN, D.R. and MILLER, R.J. 1991. Neurohormonal control of fluid and electrolyte transport in intestinal mucosa. In: *Handbook of Physiology*, Section 6, *Gastrointestinal Tract*, Vol. IV, *Intestinal Secretion and Absorption*, Chap. 24, pp. 527–589. FIELD, M. and FRIZELL, R.A. (Eds). American Physiological Society, Bethesda, Maryland.

BROWN, E.A., MARKANDU, N., SAGNELLA, G.A., JONES, B.E. and MACGREGOR, G.A. 1985. Sodium retention in nephrotic syndrome is due to an intrarenal defect. *Nephron* **39**: 290–295.

BROWN, J.J., LEVER, A.E., ROBERTSON, J.I.S. and SEMPLE, P.F. 1984. Should dietary sodium be reduced? The sceptics' position. *Q. J. Med.* **53**: 427–437.

BROWN, J.M., GROSSO, M.A. and MOORE, E.F. 1990. Hypertonic saline and dextran: impact on cardiac function in the isolated rat heart. *J. Trauma* **30**: 646–651.

BROWN, K.H. 1991. Dietary management of acute childhood diarrhoea: optimal timing of feeding and appropriate use of milks and mixed diets. *J. Pediatr.* **118**: S92–S98.

BROWN, M.A., SINOSICH, M.J., SAUNDERS, D.M. and GALLERY, E.D.M. 1986. Potassium regulation and progesterone–aldosterone interrelationships in human pregnancy: a prospective study. *Am. J. Obstet. Gynecol.* **155**: 349–353.

BROWN, M.A., ZAMMIT, V.C., MITAR, D.A. and WHITWORTH, J.A. 1992. Renal–aldosterone relationships in pregnancy-induced hypertension. *Am. J. Hyperten.* **5**: 366–371.

BROWN, P.S. and BROWN, S.C. 1987. Osmoregulatory actions of prolactin and other adenohypophysial hormones. In: *Vertebrate Endocrinology: Fundamentals and Biomedical Implications*, Vol. 2, pp. 45–84. PANG, P.K.T., SCHREIBMAN, M.P. and SAWYER, W.H. (Eds). Academic Press, San Diego.

BROWN, S.A., WALTON, C.L., CRAWFORD, P. and BAKRIS, G.L. 1993. Long term effects of antihypertensive regimen on renal hemodynamics and proteinuria. *Kidney Int.* **43**: 1210–1217.

BROWNRIGG, W. 1748. *The Art of Making Common Salt*. Davis, London. (Facsimile published by University of Michigan, Ann Arbor.)

BROZENA, S. and JESSUP, M. 1990. Pathophysiologic strategies in the management of congestive heart failure. *A. Rev. Med.* **41**: 65–74.

BRUCE, D.L. 1980. In: *Functional Toxicity of Anaesthesia*, pp. 54–64. Grune and Stratton, New York City.

BRUCE, N.G., WANNAMETHEE, G. and SHAPER, A.G. 1993. Lifestyle factors associated with geographic blood pressure variations among men and women in the U.K. *J. Human Hyperten.* **7**: 229–238.

BRUNSKILL, N. and KLAHR, S. 1993. Mechanisms of progressive renal failure. In: *Prevention of Progressive Chronic Renal Failure*, pp. 1–61. EL-NAHAS, A.M., MALLICK, N.P. and ANDERSON, S. (Eds). Oxford University Press, Oxford.

BRUNTON, L. 1990. Agents for gastric acidity and peptid ulcers. In: *Goodman and Gilman's The Pharmacological Basis of Therapeutics*, 8th Edn. GILMAN, A.G., RALL, T.W., NIES, A.S. and TAYLOR, P. (Eds). Pergamon Press, Oxford.

BUCENS, I.K. and CATTO-SMITH, A.G. 1991. Hypernatraemic dehydration after Lucozade. *Med. J. Austral.* **155**: 128–129.

BUCHANAN, T.A. 1991. Insulin resistance and hyperinsulinemia: implications for the pathogenesis and treatment of hypertension. *Sem. Nephrol.* **11**: 512–522.

BULL, N.L. and BUSS, D.H. 1980. Contributions of foods to sodium intake. *Proc. Nutr. Soc.* **39**: 30A.

BULPITT, C.J. 1986. Salt and dietary intervention. In: *Dietary Salt of Hypertension: Implications for Public Health Policies*, Vol. 5, pp. 2–3. WOODS, C. (Ed.). Royal Society of Medicine Round Table.

BULPITT, C.J. 1990. Is systolic pressure more important than diastolic pressure? *J. Human Hyperten.* **4**: 471–476.

BUNKE, M., GLEASON, J.R., BRIER, M. and SLOAN, R. 1994. Effect of erythropoietin on renal excretion of a sodium load. *Clin. Pharmacol. Ther.* **55**: 563–568.

BUNNEY, W.E. and BUNNEY, B.L.G. 1987. Lithium. In: *Psychopharmacology: the Third Generation of Progress*, pp. 553–565. MELTZER, H.Y. (Ed.). Raven Press, New York.

BURCKHARD, B.D. and BURCKHARDT, G. 1988. Cellular mechanisms of proximal tubular acidification. In: *pH Homeostasis*, pp. 233–262. HAUSSINGER, D. (Ed.). Academic Press, London.

BURCKHARDT, G. and FRIEDRICH, T. 1988. Molecular identification of the proximal tubular Na^+/H^+ exchanger. In: *Nephrology*, Vol. 1, pp. 217–230. DAVISON, A.M. (Ed.) Bailliere Tindall, London.

BURKE, L.M. and READ, R.S. 1993. Dietary supplements in sport. *Sports Med.* **15**: 43–65.

BURNIER, M., BIOLLAZ, J., MAGNIN, J.C., BIDLINGMEYER, M. and BRUNNER, H.R. 1994. Renal sodium handling in patients with untreated hypertension and white–coat hypertension. *Hypertension* **23**: 496–502.

BURSET, R.G. and WATSON, M.L. 1983. The effect of sodium restriction during gestation on offspring brain development in rats. *Am. J. Clin. Nutr.* **37**: 43–51.

BUSHINSKY, D.A. 1994. Acidosis and base. *Miner. Electrolyte Metab.* **20**: 40–52.

CAMBAR, J., CAL, J.C. and TRANCHOT, J. 1992. In: *Biological Rhythms in Clinical Laboratory Medicine*, pp. 470–482. TOUITON, Y. and HAUS, E. (Eds). Springer, Berlin.

CANNON, P.J. and MARTINEZ-MALDONADO, M. 1983. The pathogenesis of cardiac edema. *Sem. Nephrol.* **3**: 211–224.

CANNON, W.B. 1932. *The Wisdom of the Body*, pp. 91–98. Kegan Paul, Trench and Trubner, New York.

CANNON, W.B., FRASER, J. and COWELL, E.M. 1918. The preventive treatment of wound shock. *J. Am. Med. Assoc.* **70**: 618–621.

CANTIN, M. and GENEST, J. 1985. The heart as an endocrine gland. *Sci. Am.* **6**: 62–67.

CANTIN, M. and GENEST, J. 1987. The heart as an endocrine organ. *Hypertension* **10**: 118–121.

CAPPUCCIO, F.P., MARKANDU, N.D., BEYNON, G.W., SHORE, A.C., SAMPSON, B. and MACGREGOR, G.A. 1985. Lack of effect of oral magnesium on high blood pressure: a double-blind study. *Br. Med. J.* **291**: 235–238.

CAPPUCCIO, F.P., BLACKWOOD, A., SAGNELLS, G.A., MARKANDU, N.D. and MACGREGOR, G.A. 1993. Association between extracellular volume expansion and urinary calcium excretion in normal humans. *J. Hypertens.* **11**: Suppl. 5, S196–S197.

CARAMELO, C. and SCHRIER, R.W. 1990. Edema of cirrhosis and its treatment. In: *The Regulation of Sodium and Chloride Balance*, Chap. 11, pp. 353–374. SELDIN, D.W. and GIEBISCH, G (Eds). Raven Press, New York.

CARE, A.D. 1988. A fresh look at hypomagnesaemia. *Br. Vet. J.* **144**: 3–4.

CAREY, R.M., SMITH, J.R. and ORTT, R.M. 1976. Gastrointestinal control of sodium excretion in sodium-depleted, conscious rabbits. *Am. J. Physiol.* **230**: 1504–1508.

CARLSON, A.J. 1916. *The Control of Hunger in Health and Disease*, p. 15. University of Chicago Press, Chicago, IL.

CARONE, F.A. 1988. Functional changes in polycystic kidney disease are tubulo-interstitial in origin. *Sem. Nephrol.* **8**: 89–93.

CARONE, F.A., BACALLAO, R. and KANWAR, V.S. 1994. Biology of polycystic kidney disease. *Lab. Invest.* **70**: 437–448.

CARPENTER, J.A. 1956. Species differences in taste preference. *J. Comp. Physiol. Psychol.* **49**: 139–144.

CARPENTER, C.C. 1987. Introduction to symposium on oral rehydration therapy. *J. Diarrh. Dis. Res.* **5**: 252–255.

CARR, S.J. and THOMAS, T.H. 1994. Perturbation of blood cell and platelet membranes in human essential hypertension. In: *Textbook of Hypertension*, pp. 160–174. SWALES, J.D. (Ed.). Blackwell Scientific, Oxford.

CARRETERO, O.A. and SCICLI, A.G. 1994. The kallikrein–kinin system as a regulator of cardiovascular and renal function. In: *Textbook of Hypertension*, pp. 328–340. SWALES, J.D. (Ed.). Blackwell Scientific, Oxford.

CARVALHO, J.J., BARUZZI, R., HOWARD, P.F., POULTER, N., ALPERS, M.P., FRANCO, L.J., MARCOPITO, L.F., SPOONER, V.J., DYER, A.R., ELLIOTT, P., STAMLER, J. and STAMLER, R. 1989. Blood pressure in four remote populations in the Intersalt study. *Hypertension* **14**: 238–246.

CHA, S.C., ABERDEEN, G.W., MUKADDAM-DAHER, S., QUILLEN, E.W. and NUWAYHID, B.S. 1993. Tubular handling of fluid and electrolytes during ovine pregnancy. *Am. J. Physiol.* **265**: F278–F284.

CHAPMAN, A.B. and SCHRIER, R.W. 1991. Pathogenesis of hypertension in autosomal dominant polycystic kidney disease. *Sem. Nephrol.* **11**: 653–660.

CHAVEZ-NEGRETE, A., MAJLIF-CRUZ, S., FRATI-MUNARI, A., PERCHES, A. and ARGUERO, R. 1991. Treatment of hemorrhagic shock with intraosseous or intravenous infusion of hypertonic saline–dextran solution. *Eur. Surg. Res.* **23**: 123–129.

CHEUNG, C.Y. 1991. Role of endogenous atrial natriuretic factor in the regulation of fetal cardiovascular and renal function. *Am. J. Obstet. Gynecol.* **165**: 1558–1567.

CHEVALIER, R.L. 1993. Atrial natriuretic peptide in renal development. *Pediatr. Nephrol.* **7**: 652–656.

CHEVALIER, R.L. 1994. Renal disease in neonates. In: *Clinical Paediatric Nephrology*, 2nd Edn, pp. 372–386. POSTLETHWAITE, R.J. (Ed.). Butterworth-Heinemann, Oxford.

CHINET, T.C., FULTON, J.M., YANKASKAS, J.R., BOUCHER, R. and STUTTS, M.J. 1994. Mechanism of sodium hyperabsorption in cultured cystic fibrosis nasal epithelium: a patch-clamp study. *Am. J. Physiol.* **266**: C1061–C1068.

CHONKO, A.M. and GRANTHAM, J.J. 1994. Treatment of edema states. In: *Maxwell and Kleeman's Clinical Disorders of Fluid and Electrolyte Metabolism*, Chap. 19, pp. 545–582. R.G. NARINS (Ed.). McGraw-Hill, New York.

CHOWDHURY, M.R., UEMURA, N., NISHIDA, Y., MORITA, H. and HOSOMI, H. 1993. Effects of endothelins on fluid and NaCl absorption across the jejunum of anesthetized dogs. *Jap. J. Physiol.* **43**: 709–726.

CHOWDREY, H.S. and LIGHTMAN, S.L. 1989. Neuroendocrine control of blood tonicity and volume. *Bailliere's Clin. Endocrinol. Metabol.* **3**: 229–247.

CHRISTY, N.P. and SHAVER, J.C. 1974. Estrogens and the kidney. *Kidney Int.* **6**: 366–376.

CHURCHILL, S., BENGELE, H.H., MELBY, J.C. and ALEXANDER, E.A. 1981. Role of aldosterone in sodium retention of pregnancy in the rat. *Am. J. Physiol.* **240**: R175–181.

CIVETTA, J.M. 1979. A new look at the Starling equation. *Crit. Care Med.* **7**: 84–90.

CLAPHAM, D.E. 1989. Potassium and tissue excitability. In: *The Regulation of Potassium Balance*, pp. 57–88. SELDIN, D.W. and GIEBISCH, G. (Eds). Raven Press, New York.

CLEWELL, J.D. and WALKER-RENARD, P. 1994. Prostaglandins for the treatment of hepatorenal syndrome. *Ann. Pharmacother.* **28**: 54–55.

CLORE, J., SCHOOLWERTH, A. and WATLINGTON, C.O. 1992. When is cortisol a mineralocorticoid? *Kidney Int.* **42**: 1297–1308.

COBB, M. and MICHELL, A.R. 1992. Plasma electrolyte concentrations in dogs receiving diuretic therapy for cardiac failure. *J. Sm. Anim. Prac.* **33**: 526–529.

COCKROFT, J.R., CHOWIENCZYK, P.J., BENJAMIN, N. and RITTER, J.M. 1994. Preserved endothelium-dependent vasodilation in patients with essential hypertension. *New Engl. J. Med.* **330**: 1036–1040.

CODY, R.J. 1990. Atrial natriuretic factor in edematous disorders. *A. Rev. Med.* **41**: 377–382.

CODY, R.J. and PICKWORTH, K.K. 1994. Approaches to diuretic therapy and electrolyte imbalance in congestive heart failure. *Cardiol. Clin.* **12**(1), 37–50.

CODY, R.J., COUIT, A.B., SCHAER, G.L., LARAGH, J.H., SEALEY, J.E. and FELDSCHUH, J. 1986. Sodium and water balance in chronic congestive heart failure. *J. Clin. Invest.* **77**: 1441–1452.

COHEN, M.B. 1991. Etiology and mechanisms of acute infectious diarrhoea in infants in the United States. *J. Pediatr.* **118**: S34–S39.

COMA 1994. Nutritional aspects of cardiovascular disease. Committee on Medical Aspects of Food Policy. Report of the Cardiovascular Review Group. Department of Health Report on Health and Social Subjects no. 46. HMSO, London.

CONGER, J.D. and SCHRIER, R.W. 1980. Renal haemodynamics in acute renal failure. *A. Rev. Physiol.* **42**: 603–614.

CONRAD, K.P., GELLAI, M., NORTH, W.G. and VALTIN, H. 1993. Influence of oxytocin on renal hemodynamics and sodium excretion. *Ann. N. Y. Acad. Sci.* **689**: 346–362.

CONTRERAS, R.J. 1989. Differences in perinatal NaCl exposure alter blood pressure levels of adult rats. *Am. J. Physiol.* **256**: R70–R77.

CONTRERAS, R.J. 1993. High NaCl intake of rat dams alters maternal behaviour and elevates blood pressure of adult offspring. *Am. J. Physiol.* **264:** R296–304.

CONTRERAS, R.J. and BIRD, E. 1986. Perinatal sodium chloride intake modifies the fluid intake of adult rats. In: *The Physiology of Thirst and Sodium Appetite*, pp. 15–19. DE CARO, G., EPSTEIN, A.N. and MASSI, M. (Eds.). Plenum Press, New York.

CONTRERAS, R.J. and KOSTEN, T. 1981. Changes in salt intake after abdominal vagotomy: evidence for hepatic sodium receptors. *Physiol. Behav.* **26:** 575–582.

CONTRERAS, R.J. and KOSTEN, T. 1986. Peripheral gustatory mechanisms of salt intake in the rat. In: *The Physiology of Thirst and Sodium Appetite*, pp. 479–485. DE CARO, G., EPSTEIN, A.N. and MASSI, M. (Eds). Plenum Press, New York.

COOPER, S.J. 1986. Benzodiazepine and endorphinergic mechanisms in relation to salt and water intake. In: *The Physiology of Thirst and Sodium Appetite*, pp. 239–244. DE CARO, G., EPSTEIN, A.N. and MASSI, M. (Eds). Plenum Press, New York.

COOPER, S.J. and GILBERT, D.B. 1986. Dopaminergic modulation of choice in salt preference tests. In: *The Physiology of Thirst and Sodium Appetite*, pp. 453–458. DE CARO, G., EPSTEIN, A.N. and MASSI, M. (Eds). Plenum Press, New York.

CORNELISSEN, G., HAUS, E. and HALBERG, F. 1992. Chronobiologic blood pressure assessment from womb to tomb. In: *Biological Rhythms in Clinical and Laboratory Medicine*, TOUITOU, Y. and HAUS, E. (Eds). Springer, Berlin.

CORRY, D. and TUCK, M. 1991. Hypertension and diabetes. *Sem. Nephrol.* **11:** 561–570.

CORUZZI, P. 1994. Escape from the sodium-retaining effects of mineralocorticoids: is there a role for dopamine? *J. Clin. Endocrinol. Metab.* **78:** 455–458.

COSTANZO, L.S. 1988. Mechanism of action of thiazide diuretics. *Sem. Nephrol.* **8:** 234–241.

COSTANZO, L. and WINDHAGER, E.E. 1992. Renal tubular transport of calcium. In: *Handbook of Physiology, Section 8, Renal Physiology*, Vol. II, Chap. 36, pp. 1759–1805. WINDHAGER, E.E. (Ed.). American Physiological Society/Oxford University Press, New York.

COUTRY, N., FARMAN, N., BONVALET, J.P. and BLOT-CHABAUD, M. 1994. Role of cell volume variations in Na–K ATPase recruitment and/or activation in cortical collecting duct. *Am. J. Physiol.* **266:** C1342–C1349.

COWLEY, A.W. 1991. Salt and hypertension—future directions. *Hypertension* **17:** Suppl. 1, 205–210.

COWLEY, A.W. and ROMAN, R.J. 1989. Control of blood and extracellular volume. *Bailliere's Clin. Endocrinol. Metabol.* **3:** 331–369.

COWLEY, A.W., SKELTON, M.M. and MERRILL, D.C. 1986. Osmoregulation during high salt intake: relative importance of drinking and vasopressin secretion. *Am. J. Physiol.* **151:** R878–R886.

COWLEY, A.W., ROMAN, R.J., FENOY, F.J. and MATTSON, D.L. 1992. Effect of renal medullary circulation on arterial pressure. *J. Hypertens.* Suppl. **10:** S187–S193.

COWLEY, A.W., SKELTON, M.M., MERRILL, D.C., QUILLEN, E.W and SWITZER, S.J. 1983. Influence of daily sodium intake on vasopressin secretion and drinking in dogs. *Am. J. Physiol.* **245:** R860–R872.

COX, R.H., BAGSHAW, R.J. and DETWEILER, D.K. 1985. Baroreceptor reflex cardiovascular control in mongrel dogs and racing greyhounds. *Am. J. Physiol.* **249:** H655–H626.

CRANE, M.G. and HARRIS, J.J. 1978. Estrogen and hypertension: effect of discontinuing estrogens on blood pressure, exchangeable sodium, and the renin–aldosterone system. *Am. J. Med. Sci.* **276**: 33–35.

CRISPIN, S.M. and STICKLAND, N.C. 1983. Gross and microscopic anatomy of the mammalian kidney. In: *Veterinary Nephrology*, chap. 2, pp. 7–25. Hall, L.W. (Ed.). Heinemann Veterinary Books, London.

CUTLER, J.A. and BRITTAIN, E. 1990. Calcium and blood pressure: an epidemiologic perspective. *Am. J. Hyperten.* **8**: 137S–146S.

CZERWINSKI, A.W. and LACH, F.L. 1982. Renal edema. In: *Sodium, its Biological Significance*, pp. 115–135. PAPPER, S. (Ed.). CRC, Boca Raton, FL.

DAFNIS, E. and SABATINI, S. 1994. Biochemistry and pathophysiology of vanadium. *Nephron* **67**: 133–143.

DAHL L.K. 1958. Salt intake and salt need. *New Engl. J. Med.* **258**: 1152–1157, 1205–1208.

DAHL, L.K. 1968. Salt in processed baby foods. *Am. J. Clin. Nutr.* **21**: 787–792.

DAHL, L.K. 1972. Salt and hypertension. *Am. J. Clin. Nutr.* **25**: 231–244.

DAHL, L.K., HEINE, M. and TASSINARI, L. 1962. Effects of chronic excess salt ingestion: evidence that genetic factors play an important role in susceptibility to experimental hypertension. *J. Expl Med.* **115**: 1173–1190.

DAHL, L.K., KNUDSEN, K.D., HEINE, M. and LEITL, G. 1967. Effects of chronic excess salt ingestion: genetic influence on the development of salt hypertension in parabiotic rats; evidence for a humoral factor. *J. Expl Med.* **126**: 687–699.

DAHL, L.K., LEITL, G. and HEINE, M. 1972. Influence of dietary potassium and sodium/potassium molar ratios on the development of salt hypertension. *J. Expl Med.* **136**: 318–330.

DAL CANTON, A. and ANDREUCCI, V.E. 1986. Renal hemodynamics in pregnancy. In: *The Kidney in Pregnancy*, pp. 1–11. ANDREUCCI, V.E. (Ed.). Martinus Nijhoff, Boston.

DANIEL, T.O. and HENRICH, W.L. 1990. Endocrine abnormalities and fluid and electrolyte disorders. In: *Fluids and Electrolytes*, 2nd Edn, Chap. 16, pp. 830–872. KOKKO, J.P. and TANNEN, R.L. (Eds). Saunders, Philadelphia.

DANIELS, B.S. and FERRIS, T.F. 1988. The use of diuretics in nonedematous disorders. *Sem. Nephrol.* **8**: 342–353.

DANIELS, S.D., MEYER, R.A. and LOGGIE, J.M.H. 1990. Determinants of cardiac involvement in children and adolescents with essential hypertension. *Circulation* **82**: 1243–1248.

DANIELSEN, J. and BUGGY, J. 1980. Depression of *ad lib.* and angiotensin-induced sodium intake at oestrus. *Brain Res. Bull.* **5**: 501–504.

DAVIDSON, C. 1985. In: *Diuretics in Heart Failure*, pp. 51–60. RAMSAY, L.E. (Ed.). Royal Society of Medicine, London.

DAVIDSON, C. 1987. In: *Diuretics in Heart Failure*, pp. 61–65. POOLE-WILSON, P.A. (Ed.). Royal Society of Medicine, London.

DAVIDSON, E.W. and DUNN, M.J. 1987. Pathogenesis of hepatorenal syndrome. *A. Rev. Med.* **38**: 361–372.

DAVIS, R.E., SHELLEY, S., MACDONALD, G.J. and DUGGAN, K.A. 1992. The effects of a high sodium diet on the metabolism and secretion of vasoactive intestinal peptide in the rabbit. *J. Physiol.* **451**: 17–23.

DAVISON, J.M. 1983. The kidney in pregnancy: a review. *J. R. Soc. Med.* **76**: 485–501.

DAVISON, J.M. 1986. Renal function during normal pregnancy and the effect of renal disease and pre-eclampsia. In: *The Kidney in Pregnancy*, pp. 65–80. ANDREUCCI, V.E. (Ed.). Martinus Nijhoff, Boston.

DAVISON, J.M. and LINDHEIMER, M.D. 1989. Volume homeostasis and osmoregulation in human pregnancy. *Bailliere's Clin. Endocrinol. Metab.* **3**: 451–472.

DAVISON, J.M., SHIELLS, E.A., PHILLIPS, P.R. and LINDHEIMER, M.D. 1988. Serial evaluation of vasopressin release and thirst in human pregnancy. *J. Clin. Invest.* **81**: 798–806.

DAWSON, K.G. and SUTTON, R.A.L. 1986. Metabolic disorders. In: *Diuretics: Physiology, Pharmacology and Clinical Use*, Chap. 18, pp. 341–362. DIRKS, J.H. and SUTTON, R.A.L. (Eds). Saunders, Philadelphia.

DE ANGELIS, C., PERRONE, A., FERRI, C., PICCOLI, A., BELLINI C., D'AMELIO, R., SANTUCCI, A. and BALSANO, F. 1993. Oxygen administration increases plasma digoxin-ike substance and renal sodium excretion in chronic hypoxic patients. *Am. J. Nephrol.* **13**: 173–177.

DE BOLD, A.J., DE BOLD, M.L. and RUBEN DE CAMPIONE, M.M. 1988. Atrial natriuretic peptide. In: *Nephrology*, Vol. 1, pp. 109–122. DAVISON, A.M. (Ed.). Bailliere Tindall, London.

DE BONO, D. 1994. Digoxin in eurhythmic heart failure: PROVED or 'not proven'? *Lancet* **343**: 128–129.

DECHELOTTE, P., ALGAN, R., HECKETSWEILER, B. and HECKETSWEILER, P. 1991. Effects of glutamine and alanine on water and electrolyte fluxes in human jejunum during experimental hypersecretion. *Gastroenterology* **100**: A683.

DECKER, K. and DIETER, P. 1988. The stimulus-activated Na^+–H^+ exchange in macrophages, neutrophils and platelets. In: *pH Homeostasis: Mechanisms and Control*, Chap. 5, pp. 79–96. HAUSSINGER, D. (Ed.). Academic Press, London.

DECOLLOGNE, S., BERTRAND, I.B., ASCENSIO, M., DRUBAIX, I. and LELIEVRE, L.G. 1993. Na, K-ATPase and Na/Ca exchange isoforms: physiological and physiopathological relevance. *J. Cardiovasc. Pharmacol.* **22** Suppl. 2., 96–98.

DEEMS, R.O. and FRIEDMAN, M.L. 1988. Sodium chloride preference is altered in a rat model of liver disease. *Physiol. Behav.* **43**: 521–525.

DE FRONZO, R.A. 1981. The effect of insulin on renal sodium metabolism. *Diabetologica* **21**: 165–171

DE FRONZO, R.A. and SMITH, J.D. 1994. Clinical disorders of hyperkalaemia. In: *Maxwell and Kleeman's Clinical Disorders of Fluid and Electrolyte Metabolism*, 5th Edn, Chap. 23, pp. 697–754. NARINS, R.G. (Ed.). McGraw-Hill, New York.

DE JONG, P.E., NAVIS, G., HEEG, J.E. and DE ZEEUW, D. 1989. Angiotensin-converting enzyme inhibition in renal disease. In: *Cardiac and Renal Failure: an Expanding Role for ACE Inhibitors*, pp. 275–288. DOLLERY, C.T. and SHERWOOD, L.M. (Eds).

DELMEZ, J., WINDUS and KLAHR, S. 1990. Salt overload states and their treatment. In: *The Regulation of Sodium and Chloride Balance*, Chap. 13, pp. 393–419. SELDIN, D.W. and GIEBISCH, G. (Eds). Raven Press, New York.

DEL-RIO, A. and RODRIGUEZ-VILLAMIL, J.L. 1993. Metabolic effects of strict salt restriction in essential hypertensive patients. *J. Intern. Med.* **233**: 409–414.

DELVALLE, J. and YAMADA, T. 1990. The gut as an endocrine organ. *A. Rev. Med.* **41**: 447–455.

DE NICOLA, A.F., GRILLO, C., GONZALES, C. and GONZALES, S. 1992. Physiological, biochemical and molecular mechanisms of salt appetite control by mineralocorticoid action in brain. *Braz. J. Med. Biol. Res.* **25**: 1153–1162.

DE NICOLA, L., ROMANO, G., MEMOLI, B., CIANCIARUSO, B., SABATINI, M., RUSSO, D., CAGLIOTI, A., FUIANO, G., DAL CANTON, A. and CONTE, G. 1993. Extra-natriuretic effects of atrial natriuretic peptide in humans. *Kidney Int.* **43**: 307–313.

DENTON, D.A. 1956. The effect of sodium depletion on the $Na^+:K^+$ ratio of the parotid saliva of the sheep. *J. Physiol.* **131**: 516–525.

DENTON, D.A. 1957. The study of sheep with permanent unilateral parotid fistulae. *Q. J. Expl Physiol.* **42**: 72–95.

DENTON, D.A. 1965. Evolutionary aspects of the emergence of aldosterone secretion and salt appetite. *Physiol. Rev.* **45**: 245–295.

DENTON, D.A. 1973. Sodium and hypertension. In: *Mechanisms of Hypertension*, pp. 46–57. SANBHI', M.P. (Ed.). Excerpta Medica, Amsterdam.

DENTON, D.A. 1982. *The Hunger for Salt*. Springer, New York.

DENTON, D.A. and SABINE, J.R. 1961. The selective appetite for Na^+ shown by Na^+-deficient sheep. *J. Physiol.* **157**: 97–116.

DENTON, D.A., GODING, J.R. and WRIGHT, R.D. 1959. Control of adrenal secretion of electrolyte-active steroids. *Br. Med. J.* **ii**, 447–456, 522–530.

DE SANTO, N., TREVISAN, M., CAPASSO, G., GIORDANO, D.R., LATTE, M. and KROGH, V. 1988. Blood pressure and hypertension in childhood: epidemiology, diagnosis and treatment. *Kidney Int.* **34**: Suppl. 25., S115–S118.

DE SIATI, L., BALDONCINI, R., COASSIN, S., DE ANGELIS, C., FERRI, C., SANTUCCI, A., PERRONE, A. and BALSANO, F. 1993. Renal sodium excretory function during acute oxygen administration. *Respiration* **60**: 338–342.

DE SNOOK, H. 1937. Das trinkende kind in uterus. *Monatsch. Geburtch Gynak.* **105**: 88–92.

DETHIER, V.G. 1980. Biological and behavioural aspects of salt intake: a summation. In: *Biological and Behavioural Aspects of Salt Intake*, pp. 411–417. KARE, M.R., FREGLY, M.J. and BERNARD', R.A. (Eds). Academic Press, New York.

DEVLIN, T.J. and ROBERTS, W.K. 1963. Dietary maintenance requirement of sodium for lambs. *J. Anim. Sci.* **22**: 648–653.

DE WARDENER, H.E. 1978. The control of sodium excretion. *Am. J. Physiol.* **235**: F163–F173.

DE WARDENER, H.E. 1985. *The Kidney*, 5th Edn, Chaps 7, 12 and 16. Churchill Livingstone, Edinburgh.

DE WARDENER, H.E. 1988. Natriuretic factors other than atrial natriuretic peptide. In: *Nephrology*, Vol. 1, pp. 137–144. DAVISON, A.M. (Ed.). Bailliere Tindall, London.

DE WARDENER, H.E. 1990. The primary role of the kidney and salt intake in the aetiology of essential hypertension. *Clin. Sci.* **79**: 193–200; 289–297.

DE WARDENER, H.E. and CLARKSON, E.M. 1985. The concept of a natriuretic hormone. *Physiol. Rev.* **65**: 659–759.

DE WARDENER, H.E. and KAPLAN, N.M. 1993. On the assertion that a moderate restriction of sodium intake may have adverse health effects. *Am. J. Hyperten.* **6**: 810–814.

DE WARDENER, H.E. and MacGREGOR, G.A. 1980. Dahl's hypothesis that a saliuretic substance may be responsible for a sustained rise in arterial pressure: its possible role in essential hypertension. *Kidney Int.* **18**: 1–9.

DE WARDENER, H.E. and MacGREGOR, G.A. 1982. The natriuretic hormone and essential hypertension. *Lancet* **1**: 1450–1454.

DI BARTOLA, S.P. and DE MORAIS, H.S. 1992. Disorders of potassium: hypokalaemia and hyperkalaemia. In: *Fluid Therapy in Small Animal Practice*, Chap. 4, pp. 89–114. DI BARTOLA, S. (Ed.). Saunders, Philadelphia.

DI BONA, G.F. 1989. Neural control of renal tubular solute and water transport. *Miner. Electrolyte Metab.* **15**: 44–50.

DI BONA, G.F. 1990. Role of renal nerves in volume homeostasis. *Acta Physiol. Scand.* **138**: Suppl. 591, 18–27.

DI BONA, G.F. 1992. Sympathetic neural control of the kidney in hypertension. *Hypertension* **19**: Suppl. 1, I28–I35.

DIETL, P., GOOD, D. and STANTON, B. 1990. Adrenal corticosteroid action on the thick ascending limb. *Sem. Nephrol.* **10**: 350–364.

DILLON, M.J. 1991. Blood pressure measurement in childhood. In: *Handbook of Hypertension*, Vol. 14, *Blood Pressure Measurement*, Chap. 6, pp. 126–139. O'BRIEN, E. and O'MALLEY, K. (Eds). Elsevier, Amsterdam.

DI NICOLANTONIO, R., SPARGOS, S. and MORGAN, T.O. 1987. Prenatal high salt diet increases blood pressure and salt retention in the spontaneously hypertensive rat. *Clin. Expl Pharmacol. Physiol.* **14**: 233–235.

DOETSCH, G.S., GANCHROW, J.J., NELSON, L.M. and ERICKSON, R.P. 1969. Information processing in the taste system of the rat. In: *Olfaction and Taste*, pp. 492–511. PFAFFMANN, C. (Ed.). Rockefeller University Press.

DONADIO, J.V., BERGSTRAHL, E.J., OFFORD, K.P., SPENCER, D.C. and HOLLEY, K.E. 1994. A controlled trial of fish oil in IgA nephropathy. *New Engl. J. Med.* **331**: 1194–1199.

DORHOUT MEES, E.J. and KOOMANS, H.A. 1990. Pathogenesis of edema in the nephrotic syndrome. In: *The Regulation of Sodium and Chloride Balance*, pp. 321–352. SELDIN, D.W. and GIEBISCH, G. (Eds). Raven Press, New York.

DORIS, P.A. 1994. Regulation of Na, K-ATPase by endogenous ouabain-like materials. *Proc. Soc. Expl Biol. Med.* **205**: 202–212.

DORUP, I., SKAJAA, K. and THYBO, N.K. 1993. Oral magnesium supplementation restores the concentrations of magnesium, potassium and sodium–potassium pumps in skeletal muscle of patients receiving diuretic treatment. *J. Intern. Med.* **233**: 117–123.

DOUCET, A. 1988. Function and control of Na–K ATPase in single nephron segments of the mammalian kidney. *Kidney Int.* **34**: 749–760.

DOUGLAS, J.F. 1985. Acute renal failure. In: *Postgraduate Nephrology*, Chap. 7. MARSH, F. (Ed.). Heinemann, London.

DOW, S.W. and FETTMAN, K.J. 1992. Chronic renal disease and potassium depletion in cats. *Sem. Vet. Med. Surg. Small Anim.* **7**: 198–201.

DRUCKER, W.R., CHADWICK, C.D.J. and GANN, D.A. 1981. Transcapillary refill in haemorrhage and shock. *Arch. Surg.* **116**: 1344–1351.

DUHM, J. 1982. Note on the interaction of lithium ions with the transport function of the Na^+–K^+ pump. In: *Basic Mechanisms of the Action of Lithium*, pp. 21–27. EMRICH, H.M., ALDENHOFF, J.B. and LUX, H.D. (Eds). Excerpta Medica, Amsterdam.

DUMAS, P., TREMBLAY, J. and HAMET, P. 1994. Stress modulation by electrolytes in salt-sensitive spontaneously hypertensive rats. *Am. J. Med. Sci.* **307**: Suppl. 1, S130–S137.

DUNN, J.F., BANNISTER, N., KEMP, G.J. and PUBLICOVER, S.J. 1993. Sodium is elevated in mdx muscles: ionic interactions with dystrophic cells. *J. Neurol. Sci.* **114**: 76–80.

DUSTAN, H.P. and KIRK, K.A. 1989. The case for or against salt in hypertension. *Hypertension* **13**: 697–705.

DYCKNER, T. and WEBSTER P.O. 1987. Potassium/magnesium depletion in patients with cardiovascular disease. *Am. J. Med.* **82**: 11–17.

DYER, A.R., ELLIOTT, P. and SHIPLEY, M. 1994. Urinary electrolyte excretion in 24 hours and blood pressure in the INTERSALT study: II. Estimates of electrolyte–blood pressure associations corrected for regression dilution bias. *Am. J. Epidem.* **139**: 940–951.

DYSON, D.H. and PASCOE, P.J. 1990. Influence of pre-induction methoxamine, lactate Ringer solution, or hypertonic saline solution infusion or postinduction dobutamine infusion on anesthetic-induced hypotension in horses. *Am. J. Vet. Res.* **51**: 17–21.

EATON, S.B. and KONNOR, M. 1985. Paleolithic nutrition. *New Engl. J. Med.* **312**: 283–289.

EDMUNDS, M.E. and RUSSELL, G.I. 1994. Hypertension in renal failure. In: *Textbook of Hypertension*, pp. 798–810. SWALES, J.D. (Ed.). Blackwell, Oxford.

EDWARDS, D.G., KAYE, A.E. and DRUCE, E. 1989. Sources and intakes of sodium in the United Kingdom diet. *Eur. J. Clin. Nutr.* **43**: 855–861.

EKNOYAN, G. 1990. Diagnosis of disturbances. In: *The Regulation of Sodium and Chloride Balance*, Chap. 7, pp. 237–260. SELDIN, D.W. and GIEBISCH, G. (Eds). Raven Press, New York.

EKNOYAN, G. and SUKI, W.N. 1991. Renal consequences of antihypertensive therapy. *Sem. Nephrol.* **11**: 129–137.

EL-DAHR, S.S. and CHEVALIER, R.L. 1990. Special needs of the newborn infant in fluid therapy. *Pediatr. Clin. N. Am.* **37**: 323–336.

ELLIOTT, E.J., ARMITSTEAD, J.C., FARTHING, M.J. and WALKER-SMITH, J.A. 1988. Oral rehydration without bicarbonate for prevention and treatment of dehydration: a double-blind controlled trial. *Aliment. Pharmacol. Ther.* **2**: 253–262.

ELLIOTT, H.L. 1994. Interactions between antihypertensive drugs and other medications. In: *Textbook of Hypertension*, pp. 1150–1155. SWALES, J.D. (Ed.). Blackwell, Oxford.

ELLIOTT, P. 1991. Observational studies of salt and blood pressure. *Hypertension* **17**: Suppl. 1, S13–S18.

ELMER, P.J., GRIMM, R.H., FLACK, J. and LAING, B. 1991. Dietary sodium reduction for hypertension prevention and treatment. *Hypertension* **17**: Suppl. I, I182–I189.

ELMSTAHL, S. and WINGE, L. 1993. Increased sweat sodium concentration in patients with Alzheimer's disease. *Dementia* **4**: 50–53.

ELY, D.L., FOLKOW, B. and PARADISE, N. 1990. Risks associated with dietary sodium restriction in spontaneous hypertensive rat model of hypertension. *Am. J. Hyperten.* **3**: 650–660.

EPSTEIN, A.N. 1986. Hormonal synergy on the cause of salt appetite. In: *The Physiology of Thirst and Sodium Appetite*, pp. 395–403. DE CARO, G., EPSTEIN, A.N. and MASSI, M. (Ed.). Plenum, New York.

EPSTEIN, M. 1983. Renal sodium handling in cirrhosis. *Sem. Nephrol.* **3**: 225–240.

EPSTEIN, M. and PEREZ, G.O. 1994. Pathophysiology of the edema-forming states. In: *Maxwell and Kleeman's Clinical Disorders of Fluid and Electrolyte Metabolism*, 5th Edn, Chap. 18, pp. 523–544. NARINS, R.G. (Ed.). McGraw-Hill, New York.

ERNST, N.D. 1991. Health promotion roles of the Federal Government and food industry in nutrition and blood pressure. *Hypertension* **17**: Suppl. I, I196–I200.

ESKEW, G.L. 1948. *Salt, The Fifth Element*. Ferguson, Chicago, IL.

ESPINER, E.A. 1994. Physiology of natriuretic peptides. *J. Intern. Med.* **235**: 527–541.

ESPINER, E.A. and NICHOLLS, M.G. 1993. Renin and the control of aldosterone. In: *The Renin Angiotensin System*, Vol. 1, Chap. 24, pp. 1–24. ROBERTSON, J.I.S. and NICHOLLS, M.E. (Eds). Gower Medical Publishing, London.

EVANS, D.L., ROSE, R.J., KNIGHT, P.K., CLUER, D. and MANEFIELD, G.W. 1994. Some physiological responses to incremental treadmill exercise in the racing camel. *Acta Physiol. Scand.* **150:** Suppl. 617, 32–39.

FABER, M.D., KUPIN, W.L., HEIKG, C.W. and NARINS, R.G. 1994. Common fluid–electrolyte and acid–base problems in the intensive care unit. *Sem. Nephrol.* **14:** 8–23.

FAGERBERG, B., ANDERSSON, O.K., PERSSON, B., HEDNER, T., HEDNER, J. and TOWLE, A. 1985. Fluid homeostasis and haemodynamics during sodium restriction in hypertensive men. *J. Hyperten.* **3:** Suppl. 3, S327–S329.

FAGIN, J.A., IKEJIRI, K. and LEVIN, S.R. 1987. Insulinotropic effects of vanadate. *Diabetes* **36:** 1448–1452.

FALK, J.L. 1965. Limitations to the specificity of NaCl appetite in sodium depleted rats. *J. Comp. Physiol. Psychol.* **60:** 393–396.

FAN, T.-H. M., FRANTZ, R.P., ELAM, H., SAKAMOTO, S., IMAI, N. and CHANG-SENG, L. 1993. Reduction of myocardial Na–K ATPase activity and ouabain binding sites in heart failure: prevention by nadolol. *Am. J. Physiol.* **265:** H2086–H2093.

FANESTIL, D.D. 1988. Mechanism of action of aldosterone blockers. *Sem. Nephrol.* **8:** 249–263.

FANESTIL, D.D. 1994. Compartmentation of body water. In: *Maxwell and Kleeman's Clinical Disorders of Fluid and Electrolyte Metabolism*, 5th Edn, Chap. 3, pp. 3–20. NARINS, R.G. (Ed.). McGraw-Hill, New York.

FEDDE, M.R. and WOOD, S.C. 1993. Rheological characteristics of horse blood: significance during exercise. *Respir. Physiol.* **94:** 323–335.

FELD, L.G., SPRINGATE, J.E. and IZZO, J.L. 1990. Special considerations in hypertension. In: *Kidney Electrolyte Disorders*, pp. 565–599. CHAN, J.C.M. and GILL, J.R. (Eds.). Churchill Livingstone, New York.

FELDER, R.A., ROBILLARD, J., EISNER, G.M. and JOSE, P.A. 1989. Role of endogenous dopamine on renal sodium excretion. *Sem. Nephrol.* **9:** 91–93.

FERRANNINI, E. and NATALI, A. 1994. Hypertension, insulin resistance and diabetes. In: *Textbook of Hypertension*, pp. 785–797. SWALES, J.D. (Ed.). Blackwell, Oxford.

FIELD, A.C., SUTTLE, N.F. and GUNN, R.G. 1968. Seasonal changes in the composition and mineral content of the body of hill ewes. *J. Agric. Sci.* **71:** 303–310.

FIELD, K.J. and GIEBISCH, G. 1989. Mechanisms of segmental potassium reabsorption and secretion. In: *The Regulation of Potassium Balance*, pp. 139–156. SELDIN, D.W. and GIEBISCH, G. (Eds). Raven Press, New York.

FIELD, M. 1993. Intestinal electrolyte secretion. *Arch. Surg.* **128:** 273–278.

FIELD, M., RAO, M.C. and CHANG, E.B. 1989. Intestinal electrolyte transport and diarrhoeal disease. *New Engl. J. Med.* **321:** 800–806; 879–883.

FIELDMAN, N.R. 1988. The role of arginine vasopressin in the regulation of urinary output in the postoperative period. Master of Surgery Thesis, University of Cambridge.

FIELDMAN, N.R., FORSLING, M.L. and LE QUESNE, L.P. 1985. The effect of vasopressin on solute and water excretion during and after surgical operations. *Ann. Surg.* **201:** 383–390.

FINE, B.P., TY, A., LESTRANGE, N. and LEVINE, O.R. 1987. Sodium deprivation growth failure in the rat: alteration in tissue composition and fluid space. *J. Nutr.* **117:** 1623–1628.

FISCHER, G.M., COX, R.H. and DETWEILER, D.K. 1975. Altered arterial connective tissue in racing greyhound dogs. *Experientia* **31:** 1426–1427.

FISCHER, P.W.F. and GIROUX, A. 1987. Effects of dietary magnesium on sodium–potassium pump action in the heart of rats. *J. Nutr.* **117**: 2091–2095.

FISH, E.M. and MOLITORIS, B.A. 1994. Alterations in epithelial polarity and the pathogenesis of disease states. *New Engl. J. Med.* **330**: 1580–1588.

FITZSIMONS, J.T. 1979. *The Physiology of Thirst and Sodium Appetite*. Cambridge University Press, Cambridge.

FITZSIMONS, J.T. 1986. Endogenous angiotensin and sodium appetite. In: *The Physiology of Thirst and Sodium Appetite*, pp. 383–393. De CARO, G., EPSTEIN, A.N. and MASSI (Eds). Plenum Press, New York.

FITZSIMONS, J.T. 1993. Physiology and pathophysiology of thirst and sodium appetite. In: *Clinical Disturbances of Water Metabolism*, Chap. 5, pp. 65–97. SELDIN, D.W. and GIEBISCH, G. (Eds). Raven Press, New York City.

FITZSIMONS, J.T. and STRICKER, E.M. 1971. Sodium appetite and the renin–angiotensin system. *Nature* **231**: 58–60.

FLEAR, C.T.G., BHATTACHARYA, S.S. and SINGH, L.M. 1980. Solute and water exchanges between cells and extracellular fluids in health and disturbances after trauma. *J. Parent. Ent. Nutr.* **4**: 98–120.

FLYNN, F.W., SCHULKIN, J. and HAVENS, W. 1993. Sex differences in salt preference and taste reactivity in rats. *Brain Res. Bull.* **32**: 91–95.

FRANCIS, G.S. 1994. Vasodilators and inotropic agents in the treatment of congestive heart failure. *Sem. Nephrol.* **14**: 464–478.

FRANKLIN, S.S. and KLEIN, K.L. 1994. Acute renal failure: fluid and electrolyte and acid–base complications. In: *Maxwell and Kleeman's Clinical Disorders of Fluid and Electrolyte Metabolism*, 5th Edn, Chap. 38, pp. 1175–1194. NARINS, R.G. (Ed.). McGraw-Hill, New York.

FRANKMANN, S.P. and SMITH, G.P. 1993. Hepatic vagotomy does not disrupt the normal satiation of NaCl appetite. *Physiol. Behav.* **53**: 337–341.

FRANKMANN, S.P., DORSA, D.M., SAKAI, R.R. and SIMPSON, J.B. 1986. A single experience with hyperoncotic colloid dialysis persistently alters water and sodium intake. In: *The Physiology of Thirst and Sodium Appetite*, pp. 115–121. De CARO, G., EPSTEIN, A.N. and MASSI, M. (Eds). Plenum Press, New York.

FRANKMANN, S.P., ULRICH, P. and EPSTEIN, A.N. 1991. Transient and lasting effects of reproductive episodes on NaCl intake of the female rat. *Appetite* **16**: 193–204.

FRASER, C.L. and ARIEFF, A.I. 1994. Metabolic encephalopathy associated with water, electrolyte and acid–base disorders. In: *Maxwell and Kleeman's Clinical Disorders of Fluid and Electrolyte Metabolism* 5th Edn, Chap. 46, pp. 1491–1548. NARINS, R.G. (Ed.). McGraw-Hill, New York.

FRATTOLA, A., PARATI, G., CUSPIDI, C., ALBIBI, F. and MANCIA, G. 1993. Prognostic value of 24–hour blood pressure variability. *J. Hyperten.* **11**: 1133–1137.

FREGLY, M.J. 1959. Specificity of sodium chloride aversion of hypertensive rats. *Am. J. Physiol.* **196**: 1326–1330.

FREGLY, M.J. 1980. Salt and social behaviour. In: *Biological and Behavioural Aspects of Salt Intake*, pp. 3–11. KARE, K.R., FREGLY, M.J. and BERNARD, R.A. (Eds). Academic Press, New York.

FREGLY, M.J. and WALTERS, I.W. 1966. Action of mineralocorticoid on spontaneous salt appetite of adrenalectomised rats. *Physiol. Behav.* **1**: 65–74.

FREIS, E.D. 1981. Sodium deprivation as an approach to hypertension. In: *Frontiers in Hypertension Research*, pp. 43–45. LARAGH, J.H., BUHLER, F.R. and SELDIN, D.W. (Eds) Springer, New York.

FREIS, E.D. 1987. Diuretic–induced hypokalaemia. *Postgrad. Med.* **81**: 123–129.

FREIS, E.D. 1990. The cardiotoxicity of thiazide diuretics: review of the evidence. *J. Hyperten.* **8**: Suppl. 2, S22–S32.

FRENCH, M.H. 1945. Geophagia in animals. *East African Med. J.* **22**: 103–110, 152–161.

FRIED, T.A. and KUNAU, R.T. 1986. Thiazide diuretics. In: *Diuretics: Physiology, Pharmacology and Clinical Use*, Chap. 4. DIRKS, J.H. and SUTTON, R.A.L. (Eds), pp. 67–86. Saunders, Philadelphia.

FRIEDMAN, P.A. 1988. Biochemistry and pharmacology of diuretics. *Sem. Nephrol.* **8**: 198–212.

FROST, C.D., LAW, M.R. and WALD, N.J. 1991. By how much does dietary salt reduction lower blood pressure? II. Analysis of observational data within populations. *Br. Med. J.* **302**: 815–818.

FUNDER, J.W. 1990. Corticosteroid receptors and renal 11β-droxysteroid dehydrogenase activity. *Sem. Nephrol.* **10**: 311–319.

GABEL, J.C. and DRAKE, R.E. 1979. Pulmonary capillary pressure and permeability. *Crit. Care Med.* **7**: 92–98.

GALBRAITH, J.K. 1967. *The New Industrial State.* Penguin Books, Harmondsworth.

GALDZICKI, Z., COAN, E. and RAPOPORT, S.I. 1993. Cultured hippocampal neurons from trisomy 16 mouse, a model for Down's syndrome, have an abnormal action potential due to a reduced inward sodium current. *Brain Res.* **604**: 69–79.

GALEF B.G. 1986. Social interaction modifies learned aversions, sodium appetite and both palatability and handling-time induced dietary preference in rats. *J. Comp. Physiol. Psychol.* **100**: 432–439.

GALLAGHER, D.E. and O'ROURKE, M.F. 1993. What is arterial pressure? In: *Arterial Vasodilation: Mechanisms and Therapy*, pp. 134–148. O'ROURKE, M.F., SAFAR, M.E. and DZAU, V.J. (Eds). Edward Arnold, London.

GALLERY, E.D.M. 1984. Volume homeostasis in normal and hypertensive human pregnancy. *Sem. Nephrol.* **4**: 221–231.

GAMBLE, J.L. 1953. Early history of fluid replacement therapy. *Pediatrics* **11**: 554–567.

GAMBLE, J.L. 1954. *Chemical Anatomy, Physiology and Pathology of Extracellular Fluid.* Harvard University Press, Cambridge, MA.

GANGULI, M.C., SMITH, J.D. and HANSON, L.E. 1970. Sodium metabolism and its requirement during reproduction in female rats. *J. Nutr.* **99**: 225–234.

GANN, D.S. and KENNEY, P.R. 1990. Special problems of fluid and electrolyte management in surgery. In: *Kidney Electrolyte Disorders*, pp. 343–361. CHAN, J.C.M. and GILL, J.R. (Eds). Churchill Livingstone, New York City.

GANTEN, D., LINDPAWTER, K., UNGER, Th. and MULLINS, J. 1989. Application of the new biology to hypertension research: the renin–angiotensin paradigm. In: *Cardiac and Renal Failure: An Expanding Role for ACE-Inhibitors*, pp. 59–80. DOLLERY, C.T. and SHERWOOD, L.T. (Eds). Hanley and Belfus, Philadelphia.

GANZ, M.B. 1991. Regulation of intracellular pH in glomerular mesangial cells. *Sem. Nephrol.* **11**: 16–27.

GARAY, R., SENN, N. and OLLIVIER, J.-P. 1994. Erythrocyte ion transport as an indicator of sensitivity to anti–hypertensive drugs. *Am. J. Med. Sci.* **307**: Suppl. 1, S120–S125.

GARCIA, N.H. and GARVIN, J.L. 1994. Endothelin's biphasic effect on fluid absorption in the proximal straight tubule and its inhibitory cascade. *J. Clin. Invest.* **93**: 2572–2577.

GARDEMANN, A., WATANABE, Y., GROSSE, V., HESSE, S. and JUNGERMANN, K. 1992. Increases in intestinal glucose absorption and hepatic glucose uptake elicited by luminal but not vascular glutamine in the jointly perfused small intestine and liver of the rat. *Biochem. J.* **283:** 759–765.

GASKELL, C.J. 1985. Nutrition in diseases of the urinary tract in the dog and cat. *Vet. Ann.* **25:** 383–391.

GEETH, H. and SKELTY, K.T. 1987. Down's syndrome and cystic fibrosis in an Indian population. *Br. Med. J.* **294:** 156.

GEHEB, M.A., KRUSE, J.A., HAUPT, M.T., DESAI, T.K. and CARLSON, R.W. 1994. Fluid and electrolyte abnormalities in critically ill patients: fluid resuscitation, lactate metabolism and calcium metabolism. In: *Maxwell and Kleeman's Clinical Disorders of Fluid and Electrolyte Metabolism*, 5th Edn, Chap. 45, pp. 1463–1490. NARINS, R.G. (Ed.). McGraw-Hill, New York.

GELEIJNSE, J.M., WITTEMAN, J.C.M., BAK, A.A.A., DEN BREEIJEN, J.H. and GROBBE, D.E. 1994. Reduction in blood pressure with a low sodium, high potassium, high magnesium salt in older subjects with mild to moderate hypertension. *Br. Med. J.* **309:** 436–440.

GENEST, J. 1986. The atrial natriuretic factor. *Br. Heart J.* **56:** 302–306.

GENTILINI, P. and LAFFI, G. 1992. Pathophysiology and treatment of ascites and the hepatorenal syndrome. *Bailliere's Clin. Gastroenterol.* **6.3:,** 581–607.

GERBER, A.L., ARENDT, R.M. and PAUMGARTNER, G. 1987. Atrial natriuretic factor: possible implications in liver disease. *J. Hepatol.* **5:** 123–132.

GERZER, R. and DRUMMER, C. 1993. Is the renal natriuretic peptide urodilatin involved in the regulation of natriuresis? *J. Cardiovasc. Pharmacol.* **22:** Suppl. 2., 86–87.

GHIONE, S., BALZAN, S., DECOLLOGNE, S., PACI, A., PIERACCINI, L. and MONTALI, U. 1993. Endogenous digitalis-like activity in the newborn. *J. Cardiovasc. Pharmacol.* **22:** Suppl. 22, S25–S28.

GIBBONS, G.H. and DZAU, V.J. 1994. The emerging concept of vascular remodeling. *New Engl. J. Med.* **330:** 1431–1438.

GIBSON, K.J. and LUMBERS, E.R. 1992. Mechanisms by which the pregnant ewe can sustain increased salt and water supply to the fetus. *J. Physiol.* **445:** 569–579.

GICK, G.G., ISMAIL-BEIGI, F. and EDELMAN, I.S. 1988. Hormonal regulation of Na–K ATPase. In: *The Na$^+$ K$^+$ Pump: Part B; Cellular Aspects*, pp. 227–295. SKOU, J.C., NORBY, J.G., MAUNSBACH, A.B. and ESMANN, M. (Eds). Liss, New York.

GIEBISCH, G. and ARONSON, P.S. 1986. The proximal nephron. In: *Physiology of Membrane Disorders*, Chap. 37, pp. 669–700. ANDREOLI, T.E., HOFFMAN, J.F., FANESTIL, D.D. and SCHULTZ, S.G. (Eds). Plenum Press, New York.

GIEBISCH, G. and KLEIN-ROBBENHAAR, G. 1993. Recent studies on the characterization of loop diuretics. *J. Cardiovasc. Pharmacol.* **22:** Suppl. 3, S1–S10.

GIFFORD, R.W. 1986. Role of diuretics in treatment of essential hypertension. *Am. J. Cardiol.* **58:** 15A–17A.

GILL, J.R., GULLNER, H.G., LAKE, C.R. and LAKATUA, D.J. 1988. Plasma and urinary catecholamines in salt–sensitive idiopathic hypertension. *Hypertension* **11:** 312–319.

GINES, P., ABRAHAM, W.T. and SCHRIER, R.W. 1994. Vasopressin in pathophysiological states. *Sem. Nephrol.* **14:** 384–397.

GLEADHILL, A., MICHELL, A.R., MARLEN, D. and HARRIS, P.A. 1995. Effects of exercise on equine GFR.

GLYN, I.M. 1993. All hands to the sodium pump. *J. Physiol.* **462:** 1–30.

GOETZ, K.L. 1990. The tenuous relationship between atriopeptin and sodium excretion. *Acta Physiol. Scand.* **139:** Suppl. 591, 88–96.

GOETZ, K.L. 1993. Is urodilatin rather than atriopeptin the primary natriuretic peptide of the ANP family? *J. Cardiovasc. Pharmacol.* **22**: Suppl. 2, 84–85.

GOLDBERG, C.A. and SCHRIER, R.W. 1991. Hypertension in pregnancy. *Sem. Nephrol.* **11**: 576–593.

GOLDBLATT, H.E., LYNCH, J., HANZAL, F. and SUMMERVILLE, W.W. 1934. The production of persistent elevation of systolic blood pressure by means of renal ischemia. *J. Expl Med.* **59**: 347–378.

GOLIN, R. GENOVESI, S. STELLA, A. and ZANCHETTI, A. 1987. Afferent pathways of renorenal reflexes controlling sodium and water excretion in the cat. *J. Hyperten.* **5**: 417–424.

GONZALEZ-CAMPOY, J., ROMERO, J.C. and KNOX, F.G. 1989. Escape from the sodium-retaining effects of mineralocorticoids: role of ANF and intrarenal hormone systems. *Kidney Int.* **35**: 767–777.

GOODWIN, F.J. 1985. Proteinuria and nephrotic syndrome. In: *Postgraduate Nephrology.* MARSH, E. (Ed.). Heinemann, London.

GORDON, J.A. and ANDERSON, J.A. 1981. Hepatorenal syndrome. *Sem. Nephrol.* **1**: 37–42.

GOTHBER, G., LUNDIN, S., AURELL, M. and FOLKOW, B. 1983. Response to slow, graded bleeding in salt-depleted rats. *J. Hyperten.* **1**: 24–26.

GOTO, A., YAMADA, K., ISHII, M. and SUGIMOTO, T. 1990. Digitalis-like activity in human plasma: relation to blood pressure and sodium balance. *Am. J. Med.* **89**: 420–426.

GOULDING, A., GOLD, E. and CAMPBELL, A.J. 1993. Salty foods increase calcium requirements and are a potential risk factor for osteoporosis. *IDF News Nutr. Newslett.* **2.138**: 16–19.

GOW, C.B. and PHILLIPS, P.A. 1994. Epidermal growth factor as a diuretic in sheep. *J. Physiol.* **477**: 27–33.

GRACE, N.D. 1988. Effect of varying sodium and potassium intakes on sodium, potassium and magnesium contents of the ruminoreticulum and apparent absorption of magnesium in non–lactating dairy cattle. *N.Z. J. Agric. Res.* **31**: 401–407.

GRAHAM, A., HASANI, A., ALTON, E.W., MARTIN, G.P., MARRIOTT, C., HODSON, M.E., CLARKE, S.W. and GEDDES, D.M. 1993. No added benefit from nebulized amiloride in patients with cystic fibrosis. *Eur. Respir. J.* **6**: 1243–1248.

GRANGER, J.P. 1992. Pressure natriuresis: role of renal interstitial hydrostatic pressure. *Hypertension* **19**: Suppl. 1, I9–I17.

GRANGER, J.P., WEST, D. and SCOTT, J. 1994. Abnormal pressure natriuresis in the dog model of obesity-induced hypertension. *Hypertension* **23**: I8–I11.

GRAVES, S.W. and WILLIAMS, G.H. 1987. Endogenous digitalis–like natriuretic factor. *A. Rev. Med.* **38**: 433–444.

GRAY, T.C., NUNN, J.F. and UTTING, J.E. 1989. *General Anaesthesia*, pp. 234–250, 1214. Butterworth, London.

GRECO, B.A. and JACOBSON, H.R. 1990. Fluid and electrolyte problems in surgery, trauma and burns. In: *Fluids and Electrolytes*, 2nd Edn, Chap. 21, pp. 990–1023. KOKKO, J.P. and TANNEN, R.L. (Eds). Saunders, Philadelphia.

GREEN, J. 1994. The physicochemical structure of bone: cellular and noncellular elements. *Miner. Electrolyte Metab.* **20**: 7–15.

GREENOUGH, W.B. and KHIN-MAUNG, U. 1991. Cereal-based oral rehydration therapy. Strategic issues for its implementation in national diarrheal disease control programs. *J. Pediatr.* **118**: S80–S85.

GRIFFITHS, G. and WALTERS, R.K.J. 1966. The sodium and potassium content of some grass genera species and varieties. *J. Agric. Sci.* **67**: 81–89.

GRIFFITHS, J.K. and GORBACH, S.L. 1993. Other bacterial diarrhoeas. *Bailliere's Clin. Gastroenterol.* **7**: 263–305.

GRIM, C.E. and SCOGGINS, B.A. 1986. The rapid adjustment of renal sodium excretion to changes in sodium intake in sheep. *Life Sci.* **39**: 215–222.

GROBBEE, D.E. 1994. Electrolytes and hypertension: results from recent studies. *Am. J. Med. Sci.* **307**: Suppl. 1, S17–S20.

GROSSMAN, S.P. 1990. *Thirst and Sodium Appetite: Physiological Basis.* Academic Press, San Diego.

GROUTIDES, C. and MICHELL, A.R. 1990. Changes in plasma composition in calves surviving or dying from diarrhoea. *Br. Vet. J.* **146**: 205–210.

GRUBB, B.R., VICK, R.N. and BOUCHER, R.C. 1994. Hyperabsorption of Na^+ and raised Ca^{2+}–mediated Cl^- secretion in nasal epithelia of CF mice. *Am. J. Physiol.* **266**: C1478–C1483.

GRUNERT, R.R., MEYER, J.H. and PHILLIPS, P.H. 1950. The sodium and potassium requirements of the rat for growth. *J. Nutr.* **42**: 609–618.

GUARENA, C., BOERO, R., QUARELLO, F., BERTO, I., MIRACA, R., ROUX, V., IADAROLA, G. and PICCOLI, G. 1993. Abnormal erythrocyte sodium transport in patients with adult polycystic kidney and hypertension. *Arch. Mal. Coeur-Vaiss.* **86**: 1241–1243.

GULLANS, S.R. and VERBALIS, J.G. 1993. Control of brain volume during hyperosmolar and hypoosmolar conditions. *A. Rev. Med.* **44**: 289–301.

GUSTAFSSON, H. 1993. Vasomotion and underlying mechanisms in small arteries. An *in vitro* study of rat blood vessels. *Acta Physiol. Scand.* **614**: Suppl. 1–44.

GUYTON, A.C. 1992a. Pulmonary circulation, pulmonary edema, pleural fluid. In: *Textbook of Medical Physiology*, 8th Edn, Chap. 38, pp. 414–421. Saunders, Philadelphia.

GUYTON, A.C. 1992b. Kidney and fluids in pressure regulation—small volume but large pressure changes. *Hypertension* **19**: Suppl. 1, 12–18.

HAAS, J.A. and GLICK, S. 1978. Radioimmunoassayable plasma vasopressin associated with surgery. *Arch. Surg.* **113**: 597–600.

HAAS, J.A. and KNOX, F.G. 1990. Mechanisms for escape from the salt–retaining effects of mineralocorticoids: role of deep nephrons. *Sem. Nephrol.* **10**: 380–387.

HACKENTHAL, E. and NOBILING, R. 1994. Renin secretion and its regulation. In: *Textbook of Hypertension*, Chap.11, pp. 232–244. SWALES, J.D. (Ed.). Blackwell Scientific, Oxford.

HADDY, F.J. 1980. Mechanisms, prevention and therapy of sodium-dependent hypertension. *Am. J. Med.* **69**: 334–344.

HADDY, F.J. 1988. Ionic control of vascular smooth muscle cells. *Kidney Int.* **34**: Suppl. 25, S2–S8.

HADDY, F., PAMMANI, M. and CLOUGH, D. 1978. The sodium–potassium pump in volume-expanded hypertension. *Clin. Expl Hyperten.* **1**: 295–336.

HALES, S. 1733. *Statistical Essays. Vol. II: Haemostatics*, pp. 31–47. Experiments VII and VIII. Royal Society, London.

HALL, J.E. 1986. Control of sodium excretion by angiotensin. II. Intrarenal mechanisms and blood pressure regulation. *Am. J. Physiol.* **250**: R960–R972.

HALL, J.E. and GRANGER, J.P. 1994. Role of sodium and fluid excretion in hypertension. In: *Textbook of Hypertension*, pp. 360–387. SWALES, J.D. (Ed.). Blackwell Scientific, Oxford.

HALL, J.E., GUYTON, A.C. and MIZELLE, H.L. 1990. Role of the renin–angiotensin system in control of sodium excretion and arterial pressure. *Acta Physiol. Scand.* **139**: Suppl. 591, 48–62.

HALL, J.E., BRANDS, M.W., HILDEBRANDT, D.A. and MIZELLE, H.L. 1992. Obesity-associated hypertension: hyperinsulinemia and renal mechanisms. *Hypertension* **19**: Suppl. 1, I45–I55.

HALM, D.R. and FRIZELL, R.A. 1991. Ion transport across the large intestine. In: *Handbook of Physiology*, Section 6, *Gastrointestinal Tract*, Vol. IV, Intestinal Secretion and Absorption, Chap. 8, pp. 257–273. FIELD, M. and FRIZELL, R.A. (Eds). American Physiological Society, Bethesda, Maryland.

HALPERI, M.L., CARLISLE, E.J.F., DONNELLY, S., KAMEL, K.S. and VASUVATTAKUL, S. 1994. Renal tubular acidosis. In: *Maxwell and Kleeman's Clinical Disorders of Fluid and Electrolyte Metabolism*, 5th Edn, Chap. 27, pp. 875–910. NARINS, R.G. (Eds). McGraw-Hill, New York.

HAMET, P., MONGEAU, E., LAMBERT, J., BELLAVANCE, F., DAIGNAULT-GELINAS, H., LEDOUX, M. and WHISSELL-CAMBIOTTI, L. 1991. Interactions among calcium, sodium and alcohol intake in determinants of blood pressure. *Hypertension* **17**: Suppl. I, 150–154.

HAMLIN, M.N., WEBB, R.C., LING, W.D. and BOHR, D.F. 1988. Parallel effects of DOCA on salt appetite, thirst and blood pressure in sheep. *Expl Biol. Med.* **188**: 46–51.

HAMLYN, J.M., RINGEL, R., SCHAEFFER, J., LEVINSON, P.D., HAMILTON, P.D., KOWARSKI, A.A. and BLAUSTEIN, M.P. 1982. A circulating inhibitor of (Na$^+$ + K$^+$) ATPase associated with essential hypertension. *Nature* **300**: 650–652.

HAMPTON, J. 1987. In: *Diuretics in Heart Failure*, pp. 11–14, 35. POOLE-WILSON, P.A. (Ed.). Royal Society of Medicine, London.

HANSSON, G.C., MU, J.Y. and LUNDGREN, O. 1993. An intestinal natriuretic factor. *J. Cardiovasc. Pharmacol.* **22**: Suppl. 2, 60–62.

HARGREAVES, M., COSTILL, D., BURKE, L., McCONNELL, G. and FEBBRAIO, M. 1994. Influence of sodium on glucose bioavailability during exercise. *Med. Sci. Sports Exerc.* **26**: 365–368.

HARLAN, W.R., OBERMAN, A. and MITCHELL, R.E. 1973. A 30-year study of blood pressure in a white male cohort. In: *Hypertension: Mechanisms and Management.* ONESTI, G., KIM, E.U. and MOYER, J.H. (Eds). Grune and Stratton, New York.

HARMSEN, E. and LEENEN, F.H. 1992. Dietary sodium induced cardiac hypertrophy. *Can. J. Physiol. Pharmacol.* **70**: 580–586.

HARRIMAN, A.E. and MACLEOD, R.B. 1953. Discriminative thresholds for salt of normal and adrenalectomised rats. *Am. J. Psychol.* **66**: 465–471.

HARRIS, D.J., ALLEN, D.J. and CAPLE, I.W. 1986. Effects of low sodium nutrition on fertility of dairy cows. *Proc. Nutr. Soc. Austral.* **11**: 92.

HARSHFIELD, G.A., PULLIAM, D.A., SOMES, G.W. and ALPERT, B.S. 1993. Ambulatory blood pressure patterns in youth. *Am. J. Hyperten.* **6**: 968–973.

HASSAN, M.O., AL-SHAFIE, O.T. and JOHNSTON, W.J. 1993. Loss of the nocturnal dip and increased variability of blood pressure in normotensive patients with noninsulin-dependent diabetes mellitus. *Clin. Physiol.* **13**: 519–523.

HATTON, D.C. and McCARRON, D.A. 1994. Dietary calcium and blood pressure in experimental models of hypertension. A review. *Hypertension* **23**: 513–530.

HAUPERT, G.T. 1988. Physiological inhibitors of Na-K ATPase: concept and status. In: *The Na$^+$ K$^+$ Pump: Part B, Cellular Aspects*, pp. 293–320. SKOU, J.C., NORBY, J.C., MAUNSBACH, A.B. and ESMANN, M. (Eds). Alan Liss, New York.

HAYCOCK, G.B. 1993. The influence of sodium on growth in infancy. *Pediatr. Nephrol.* **7**: 871–875.

HAYCRAFT, J.B. 1886. Taste. *Brain* **10**: 45–63.

HAYNES, W.G., NOON, J.P., WALKER, B.R. and WEBB, D.J. 1993. Inhibition of nitric oxide synthesis increases blood pressure in healthy humans. *J. Hyperten.* **11**: 1375–1380.

HEILMANN, L., VON TEMPELHOFF, G.F. and ULRICH, S. 1993. The Na/K co-transport system in erythrocytes from pregnant patients. *Arch. Gynecol. Obstet.* **253**: 167–174.

HELLER, J., HORACEK, V. and KAMARADOVA, S. 1988. Efferent arteriole: the main target for some renal vasoconstrictors. In: *Nephrology*, Vol. 1, pp. 75–86. DAVISON, A.M. (Ed.). Bailliere Tindall, London.

HENKIN, R.I. 1980. Salt taste and salt preference in normal and hypertensive rats and humans. In: *Biological and Behavioural Aspects of Salt Intake*, pp. 367–396. KARE, M.R., FREGLY, M.J. and BERNARD, R.A. (Eds). Academic Press, New York.

HENKIN, R.I. and SOLOMON, D.H. 1962. Salt taste threshold in adrenal insufficiency in man. *J. Clin. Invest.* **42**: 727–735.

HENKIN, R.I., GILL, J.R. and BARTTER, F.C. 1963. Studies on taste thresholds in normal man and in patients with adrenal cortical insufficiency. *J. Clin. Invest.* **42**: 727–735.

HENRIKSEN, J.H. and LARSEN, H.R. 1994. Hepatorenal disorders: role of the sympathetic nervous system. *Sem. Liver Dis.* **14**: 35–43.

HERIN, P. and APERIA, A. 1994. Neonatal kidney, fluids and electrolytes. *Current Opin. Pediatr.* **6**: 154–157.

HERMANN, G.E., KOHLERMAN, N.H. and ROGERS, R.C. 1983. Hepatic-vagal and gustatory interactions in the brainstem of the rat. *J. Auton. Nervous Syst.* **9**: 477–495.

HERTZ, L. 1977. Drug induced alteration of ion distribution at the cellular level of the central nervous system. *Pharmacol. Rev.* **29**: 33–65.

HILTON, P.J. 1986. Cellular sodium transport in essential hypertension. *New Engl. J. Med.* **314**: 222–226.

HIRANO, J. 1986. The spectrum of prolactin action in teleosts. In: *Comparative Endocrinology: Developments and Directions*, pp. 53–74. RALPH, C.L. (Ed.). Liss, New York.

HOFFMAN, J.F. 1986. Active transport of Na^+ and K^+ by red blood cells. In: *Physiology of Membrane Disorders*, 2nd Edn, Chap. 13, pp. 221–234. ANDREOLI, T.E., HOFFMAN, J.F., FANESTIL, D.D. and SCHULTZ, S.G. (Eds). Plenum Press, New York.

HOFMAN, A., HAZEBROEK, A. and VALKENBURG, H.A. 1983. A randomized trial of sodium intake and blood pressure in newborn infants. *J. Am. Med. Assoc.* **250**: 370–373.

HOLLENBERG, N.K. 1980. Set point for sodium homeostasis: surfeit, deficit and their implications. *Kidney Int.* **17**: 423–429.

HOLLENBERG, N.K. 1982. Surfeit, deficit and the set point for sodium homeostasis. *Kidney Int.* **21**: 884–885.

HOLLENBERG, N.K. 1983. The relationships between sodium intake and 'the state of sodium balance'. *Sem. Nephrol.* **3**: 171–179.

HOLLENBERG, N.K. 1987. Cardiovascular therapeutics in the 1980s. *Am. J. Med.* **82**: 1–3.

HOLLENBERG, N.K. and SCHULMAN, G. 1985. Renal perfusion and function in sodium–retaining states. In: *The Kidney: Physiology and Pathophysiology*, Vol.

1, pp. 1119–1162. SELDIN, D.W. and GIEBISCH, G. (Eds). Raven Press, New York.

HOLLENBERG, N.K., INGELFINGER, J.R. and DZAU, V.J. 1993. The renin–angiotensin system. In: *Maxwell and Kleeman's Clinical Disorders of Fluid and Electrolyte Metabolism*, 5th Edn, Chap. 16, pp. 477–492. NARINS, R.G. (Ed.). McGraw-Hill, New York.

HOLLIFIELD, J.W. 1987. Magnesium depletion, diuretics and arrhythmias. *Am. J. Med.* **82**: 30–37.

HOLLIFIELD, J.W. 1989. Thiazide treatment: effect on magnesium and ventricular ectopics. *Am. J. Cardiol.* **63**: 22G–25G.

HORROBIN, D.F., BURSTYN, P.G., LLOYD, I.J., DURKIN, N., LIPTON, A. and MUIRUBI, K.L. 1971. Action of prolactin on human renal function. *Lancet* **2**: 352–354.

HORVATH, K., HILL, I.D., DEVARAJAN, P., MEHTA, D., THOMAS, S.C., LU, R.B. and LEBENTHAL, E. 1994. Short term effect of epidermal growth factor EGF on sodium and glucose cotransport of isolated jejunal epithelial cells. *Biochim. Biophys. Acta* **1222**: 215–222.

HOWARD, R.L., BICHET, D.G. and SCHRIER, R.W. 1993. Pathogenesis of hypernatremia and polyuric states. In: *Clinical Disturbances of Water Metabolism*, Chap. 10, pp. 189–209. SELDIN, D.W. and GIEBISCH, G. (Eds). Raven Press, New York.

HULTER, H.N. 1994. Adrenal steroid hormones. In: *Maxwell and Kleeman's Clinical Disorders of Fluid and Electrolyte Metabolism*, 5th Edn, Chap. 14, pp. 399–441. NARINS, R.G. (Ed.). McGraw-Hill, New York.

HUMES, H.D. 1984. Regulation of intracellular sodium. *Sem. Nephrol.* **4**: 117–133.

HUMES, H.D. and WEINBERG, J.M. 1986. Toxic nephropathies. In: *The Kidney*, 3rd Edn, Chap. 34. BRENNER, B.M. and RECTOR, F.C. (Eds). Saunders, Philadelphia.

HUMPHREYS, M.H. 1994. Mechanisms and management of nephrotic edema. *Kidney Int.* **45**: 266–281.

HUMPHREYS, M.H. and RECTOR, F.C. 1985. Pathophysiology of edema formation. In: *The Kidney: Physiology and Pathophysiology*, Vol. 1, pp. 1163–1179. SELDIN, D.W. and GIEBISCH, G. (Eds). Raven Press, New York.

HUNT, J.B., THILLAINAVAGAM, A.U., SALIM, A.F.M., CARNABY, S., ELLIOTT, E.J. and FARTHING, M.J.C. 1992. Water and solute absorption from a new hypotonic oral rehydration solution: evaluation in human and animal perfusion models. *Gut* **33**: 1652–1659.

HUNT, J.C. 1983. Sodium intake and hypertension: a cause for concern. *Ann. Intern. Med.* **98**: 724–728.

HUNYER, S.N. 1991. Blood pressure measurement in clinical practice. In: *Handbook of Hypertension*, Vol. 14, *Blood Pressure Measurement*, Chap. 4, pp. 95–11. O'BRIEN, E. and O'MALLEY, K. (Eds). Elsevier, Amsterdam.

HURA, C.E., KUNAU, R.T. and STEIN, J.H. 1988. Use of diuretics in salt–retaining states. *Sem. Nephrol.* **8**: 318–332.

HYMANS, D.E. 1986. Diuretic therapy in the elderly. *Drugs* **31**: Suppl. 4, 138–153.

IACONO, J. and DOUGHERTY, R. 1990. Blood pressure and fat intake. In: *Hypertension: Pathophysiology, Diagnosis and Management*, p. 257. BRENNER, B. and LARAGH, J. (Eds). Raven Press, New York.

ICHIHARA, A., SUZUKI, H. and SARUTA, T. 1993. Effects of magnesium on the renin–angiotensin–aldosterone system in human subjects. *J. Lab. Clin. Med.* **122**: 432–440.

IIMURA, O. and SHIMAMOTO, K. 1993. Salt and hypertension: water–sodium handling in essential hypertension. *Ann. N. Y. Acad. Sci.* **676**: 105–121.

IKENAGA, H., SUZUKI, H., ISHII, N., ITOH, H. and SARUTA, T. 1993. Role of nitric oxide on pressure-natriuresis in Wistar–Kyoti and spontaneously hypertensive rats. *Kidney Int.* **43**: 205–211.

ILLNER, H. 1984. Changes in red blood cell transport in shock. In: *Shock and Related Topics*, pp. 25–43. SHIRES, G.T. (Ed.). Churchill Livingstone, New York.

IMBS, J.L., SCHMIDT, M. and GIESEN-CROUSE, E. 1987. Pharmacology of loop diuretics: state of the art. *Adv. Nephrol.* **16**: 137–158.

IMURA, H. and NAKAO, K. 1989. Atrial natriuretic peptide in heart failure: altered gene expression and processing and possible pathophysiological significance. In: *Cardiac and Renal Failure: an Expanding Role for ACE Inhibitors*, pp. 127–137. DOLLERY, C.T. and SHERWOOD, L.M. (Eds). Hanley and Belfus, Philadelphia.

IMURA, H., NAKAO, K. and ITOH, H. 1992. The natriuretic peptide system in the brain: implications in the central control of cardiovascular and neurendocrine functions. *Front. Neuroendocrinol.* **13**: 217–249.

ING, R.D., NAZEERI, M.N., ZELDES, S., DULCHAVSKY, S.A. and DIEBEL, L.N. 1994. Hypertonic saline/dextran improves septic myocardial performance. *Ann. Surg.* **60**: 505–508.

INOUE, Y., GRANT, J.P. and SNYDER, P.J. 1993. Effect of glutamine-supplemented total parenteral nutrition on recovery of the small intestine after starvation atrophy. *J. Parenter. Enteral Nutr.* **17**: 165–170.

ISEKI, K., MASSRY, S.G. and CAMPESE, 1986. Effects of hypercalcaemia and parathyroid hormone on blood pressure in normal and renal–failure rats. *Am. J. Physiol.* **250**: F924–929.

ISHIKAWA, I. 1991. Acquired cystic disease: mechanisms and manifestations. *Sem. Nephrol.* **11**: 671–684.

ISHIKAZA, Y., YAMAMOTO, Y., TANAKA, M., KATO, F., YOKOTA, N., KITAMURA, K. *et al.* 1993. Molecular forms of human brain natriuretic peptide BNP. in plasma of patients on hemodialysis. *J. Am. Soc. Nephrol.* **4**: 440.

ISMAIL-BEIGI, F. 1992. Regulation of Na–K ATPase expression by thyroid hormone. *Sem. Nephrol.* **12**: 44–88.

JACOBS, L.M. 1994. Timing of fluid resuscitation in trauma. *New Engl. J. Med.* **331**: 1153–1154.

JAMISON, R.L., ROY, D.R. and LAYTON, H.E. 1993. Countercurrent mechanism and its regulation. In: *Clinical Disturbances of Water Metabolism*, Chap. 7, pp. 119–156. SELDIN, D.W. and GIEBISCH, G. (Eds). Raven Press, New York.

JANSSEN, B.J.A. and SMITS, J.F.M. 1989. Renal nerves in hypertension. *Miner. Electrolyte Metab.* **15**: 74–82.

JENKINS, H.R., SCHNACKENBERG, U. and MILLA, P.J. 1993. *In vitro* studies of sodium transport in human infant colon: the influence of acetate. *Pediatr. Res.* **34**: 666–669.

JIANG, C., FINKBEINER, W.D., WIDDICOMBE, J.H., McCRAY, P.B. and MILLER, S.S. 1993. Altered fluid transport across airway epithelium in cystic fibrosis. *Science* **262**: 424–427.

JOHNSON, A.K. 1985. The periventricular anteroventral third ventricle AV3V.: its relationship with the subfornical organ and neural systems involved in maintaining body fluid homeostasis. *Brain Res. Bull.* **15**: 595–601.

JOHNSON, J.A. and DAVIS, J.O. 1976. The effect of estrogens on renal sodium excretion in the dog. In: *Hypertension in Pregnancy*, pp. 239–248. LINDHEIMER, M.D., KATZ, A.I. and ZUSPAN, F.P. (Eds). John Wiley, New York.

JOHNSON, J.A., DAVIS, J.O., BAUMBER, J.S. and SCHNEIDER, E.G. 1970. Effects of estrogens and progesterone on electrolyte balances in normal dogs. *Am. J. Physiol.* **219**: 1691–1697.

JOHNSON, J.A., DAVIS, J.O., BROWN, P.R., WHEELER, P.D. and WITTY, R.T. 1972. Effects of estradiol on sodium and potassium balances in adrenalectomized dogs. *Am. J. Physiol.* **223**: 194–197.

JOHNSTON, C.J., ANOLDA, L.F., TSINDELL, K.T., PHILLIPS, P.A. and HODSMAN, G.P. 1987. Response of vasoactive hormones in congestive heart failure. *Can. J. Phys. Pharmacol.* **65**: 1706–1711.

JORDAN, P.A., BOTKIN, D.B., DOMINISKI, A.S., LOWENDORF, H.S. and BELOVSKY, G.E. 1973. Sodium as a critical nutrient for the moose of Isle Royale. In: *Proc. N. Am. Moose Conf. Workshop.* **9**: p.13. [Discussion by Denton (1982), pp. 37–40.]

JOSE, P.A., STEWART, C.L., TINA, L.V. and CALIAGNO, P.L. 1987. Renal disease. In: *Neonatology*, 3rd Edn, Chap. 35, pp. 795–849. AVERY, G.B. (Ed.). Lippincott, Philadelphia.

JOYCE, J.P. and RATTRAY, P.V. 1970. The nutritive value of white clover and perennial ryegrass. II. Intake and utilisation of sulphur, potassium and sodium. *N. Z. J. Agric. Res.* **13**: 792–799.

KAEHNY, W.D. and EVERSON, G.T. 1991. Extrarenal manifestations of autosomal dominant polycystic kidney disease. *Sem. Nephrol.* **11**: 661–670.

KAGAWA, K., SUZUKI, S., MATSUSHITA, K., UEMURA, N., MORITA, H. and HOSOMI, H. 1994. Relationship between the suppressive actions on intestinal absorption and on cGMP production for the natriuretic peptide family in dogs. *Clin. Expl Pharmacol. Physiol.* **21**: 83–92.

KAHN, A.M. 1988. The effect of diuretics on the renal handling of urate. *Sem. Nephrol.* **8**: 305–314.

KAIRANE, C., ZILMER, M., MUTT, V. and SILLARD, R. 1994. Activation of Na, K-ATPase by an endogenous peptide, PEC–60. *FEBS Lett.* **345**: 1–4.

KAPLAN, N.M. 1986. Hypertension. In: *Diuretics: Physiology, Pharmacology and Clinical Use*, Chap. 11, pp. 207–236. DIRKS, J.H. and SUTTON, R.A.L. (Eds). Saunders, Philadelphia.

KAPLAN, N.M. 1988. Calcium and potassium in the treatment of essential hypertension. *Sem. Nephrol.* **8**: 176–184.

KAPLAN, N.M. 1990. Sodium handling in hypertensive states. In: *The Regulation of Sodium and Chloride Balance*, Chap. 16, pp. 457–480. SELDIN, D.W. and GIEBISCH, G. (Eds). Raven Press, New York.

KATZ, A.I. 1988. Role of Na–K ATPase in kidney function. In: *The $Na^+ K^+$ Pump: Part B, Cellular Aspects*, pp. 207–232. SKOU, J.C., NORBY, J.C., MAUNSBACH, A.B. and ESMANN, M. (Eds). Liss, New York.

KATZ, A.I. 1990. Corticosteroid regulation of Na–K ATPase along the mammalian nephron. *Sem. Nephrol.* **10**: 388–399.

KATZ, A.I. and GENANT, H.K. 1971. Effect of extracellular volume expansion on renal cortical and medullary Na–K ATPase. *Pflugers Archiv.* **330**: 136–148.

KATZ, A.I. and LINDHEIMER, M.D. 1973. Renal sodium – and potassium – activated adenosine triphosphatase and sodium reabsorption in the hypothyroid rat. *J. Clin. Invest.* **52**: 796–804.

KATZ, A.I. and LINDHEIMER, M.D. 1977. Actions of hormones on the kidney. *A. Rev. Physiol.* **39**: 97–133.

KATZ, D.P., SKEIE, B., ATTELIS, N., KUETAN, V. and ASHKANAZI, J. 1993. Nutritional support. In: *Textbook of Trauma, Anaesthesia and Critical Care*, Chap. 78, pp. 930–957. GRADE, C.M. (Ed.). Mosby, St Louis.

KATZ, F.H. and KAPPAS, A. 1967. The effects of estradiol and estrol on plasma levels of cortisol and thyroid hormone binding globulins and on aldosterone and cortisol secretion rates in man. *J. Clin. Invest.* **46**: 1768–1777.

KATZ, J.I., SKOM, H.J. and WAKERLIN, G.E. 1957. Pathogenesis of spontaneous and pyelonephritic hypertension in the dog. *Circ. Res.* **5**: 137–143.

KATZ, M.A. 1980. Interstitial space—the forgotten organ. *Med. Hypotheses* **6**: 885–898.

KAUNITZ, H. 1956. Causes and consequences of salt consumption. *Nature* **178**: 1141–1144.

KAY, R.N.B. 1960. The rate of flow and composition of various salivary secretions in sheep and calves. *J.Physiol.* **150**: 515–537.

KEENAN, B.S., BUZEK, S.W. and GARZA, C. 1983. Cortisol and its possible role in regulation of sodium and potassium in human milk. *Am. J. Physiol.* **244**: E253–E261.

KELLER-WOOD, M. 1994. Vasopressin response to hyperosmolality and hypotension during ovine pregnancy. *Am. J. Physiol.* **266**: R188–R193.

KELLY, R.A. 1987. Endogenous cardiac glycoside-like compounds. *Hypertension* **10**: Suppl. 1, I87–I92.

KENYON, C.J. and JARDINE, A.G. 1989. Atrial natriuretic peptide: water and electrolyte homeostasis. *Bailliere's Clin. Endocrinol. Metab.* **3**: 431–450.

KINCAID-SMITH, P. and WHITWORTH, J.A. 1988. Pathogenesis of hypertension in chronic renal disease. *Sem. Nephrol.* **8**: 155–163.

KINSELLA, J.L. 1990. Action of glucocorticoids on proximal tubule transport system. *Sem. Nephrol.* **10**: 330–338.

KIRCHNER, K.A. and STEIN, J.H. 1994. Sodium metabolism. In: *Maxwell and Kleeman's Clinical Disorders of Fluid and Electrolyte Metabolism*, 5th Edn, Chap. 3, 45–80. NARINS, R.G. (Ed.). McGraw-Hill, New York.

KIRKMAN, E. and LITTLE, R. 1994. Cardiovascular regulation during hypovolaemic shock – central integration. In: *Blood Loss and Shock*, pp. 61–75. SECHER, N.H., PAWELCZYK, J.A. and LUDBROOK, J. (Eds). Edward Arnold, London.

KISHIMOTO, H., TSUMURA, K., FUJIOKA, S., UCHIMOTO, S. and MORII, H. 1993. Chronic parathyroid hormone administration reverses the antihypertensive effect of calcium loading in young spontaneously hypertensive rats. *Am. J. Hyperten.* **6**: 234–240.

KJELDSEN, K., EVERTS, M.E. and CLAUSEN, T. 1986. Effects of semi–starvation and potassium deficiency on the concentration of ouabain binding sites and sodium and potassium content in rat skeletal muscle. *Br. J. Nutr.* **56**: 519–532.

KLEINKNECHT, D. 1980. Pathogenesis of acute, reversible renal failure. In: *Acute Renal Failure*, pp. 1–13. CHAPMAN, A. (Ed.). Churchill Livingstone, Edinburgh.

KLEINKNECHT, D. 1993. Diseases of the kidney caused by non-steroidal anti–inflammatory drugs. In: *Analgesic and NSAID-Induced Kidney Disease*, Chap. 14, pp. 160–173. STEWART, J.H. (Ed.). Oxford University Press, Oxford.

KLEMM, S.A., GORDAN, R.D., TUNNY, T.J. and THOMPSON, R.E. 1991. The syndrome of hypertension and hyperkalaemia with normal GFR Gordon's syndrome.: is there increased proximal reabsorption? *Clin. Invest. Med.* **14**: 551–558.

KLEYMAN, T.R. and CRAGOE, E.J. 1988. The mechanism of action of amiloride. *Sem. Nephrol.* **8**: 242–248.

KNEPPER, M.A., NIELSEN, S., CHOU, C.-L. and DI GIOVANNI, S.R. 1994. Mechanism of vasopressin action in the renal collecting duct. *Sem. Nephrol.* **14**: 302–321.

KNIGHT, P.K., CLUER, D., MANEFIELD, G.W. and GORDE, A.V. 1994a. Haematology in the racing camel at rest: seasonal and training variations. *Acta Physiol. Scand.* **150**: Suppl. 617, 19–24.

KNIGHT, P.K., ROSE, R.J., EVANS, D.L., CLUER, D., HENCKEL, P. and SALTIN, B. 1994b. Metabolic responses to maximal intensity exercise in the racing camel. *Acta Physiol. Scand.* **150**: Suppl. 617, 61–77.

KNOCHEL, J.P. 1989. Clinical expression of potassium disturbances. In: *The Regulation of Potassium Balance*, Chap. 9, pp. 207–240. SELDIN, D.W. and GIEBISCH, G. (Eds). Raven Press, New York.

KNOCHEL, J.P. and READ, G. 1994. Disorders of heat regulation. In: *Maxwell and Kleeman's Clinical Disorders of Fluid and Electrolyte Metabolism*, 5th Edn, Chap. 47, pp. 1549–1590. NARINS, R.G. (Ed.). McGraw-Hill, New York.

KNOX, F.G. and GRANGER, J.P. 1992. Control of sodium excretion: an integrative approach. In: *Handbook of Physiology*, Section 8, *Renal Physiology*, Vol. I, Chap. 21. WINDHAGER, E.E. (Ed.), pp. 927–967. American Physiological Society/Oxford University Press, New York.

KNUTSEN, R., BOHMER, T. and FALCH, J. 1994. Invravenous theophylline–induced excretion of calcium, magnesium and sodium in patients with recurrent asthmatic attacks. *Scand. J. Clin. Lab. Invest.* **54**: 119–125.

KO, T.C., BEAUCHAMP, D., TOWNSEND, C.M. and THOMPSON, J.C. 1993. Glutamine is essential for epidermal growth factor-stimulated intestinal cell proliferation. *Surgery* **114**: 147–154.

KOBAYASHI, D.L., PETERSON, M.E., GRAVES, T.K., LESSER, M. and NICHOLS, C.E. 1990. Hypertension in cats with chronic renal failure or hyperthyroidism. *J. Vet. Int. Med.* **4**: 58–62.

KOBRIN, S.M. and GOLDFARB, S. 1990. Magnesium deficiency. *Sem. Nephrol.* **10**: 525–535.

KOEHLER, E.M., McLEMORE, G.L., TANG, W. and SUMMY-LONG, J.Y. 1993. Osmoregulation of the magnocellular system during pregnancy and lactation. *Am. J. Physiol.* **264**: R555–R560.

KOEPKE, J.R. 1989. Effect of environmental stress on neural control of renal function. Miner. *Electrolyte Metab.* **15**: 83–87.

KOEPPEN, B.M. 1990. Mechanisms of segmental sodium and chloride reabsorption. In: *The Regulation of Sodium and Chloride Balance*, pp. 59–104 (Ed.). SELDIN, D.W. and GIEBISCH, G. (Eds). Raven Press, New York.

KOEPSELL, H. and SPANGENBER, J. 1994. Function and presumed molecular structure of Na^+–D-glucose cotransport system. *J. Membr. Biol.* **138**: 1–11.

KOH, S.D. and TEITELBAUM, P. 1961. Absolute behavioural taste thresholds in the rat. *J. Comp. Physiol. Psychol.* **54**: 223–229.

KOOMANS, H.A., GEERS, A.B., MEIRACKER, A.H.D., ROOS, P. and DORHOUT MEES, E.J. 1984. Effects of plasma volume expansion on renal salt handling in patients with the nephrotic syndrome. *Am. J. Nephrol.* **4**: 227–234.

KOPP, U.C. and DI BONA, G.F. 1993. Neural regulation of renin secretion. *Sem. Nephrol.* **13**: 543–551.

KORSGAARD, N., AALKJAER, C., HEAGERTY, A.M., IZZARD, A.S. and MULVANY, M.J. 1993. Histology of subcutaneous small arteries from patients with essential hypertension. *Hypertension* **22**: 523–526.

KOSHIDA, R., SAKAZUME, S., HARUYAMA, H., OKUDA, N., OHAMA, K. and ASANO, S. 1994. A case of pseudo-Bartter's syndrome due to intestinal malrotation. *Acta Pediatr. Jpn* **26**: 107–111.

KOTCHEN, T.A. 1983. Pressor function of the kidney: the renin–angiotensin system. *Sem. Nephrol.* **3**: 4–13.

KRAMER, H.J. 1987. In: *Diuretics in Heart Failure*, pp. 37–44. POOLE-WILSON, P.A. (Ed.). Royal Society of Medicine, London.

KRAUSZ, M.M., LANDAU, E.H., KLIN, B. and GROSS, D. 1992. Hypertonic saline treatment of uncontrolled haemorrhagic shock at different periods from bleeding. *Arch. Surg.* **127**: 93–96.

KREIMEIER, U., BRUCKNER, U.B., NIEMCZYK, S. and MESSMER, K. 1990. Hyperosmotic saline dextran for resuscitation from traumatic–haemorrhagic hypotension: effect on regional blood flow. *Circ. Shock* **32**: 83–99.

KREIMEIER, U., RUIZ-MORALES, M. and MESSMER, K. 1993. Comparison of the effects of volume resuscitation with dextran 60 vs Ringer's lactate on central hemodynamics, regional blood flow, pulmonary function and blood composition during hyperdynamic endotoxaemia. *Circ. Shock* **39**: 89–99.

KRIEGER, J.E., ROMAN, R.J. and COWLEY, A.W. 1991. Hemodynamics and blood volume in angiotensin II salt dependent hypertension in dogs. *Am. J. Physiol.* **257**: H1402–H1412.

KRISHNA, G.G. 1990. Hypokalemic states: current clinical issues. *Sem. Nephrol.* **10**: 515–524.

KRISHNA, G.G. 1994. Role of potassium in the pathogenesis of hypertension. *Am. J. Med. Sci.* **307**: Suppl. 1, S21–S25.

KRISHNA, G.G., SHULMAN, M.D. and NARINS, R.G. 1988. Clinical use of potassium-sparing diuretics. *Sem. Nephrol.* **8**: 354–364.

KRISHNA, G.G., STEIGERWALT, S.P., PIKUS, R., KAUL, R., NARINS, R.G., RAYMOND, K.H. and KUNAU, R.T. 1994. Hypokalaemic states. In: *Maxwell and Kleeman's Clinical Disorders of Fluid and Electrolyte Metabolism*, 5th Edn, Chap. 22, pp. 659–696. NARINS, R.G. (Ed.). McGraw-Hill, New York.

KULLAMA, L.K., AGNEW, C.L., DAY, L., ERVIN, M.G. and ROSS, M.G. 1994. Ovine fetal swallowing and renal responses to oligohydroamnios. *Am. J. Physiol.* **266**: R972–R978.

KUROKAWA, K., FINE, L.G. and KLAHR, S. 1982. Renal metabolism in obstructive nephropathy. *Sem. Nephrol.* **2**: 31–40.

KUROKAWA, K., OKUDA, T., MATSUNGA, H., CHANG, H. and YAMASHITA, N. 1988. Calcium-activated chloride conductance in mesangial cells: its regulation by vasoactive peptides and growth-promoting factors. In: *Nephrology*, Vol. 1, pp. 541–551. DAVISON, A.M. (Ed.). Bailliere Tindall, London.

KURTZ, T.W., AL-BANDES, H.A. and MORRIS, R.C. 1987. Salt-sensitive essential hypertension in men: is the sodium ion alone important? *New Engl. J. Med.* **317**: 1043–1048.

KURTZMAN, N.A. 1985. Salt, hypertension and confusion. *Sem. Nephrol.* **5**: 1.

LAFFI, G., LA VILLA, G. and GENTILINI, P. 1994. Pathogenesis and management of the hepatorenal syndrome. *Sem. Liver Dis.* **14**: 71–81.

LAHERA, V., SALAZAR, J., SALOM, M.G. and RAMERO, J.C. 1992. Deficient production of nitric oxide induces volume-dependent hypertension. *J. Hyperten.* **10**: S173–S177.

LAJOIE, G., LASZIK, Z., NADASDY, T. and SILVA, F.G. 1994. The renal–cardiac connection: renal parenchymatosis alterations in patients with heart disease. *Sem. Nephrol.* **14**: 441–463.

LANDSBERG, L. 1992. Hyperinsulinemia: possible role in obesity–induced hypertension. *Hypertension* **19**: Suppl. 1., I61–I66.

LANG, F., GEROK, W. and HAUSSINGER, D. 1993. New clues to the pathophysiology of hepatorenal failure. *Clin. Invest.* **71**: 93–97.

LANGFORD, H.G. 1991. Sodium–potassium interaction in hypertension and hypertensive cardiovascular disease. *Hypertension* **17**: Suppl. I, 155–157.

LANT, A. 1986. Diuretic drugs: progress in clinical pharmacology. *Drugs* **31**: Suppl. 4, 40–55.

LA POINTE, M.S. and BATTLE, D.C. 1994. Na^+/H^+ exchange and vascular smooth muscle proliferation. *Am. J. Med. Sci.* **307**: Suppl. 1, S9–S16.

LARAGH, J.H. 1989. Nephron heterogeneity: clue to the pathogenesis of essential hypertension and effectiveness of angiotensin–converting enzyme inhibitor treatment. *Am. J. Med.* **87**: Suppl. 68, 68-2S–68-14S.

LARAGH, J.H. and ATLAS, S.A. 1988. Atrial natriuretic hormone: a regulator of blood pressure and volume homeostasis. *Kidney Int.* **34**: Suppl. 25, S64–S71.

LARAGH, J.H. and PECKER, M.S. 1983. Dietary salt and hypertension: some myths, hopes and truths. *Ann. Intern. Med.* **89**: 735–742.

LARAGH, J.H. and SEELEY, J.E. 1990. The renin system and its pathophysiology in disease. In: *The Regulation of Sodium and Chloride Balance*, Chap. 6, pp. 195–236. SELDIN, D.W. and GIEBISCH, G. (Eds). Raven Press, New York.

LARAGH, J.H. and SEELEY, J.E. 1992. Renin–angiotensin–aldosterone system and the renal regulation of sodium, potassium and blood pressure homeostasis. In: *Handbook of Physiology*, Section 8 *Renal Physiology*, Vol. II, pp. 1430–1518. WINDHAGER, E.E. (Ed.). American Physiological Society/Oxford University Press, New York.

LARSEN, L.A. 1982. Salt/sodium in processed food: a regulatory overview. In: *Sodium Intake—Dietary Concerns*, pp. 83–93. FREEMAN, T.M. and GREGG, O.W. (Eds). American Association of Cereal Chemists, St Paul, MN.

LAREDO, J., HAMILTON, B.P. and HAMLYN, J.M. 1994. Ouabain is secreted by bovine adrenocortical cells. *Endocrinology* **135**: 794–797.

LASKI, M.E. 1986. Diuretics: mechanism of action and therapy. *Sem. Nephrol.* **6**: 210–223.

LASSITER, W.E. 1990. Regulation of sodium chloride distribution within extracellular space. In: *The Regulation of Sodium and Chloride Balance*, pp. 23–58. SELDIN, D.W. and GIEBISCH, G. (Eds). Raven Press, New York.

LA VILLA, G., ASHBERT, M., JIMENEZ, W., GINES, P., Claria, P., Claria, J. *et al.* 1990. Natriuretic hormone activity in the urine of cirrhotic patients. *Hepatology* **12**: 467–475.

LAW, M.R., FROST, C.D. and WALD, N.J. 1991a. By how much does dietary salt reduction lower blood pressure? I. Analysis of observational data among populations. *Br. Med. J.* **302**: 811–815.

LAW, M.R., FROST, C.D. and WALD, N.J. 1991b. Analysis of data from trials of salt reduction. *Br. Med. J.* **302**: 819–824.

LAW, M.R., FROST, C.D. and WALD, N.J. 1991c. Dietary salt and blood pressure. *J. Hyperten.* **9**: Suppl. 6, S37–S41.

LEAF, A., MACKNIGHT, A.D.C., CHEUNG, J.Y. and BONVENTRE, J.V. 1986. The cellular basis of ischemic acute renal failure. In: *Physiology of Membrane Disorders*, 2nd Edn, Chap. 42, pp. 769–784. ANDREOLI, T.E., HOFFMAN, J.F., FANESTIL, D.D. and SCHULTZ, S.G. (Eds). Plenum Press, New York.

LECHENE, C. 1988. The physiological role of the Na–K pump. In: *The Na$^+$ K$^+$ Pump: Part B, Cellular Aspects*, pp. 171–194. SKOU, J.C., NORBY, J.C., MAUNSBACH, A.B. and ESMANN, M. (Eds). Liss, New York.

LEDINGHAM, J.M. 1985. Hypertension and the kidney. In: *Postgraduate Nephrology*, pp. 325–353. MARSH, F. (Ed.). Heinemann Medical, London.

LEE, K.O., TAYLOR, F.A., OH, V.M.S., CHEAH, J.S. and AW, S.F. 1991. Hyperinsulinaemia in thyrotoxic hypokalaemic periodic paralysis. *Lancet* 337: 1063–1064.

LEE, S.H., ELIAS, P.M., FEINGOLD, K.R. and MAURO, T. 1994. A role for ions in barrier recovery after acute perturbation. *J. Invest. Dermatol.* 102: 976–979.

LE FANU, J. 1994. Why blood pressure is a mystery. *The Times*, 17 November, p.17.

LEFFLER, L.W., HESSLER, J.R., GREEN, R.S. and FLETCHER, A.M. 1986. Effects of sodium chloride on pregnant sheep with reduced ureteroplacental perfusion pressure. *Hypertension* 8: 62–65.

LEON, A., BAIN, S.A.F. and LEVICK, W.R. 1992. Hypokalaemic episodic polymyopathy in cats fed a vegetarian diet. *Aust. Vet. J.* 69: 249–254.

LE QUESNE, L.P., COCHRANE, J.P.S. and FIELDMAN, N.R. 1985. The effect of vasopressin on solute and water excretion during and after surgical operation. *Ann. Surg.* 201: 383–390.

L'ESTRANGE, J.L. and AXFORD, R.F.E. 1966. Mineral balance studies on lactating ewes with particular reference to the metabolism of magnesium. *J. Agric. Sci.* 67: 295–304.

LEVENS, N.R., MARRISCOTTI, S.P., PEACH, M.J. *et al.* 1984. Angiotensin II mediates increased small intestinal fluid absorption with extracellular volume depletion in the rat. *Endocrinology* 114: 1692–1701.

LEVENSON, R. 1994. Isoforms of the Na, K-ATPase: Family members in search of function. *Rev. Physiol. Biochem. Pharmacol.* 123: 1–45.

LEVIN, M.L. 1988. Patterns of tubulo-interstitial damage associated with non–steroidal anti-inflammatory drugs. *Sem. Nephrol.* 8: 55–61.

LEVY, M. 1978. Kidney in liver disease. In: *Sodium and Water Homeostasis*, pp. 737–753. BRENNER, B.M. and STEIN, J.H., (Eds). Churchill Livingstone, London.

LEVY, M. 1993. Hepatorenal syndrome. *Kidney Int.* 43: 737–753.

LEVY, M. 1994. Pathogenesis of sodium retention in early cirrhosis of the liver: evidence for vascular overfilling. *Sem. Liver Dis.* 14: 4–13.

LEWIS, E.J., HUNSICKER, L.G., BAIN, R.P. and ROHDE, R.D. 1993. The effect of angiotensin-converting-enzyme inhibition on diabetic nephropathy. *New Engl. J. Med.* 329: 1456–1462.

LICATA, G., VOLPE, M., SCAGLIONE, R. and RUBATTU, S. 1994. Salt regulating hormones in young normotensive obese subjects. Effects of saline load. *Hypertension* 23: Suppl. 1, I20–I24.

LICHARDUS, B., FOLDES, O., STYKE, J., ZEMANKOVA, A. and KOVACS, L. 1993. On the role of digoxin-like substances, ANP and AVP in natriuresis induced by hypertonic saline infusion in dogs. *J. Cardiovasc. Pharmacol.* 22: Suppl. 2, S82–S83.

LIDDLE, G.W. and LIDDLE, R.A. 1981. Endocrinology. In: *Pathophysiology*, Chap. 7, pp. 653–754. SMITH, L.H. and THIER, S.O. (Eds). Saunders, Philadelphia.

LIEBERMAN, E. 1983. Pediatric renal hypertension. *Sem. Nephrol.* 3: 65–72.

LIEBERTHAL, W., and LEVINSKY, N.G. 1990. Treatment of acute tubular necrosis. *Sem. Nephrol.* 10: 571–583.

LIMA, A.A., SOARES, A.M., FREIRE-JUNIOR, J.E. and GUERRANT, R.L. 1992. Cotransport of sodium with glutamine, alanine and glucose in the isolated ileal mucosa. *Braz. J. Med. Biol. Res.* 25: 637–640.

LINAS, S.L. and BERL, T. 1989. Clinical diagnosis of abnormal potassium balance. In: *The Regulation of Potassium Balance*, Chap. 8, pp. 177–206. SELDIN, D.W. and GIEBISCH, G. (Eds). Raven Press, New York.

LINJEN, P., PETROV, V. and AMERY, A. 1994. Relationship between erythrocyte cation transport systems and membrane and plasma lipids in healthy men. *Am. J. Med. Sci.* **307**: Suppl. 1, S146–S149.

LIND, L., RIDEFELT, P., RASTAD, J., AKERSTROM, G. and LJUNGHALL, S. 1994. Relationship between abnormal regulation of cytoplasmic calcium and elevated blood pressure in patients with primary hyperparathyroidism. *J. Human Hyperten.* **8**: 113–118.

LINDHEIMER, M.D. 1993. Hypertension in pregnancy. *Hypertension* **22**: 127–137.

LINDHEIMER, M.D. and KATZ, A.I. 1980. Kidney in pregnancy. *Kidney Int.* **18**: 147–151.

LINDHEIMER, M.D. and KATZ, A.I. 1986. The kidney in pregnancy. In: *The Kidney*, 3rd Edn, pp. 1253–1297. BRENNER, B.M. and RECTOR, F.C. (Eds). Saunders, Philadelphia.

LINDHEIMER, M.D., BARRON, W.M., DURR, J. and DAVISON, J.M. 1986. Water homeostasis and vasopressin secretion during gestation. *Adv. Nephrol.* **15**: 1–24.

LINGREL, J.B., VAN HUYSSE, J., O'BRIEN, W., JEWELL-MOTZ, E., ASKEW, A. and SCHULTHEIS, P. 1994. Structure–function studies of the Na, K-ATPase. *Kidney Int.* **45**: Suppl. 44, S32–S39.

LINZELL, J.L. AND PEAKER, M. 1972. Day to day variations in milk composition in the goat and cow as a guide to the detection of clinical mastitis. *Br. Vet. J.* **128**: 284–295.

LO, C.S. and KLEIN, L.F. 1992. Thyroidal and steroidal regulation of Na^+, K^+ ATPase. *Sem. Nephrol.* **12**: 62–66.

LORENZ, J.N., GREENBERG, S.G. and BRIGGS, J.P. 1993. The macula densa mechanism for control of renin secretion. *Sem. Nephrol.* **13**: 531–542.

LOUIS, W.J., TABEI, R. and SPECTOR, S. 1971. Effects of sodium intake on inherited hypertension in the rat. *Lancet* ii: 1283–1286.

LOWE, S.A., MACDONALD, G.J. and BROWN, M.A. 1992a. Acute and chronic regulation of atrial natriuretic peptide in human pregnancy: a longitudinal study. *J. Hypertens.* **10**: 821–829.

LOWE, S.A., MACDONALD, G.J. and BROWN, M.A. 1992b. Atrial natriuretic peptide in pregnancy: response to oral sodium supplementation. *Clin. Expl Pharmacol. Physiol.* **19**: 607–612.

LUDAN, A.C. 1988. Current management of acute diarrhoeas—use and abuse of drug therapy. *Drugs* **36** Suppl. 4, 18–25.

LUDENS, J.H., CLARK, M.A., KOLBASA, K.P. and HAMLYN, J.M. 1993. Digitalis–like factor and ouabain-like compound in plasma of volume–expanded dogs. *J. Cardiovasc. Pharmacol.* **22**: Suppl. 2, S38–S41.

LUFT, F.C., ZEMEL, M.B., SOWERS, J.A., FINEBERG, N.S. and WEINBERGER, M.H. 1990. Sodium bicarbonate and sodium chloride: effects on blood pressure and electrolyte homeostasis in normal and hypertensive men. *J. Hyperten.* **8**: 663–670.

LUFT, F.C., MILLER, J.Z., GRIM, C.E., FINEBERG, N.S., CHRISTIAN, J.C., DAUGHERTY, S.A. and WEINBERGER, M.H. 1991. Salt sensitivity and resistance of blood pressure: age and race as factors in physiological responses. *Hypertension* **17**: Suppl. 1, I102–I108.

LUMBERS, E.R. 1993. Renin, uterus and amniotic fluid. In: *The Renin Angiotensin System*, Chap. 45, pp. 45.1–45.12. ROBERTSON, J.I.S. and NICHOLLS, M.G. (Eds). Gower Medical Publishing, London.

LUSCHER, T.F. 1994. The endothelium and cardiovascular disease—a complex relation. *New Engl. J. Med.* **330**: 1080–1083.

MAACK, T. 1988. Functional properties of atrial natriuretic peptide and its receptors. In: *Nephrology*, Vol. 1, pp. 123–136. DAVISON, A.M. (Ed.). Bailliere Tindall, London.

MAACK, T., MARION, D.N., CAMARGO, M.J.F., KLEINERT, H., VAUGHAN, E.D. and ATLAS, S.A. 1984. Effects of auriculin (atrial natriuretic factor) on blood pressure, renal function, and the renin–aldosterone system in dogs. *Am. J. Med.* **77**: 1069–1075.

MCBURNIE, M., DENTON, D.A. and TARJAN, E. 1988. Influence of pregnancy and lactation on Na appetite of BALB/C mice. *Am. J. Physiol.* **255**: R1020–1024.

MCCANCE, R.A. 1936. Experimental sodium chloride deficiency in man. *Proc. R. Soc. Lond.* **119B**:, 245–268.

MCCANCE, R.A. 1938. Effects of sodium deficiency on ECF, sweat, saliva, gastric juice and CSF. *J. Physiol.* **92**: 208–216.

MCCARRON, D.A. 1989. Calcium metabolism and hypertension. *Kidney Int.* **35**: 717–736.

MCCARRON, D.A. and MORRIS, C.D. 1986. Epidemiological evidence associating dietary calcium and calcium metabolism with blood pressure. *Am. J. Nephrol.* **6**: Suppl. 1, 3–9.

MCCARRON, D.A., MORRIS, C.D., HENRY, H.J. and STANTON, J.L. 1984. Blood pressure and nutrient intake in the United States. *Science* **224**: 1392–1397.

MCCONNELL, S.D. and HENKIN, R.I. 1973. Increased preference for Na$^+$ and K$^+$ salts in spontaneously hypertensive SH. rats. *Proc. Soc. Expl Biol. Med.* **143**: 185–188.

MCDONOUGH, A.A., TANG, M.J. and LESCALE-MATYS, L. 1990. Ionic regulation of the biosynthesis of Na–K ATPase subunits. *Sem. Nephrol.* **10**: 400–409.

MCDONOUGH, A.A., AZUMA, LESCALE-MATYS, L., TANG, M.J., NAKHOUL, F., HENSLEY, C.B. and KOMATSU', Y. 1992. Physiological rationale for multiple sodium pump isoforms. *Ann. N.Y. Acad. Sci.* **671**: 156–169.

MCDOUGALL, J.G., COGHLAN, J.P., SCOGGINS, B.A. and WRIGHT, R.D. 1974. Effect of sodium depletion on bone sodium and total exchangeable sodium in sheep. *Am. J. Vet. Res.* **35**: 923–929.

MACGREGOR, G.A. 1977. High blood pressure and renal disease. *Br. Med. J.* **2**: 624.

MACGREGOR, G.A. and CAPPUCCIO, F.P. 1993. The kidney and essential hypertension: a link to osteoporosis. *J. Hyperten.* **11**: 781–785.

MCKEEVER, K.H., HINCHCLIFF, K.W., SCHMALL, L.M. 1991. Renal tubular function in horses during submaximal exercise. *Am. J. Physiol.* **161**: R553–R560.

MCKEEVER, K.H., HINCHCLIFF, K.W., REED, S.M. and ROBERTSON, J.T. 1993. Plasma constituents during incremental treadmill exercise in intact and splenectomised horses. *Eq. Vet. J.* **25**: 233–236.

MCKINLEY, M.J. 1992. Common aspects of the cerebral regulation of thirst and renal sodium excretion. *Kidney Int.* **37**: S102–S106.

MCKINLEY, M.J., DENTON, D.A., NELSON, J.F. and WEISINGER, R.S. 1983. Rehydration induces sodium depletion in rats, rabbits and sheep. *Am. J. Physiol.* **245**: R287–R292.

MCKINLEY, M.J., DENTON, D.A., COGHLAN, J.P., HARVEY, R.B., MCDOUGALL, J.G., RUNDGREN, M., SCOGGINS, B.A. and WEISINGER, R.S. 1986. Cerebral osmoregulation of renal sodium excretion — a response analogous to thirst and vasopressin release. *Can. J. Physiol. Pharmacol.* **65**: 1724–1729.

MACKNIGHT, A.D.C. and LEAF, A. 1986. Regulation of cellular volume. In: *Physiology of Membrane Disorders*, 2nd Edn, Chap. 19, pp. 311–328. T.E. ANDREOLI, J.C. HOFFMANN, D.D. FANESTIL and S.G. SCHULZ (Eds). Plenum Press, New York.

MACKNIGHT, D.C., GRANTHAM, J. and LEAF, A. 1993. Physiologic responses to changes in extracellular osmolality. In: *Clinical Disturbances of Water Metabolism*, Chap. 3, pp.131–149. SELDIN, D.W. and GIEBISCH, G. (Eds). Raven Press, New York.

MCPARLAND, B.E., GOULDING, A. and CAMPBELL, A.J. 1989. Dietary salt affects biochemical markers of resorption and formation of bone in elderly women. *Br. Med. J.* **299**: 834–835.

MACPHAIL, S., THOMAS, T.H., WILKINSON, R., DAVISON, J.M. and DUNLOP, W. 1992. Pregnancy induced hypertension and sodium pump function in erythrocytes. *Br. J. Obstet. Gynecol.* **99**: 803–807.

MACPHAIL, S., THOMAS, T.H., WILKINSON, R., DAVISON, J.M. and DUNLOP, W. 1993. Erythrocyte sodium lithium countertransport in normal and hypertensive pregnancy in relation to haemodynamic changes. *Br. J. Obstet. Gynecol.* **100**: 673–678.

MCQUARRIE, I., THOMPSON, W.H. and ANDERSON, J.A. 1936. Effects of excessive ingestion of sodium and potassium salts on carbohydrate metabolism and blood pressure in diabetic children. *J. Nutr.* **11**: 77–101.

MCSWEENEY, C.S. and CROSS, R.B. 1992. Effects of Na intake on Na conservation, digestion and mineral metabolism in growing ruminants fed *Stylosanthes hamata* cv. Verano. *Small Ruminant Research* **7**: 299–313.

MADARA, J.L. 1991. Functional morphology of epithelium of the small intestine. In: *Handbook of Physiology*, Section 6, *Gastrointestinal Tract*, Vol. IV. Intestinal Secretion and Absorption, Chap. 3, pp. 83–119. American Physiological Society, Bethesda, MD

MADDOX, D.A., DEEN, W.M. and GENNARI, F.J. 1987. Control of bicarbonate and fluid reabsorption in the proximal convoluted tubule. *Sem. Nephrol.* **7**: 72–81.

MAESAKA, J.K., WOLF, K.G., PICCIONE, J.M. and MA, C.M. 1993. Hypouremia, abnormal tubular urate transport and plasma natriuretic factors in patients with Alzheimer's disease. *J. Am. Geriatr. Soc.* **41**: 501–506.

MAGONEY, L.T. and LAUER, R.M. 1989. Consistency of blood pressure levels in children. *Sem. Nephrol.* **9**: 230–235.

MAJID, D.S., WILLIAMS, A., KADOWITZ, P.J. and NAVAR, L.G. 1993a. Renal responses to intra-arterial administration of nitric oxide donor in dogs. *Hypertension* **22**: 535–541.

MAJID, D.S., WILLIAMS, A. and NAVAR, L.G. 1993b. Inhibition of nitric oxide synthesis attenuates pressure induced natriuretic responses in anesthetized dogs. *Am. J. Physiol.* **264**: F79–F87.

MALCOLM, A.D. 1986. Digitalis and diuretics: still the standard therapy? *Ann. Med. Int.* **137**: 190–192.

MANDEL, L.J. 1986. Bioenergetics of membrane transport processes. In: *Physiology of Membrane Disorders*, 2nd Edn, Chap. 18, pp. 295–310. ANDREOLI, T.E., HOFFMAN, J.F., FANESTIL, D.D. and SCHULTZ, S.G. (Eds). Plenum Press, New York.

MANGOS, J.A. 1986. Cystic fibrosis. In: *Physiology of Membrane Disorders*, 2nd Edn, Chap. 50, pp. 907–918. ANDREOLI, T.E., HOFFMANN, J.F., FANESTIL, D.D. and SCHULTZ, S.G. (Eds). Plenum Press, New York.

MANNING, R.D. 1987. Effects of hypoproteinemia on renal hemodynamics, arterial pressure and fluid volume. *Am. J. Physiol.* **252**: F91–F98.

MARSDEN, P.A. and BRENNER, B.M. 1991. Nitric oxide and endothelins: novel autocrine/paracrine regulators of the circulation. *Sem. Nephrol.* **11**: 169–185.

MARSH, A.C. 1982. Effect of processing on sodium content of foods. In: *Sodium Intake – Dietary Concerns*, pp. 116–121. FREEMAN, T.M. and GREGG, O.W. (Eds). American Association of Cereal Chemists, St Paul, MN.

MARTIN, B.J. and MILLIGAN, K. 1987. Diuretic–associated hypomagnesaemia in the elderly. *Arch. Intern. Med.* **147**: 1768–1771.

MARTINEZ-MALDONADO, M. and BENABE, J.C. 1990. Diuretics: primary and secondary effects. In: *The Regulation of Sodium and Chloride Balance*, Chap. 17, pp. 481–502. SELDIN, D.W. and GIEBISCH, G. (Eds). Raven Press, New York.

MARVER, D. 1992. Regulation of Na$^+$ K$^+$ ATPase by aldosterone. *Sem. Nephrol.* **13**: 56–61.

MATON, P.N. 1994. Expanding uses of octreotide. *Bailliere's Clin. Gastroenterol.* **8**: 321–337.

MATSUSHITA, M., NISHIDA, Y., HOSOMI, H. and TANAKA, S. 1991. Effects of atrial natriuretic peptide on water and NaCl absorption across the intestine. *Am. J. Physiol.* **206**: R6–R12.

MATTSON, D.L., NAKANISHI, K., PAPANEK, P.E. and COWLEY, A.W. 1994. Effect of chronic renal medullary nitric oxide inhibition on blood pressure. *Am. J. Physiol.* **266**: H1918–H1926.

MAUGHAN, R.J. 1994. Fluid and electrolyte loss and replacement in exercise. In: *Oxford Textbook of Sports Medicine*, pp. 82–93. HARRIES, M., WILLIAMS, C., STANISH, W.D. and MICHELI', L.J. (Eds). Oxford Medical Publications, Oxford.

MAVICHAK, V. and SUTTON, R.A.L. 1986. Osmotic diuresis and aquaretics. In: *Diuretics: Physiology, Pharmacology and Clinical Use*, Chap. 12, pp. 29–48. DIRKS, J.H. and SUTTON, R.A.L. (Eds). Saunders, Philadelphia.

MAXWELL, M.H., WAKS, A.U. and KRISHNA, G.G. 1994. Electrolytes in the pathogenesis of hypertension. In: *Maxwell and Kleeman's Clinical Disorders of Fluid and Electrolyte Metabolism*, 5th Edn, Chap. 48, pp. 1591–1617. NARINS, R.G. (Ed.). McGraw-Hill, New York.

MAY, C.N. and PEART, S. 1989. Investigation of the mechanisms by which angiotensin and isoprenaline alter calcium flux in juxtaglomerular cells. In: *Concepts in Hypertension*, pp. 31–37. MATHIAS, C.J. and SEVER, P.S. (Eds). Springer, New York.

MAYER, J. 1969. Hypertension, salt intake and the infant. *Postgrad. Med.* **45**: 229–236.

MCGARVEY, S.T., ZINNER, S.H., WILLETT, W.C. and ROSNER, B. 1991. Maternal prenatal dietary potassium, calcium, magnesium and infant blood pressure. *Hypertension* **17**: 218–224.

MEDFORD, R.M. 1993. Digitalis and the Na–K ATPase. *Heart Dis. Stroke* **2**: 250–255.

MENCKEN, H.L. 1919. Exeunt omnes. In: *The Vintage Mencken*, pp. 233–240. COOKE, A. (Ed.) (1955). Vintage Books Random House, New York.

MENDEZ, R.E. and BRENNER, B.M. 1990. Glomerulotubular balance and the regulation of sodium excretion by intrarenal haemodynamics. In: *The Regulation of Sodium and Chloride Balance*, pp. 105–132. SELDIN, D.W. and GIEBISCH, G. (Eds). Raven Press, New York.

MERRILL, D.C., SELTONN, M.M. and COWLEY, A.W. 1987. Angiotensin II sensitization of aldosterone responsiveness to plasma sodium in conscious dogs. *Am. J. Physiol.* **253**: R832–R837.

MERRILL, D.C., EBERT, E.J., SKELTON, M.M. and COWLEY, A.W. 1989. Effect of plasma sodium on aldosterone secretion during angiotensin II stimulation in normal humans. *Hypertension* **14**: 164–169.

MEYER, H. 1980. Na-stoftwechsel und Na-bedarf des Pferdes. *Ubers. Tierernahr* **8**: 37–64.

MEYER-LEHNERT, H. and SCHRIER, R.W. 1990. Hyponatraemia: diagnosis and treatment. In: *The Regulation of Sodium and Chloride Balance*, pp. 433–456. SELDIN, D.W. and GIEBISCH, G. (Eds). Raven Press, New York.

MICHAUD, L. and ELVEHJEM, C.A. 1944. The nutritional requirements of the dog. *N. Am. Vet.* **25**: 657–666.

MICHELL, A.R. 1969. A study of salt appetite in sheep. Thesis presented for the degree of Doctor of Philosophy in the University of London.

MICHELL, A.R. 1970. Biochemical observations on spontaneous salt appetite in sheep. *Proc. 78th Annual Convention, American Psychological Association*, pp. 211–212.

MICHELL, A.R. 1971. A biochemical basis for salt appetite. *Proc. International Union for Physiological Sciences*, Munich, Vol. IX, p.390.

MICHELL, A.R. 1973. The biochemical basis of salt appetite: preliminary examination of a hypothesis. *J. Physiol.* **234**: 114–115P.

MICHELL, A.R. 1974a. Body fluids and diarrhoea: dynamics of dysfunction. *Vet. Rec.* **94**: 311–315.

MICHELL, A.R. 1974b. Relationships between sodium transport and sodium appetite. *Proc. International Union for Physiological Sciences*, New Delhi, Vol. XI, p.231.

MICHELL, A.R. 1975a. Salt appetite; its physiological basis and significance. In *Olfaction and Taste*, pp. 261–265. DENTON, D.A. and COGHLAN, J. (Eds). Academic Press, New York.

MICHELL, A.R. 1975b. Changes of sodium appetite during the estrous cycle of sheep. *Physiol. Behav.* **14**: 223–226.

MICHELL, A.R. 1976a. Relationships between individual differences in salt appetite of sheep and their plasma electrolyte status. *Physiol. Behav.* **17**: 216–220.

MICHELL, A.R. 1976b. Skeletal sodium, a missing element in hypertension and salt excretion. *Perspect. Biol. Med.* **20**: 27–33.

MICHELL, A.R. 1977. Acid–base status, salivary Na/K and electrolyte excretion in sheep with differing sodium appetite. *Br. Vet. J.* **133**: 245–257.

MICHELL, A.R. 1978a. Salt appetite, sodium metabolism and hypertension: a deviation of perspective. *Perspect. Biol. Med.* **21**: 335–347.

MICHELL, A.R. 1978b. Sodium need and sodium appetite during the estrous cycle of sheep. *Physiol. Behav.* **21**: 519–523.

MICHELL, A.R. 1979a. Biochemistry and behaviour: systemic aspects of neurological disturbances. *J. Small Animal Pract.* **20**: 645–649.

MICHELL, A.R. 1979b. Water and electrolyte excretion during the oestrous cycle in sheep. *Q. J. Expl Physiol.* **64**:79–88.

MICHELL, A.R. 1979c. Sodium transport and salt appetite; the effect of DPH on sodium preference and electrolyte balance in rats. *Chem. Senses Flav.* **4**: 231–240.

MICHELL, A.R. 1979d. Drugs and renal function. *J. Vet. Pharmacol. Ther.* **2**: 5–20.

MICHELL, A.R. 1979e. The pathophysiological basis of fluid therapy in animals. *Vet. Rec.* **104**: 542–548; 572–575.

MICHELL, A.R. 1980a. Sodium appetite and the oestrous cycle in sheep: effect of oestrogen, progesterone and changes in food intake. *Q. J. Expl Physiol.* **65**: 27–36.

MICHELL, A.R. 1980b. Salt and hypertension. *Lancet* **1**: 1358.

MICHELL, A.R. 1981. Fluid and electrolyte excretion in sheep: the effect of reduced food and water intake with oestrus. *Q. J. Expl Physiol.* **66**: 515–521.

MICHELL, A.R. 1982a. Current concepts of fever. *J. Small Animal Pract.* **23**: 185–193.

MICHELL, A.R. 1982b. Plasma sodium concentration and dehydration. *Vet. Rec.* **110**: 457.

MICHELL, A.R. 1983a. Renal Physiology. In *Veterinary Nephrology*, Chap. 4, pp. 57–70. HALL, L.W. (Ed.). Heinemann, London.

MICHELL, A.R. 1983b. Fluid therapy for alimentary disease: origins and objectives. *Ann. Rech. Vét.* **14**: 198–204.

MICHELL, A.R. 1983c. Understanding fluid therapy. *Irish Vet. J.* **37**: 94–103.

MICHELL, A.R. 1984a. Sums and assumptions about salt. *Perspect. Biol. Med.* **27**: 221–233.

MICHELL, A.R. 1984b. Salt and hypertension. *Lancet* **2**: 634.

MICHELL, A.R. 1985a. Sodium and research in farm animals; problems of requirement, deficit and excess. *Outlook on Agriculture*, Vol. 14, pp. 179–182. Pergamon, Oxford.

MICHELL, A.R. 1985b. Sodium in health and disease; a comparative review with the emphasis on herbivores. *Vet. Rec.* **116**: 653–657.

MICHELL, A.R. 1985c. What is shock? *J. Small Animal Prac.* **26**: 719–738.

MICHELL, A.R. 1986. The gut, the unobtrusive regulator of sodium balance. *Perspect. Biol. Med.* **29**: 203–213.

MICHELL, A.R. 1988a. Drips, drinks and drenches; what matters in fluid therapy. *Irish Vet. J.* **42**: 17–22.

MICHELL, A.R. 1988b. Intake and preference for salt in man and animals. *Ann. N.Y. Acad. Sci.* **510**: 138–139.

MICHELL, A.R. 1988c. Renal function, renal damage and renal failure. In: *Renal Disease in Dogs and Cats: Comparative and Clinical Aspects*, pp. 5–29. MICHELL, A.R. (Ed.). Blackwell, Oxford.

MICHELL, A.R. 1988d. Diuretics and cardiovascular disease. *J. Vet. Pharmacol. Ther.* **11**: 246–253.

MICHELL, A.R. 1988e. What is stress: the physiology of malaise and malingering. In: *Animal Disease: a Welfare Problem*, pp. 8–21. GIBSON, T.E. (Ed.). B.V.A. Animal Welfare Foundation, London.

MICHELL, A.R. 1989a. Oral and parenteral rehydration therapy. *In Practice* **11**: 96–99.

MICHELL, A.R. 1989b. Shock in companion animals. *Vet. A.* **29**: Bailliere, London, pp. 48–58.

MICHELL, A.R. 1989c. Salt intake, animal health and hypertension: should sleeping dogs lie? In: *Recent Advances in Dog and Cat Nutrition*, pp. 275–292. RIVERS, J. and BURGER, J. (Eds). Cambridge University Press, Cambridge.

MICHELL, A.R. 1989d. Physiological aspects of the requirement for sodium in mammals. *Nutr. Res. Rev.* **2**: 149–160.

MICHELL, A.R. 1989e. Historical, comparative and pathophysiological aspects of urolithiasis. *J. R. Soc. Med.* **82**: 669–672.

MICHELL, A.R. 1991. Regulation of salt and water balance. *J. Small Animal Pract.* **32**: 135–145.

MICHELL, A.R. 1992a. Sodium preference in sheep excreting sodium predominantly in urine or faeces. *Physiol. Behav.* **52**: 285–286.

MICHELL, A.R. 1992b. Sodium in health and disease: what can we learn from animals? In: *The Advancement of Veterinary Science*, Vol. 5. MICHELL, A.R. (Ed.). Commonwealth Agricultural Bureaux, Slough.

MICHELL, A.R. 1992c. Hypocalcaemia: new solutions for old bottlenecks. *Br. Vet. J.* **148**: 271–273.

MICHELL, A.R. 1993. Hypertension in companion animals. *Vet. A.* **33**: 11–23.

MICHELL, A.R. 1994a. Salt, hypertension and renal disease: comparative medicine, models and real diseases. *Postgrad. Med. J.* **70**: 686–694.

MICHELL, A.R. 1994b. Salt, water and survival; acid tests and basic advances in fluid therapy. *Irish Vet. J.* **47**: 3–8.

MICHELL, A.R. 1994c. Fluid therapy; practical principles. *J. Small Animal Pract.* **35**: 559–565.

MICHELL, A.R. 1994d. What is the importance of salt appetite? *Perspect. Biol. Med.* **37**: 473–485.

MICHELL, A.R. 1994e. Physiological role of sodium in animals. In: *Sodium in Agriculture*, Chap. 7, pp. 91–106. PHILLIPS, J.C.J. and CHIY, P.C. (Ed.). Chalcombe Publications.

MICHELL, A.R. 1994f. The comparative clinical nutrition of sodium intake: lessons from animals. *J. Nutr. Med.* **4**: 363–370.

MICHELL, A.R. 1995a. Effective blood volume: An effective concept or a modern myth? *Perspec. Biol. Med.* In Press.

MICHELL, A.R. 1995b. Progression of chronic renal disease; have we progressed? In: *Veterinary Annual*, Vol. 35. RAW, M.E. and PARKINSON, T.J. (Ed.). Blackwell Scientific, Oxford.

MICHELL, A.R. and BODEY, A.R. 1994. Canine hypertension. *Proc. American College of Veterinary Internal Medicine*, Colorado, pp. 502–505.

MICHELL, A.R. and MOSS, P. 1988. Salt appetite during pregnancy in sheep. *Physiol. Behav.* **42**: 491–493.

MICHELL, A.R. and MOSS, P. 1992. Differences between sheep excreting sodium predominantly in their urine or in their faeces: the effect of changes in sodium intake. *Expl Physiol.* **77**: 785–791.

MICHELL, A.R. and MOSS, P. 1995. Responses to reduced water intake, including dehydration natriuresis, in sheep excreting sodium predominantly in urine or in faeces. *Expl Physiol.*, **80**, 265–274.

MICHELL, A.R. and NOAKES, D.E. 1985. The effect of oestrogen and progesterone on sodium, potassium and water balance in sheep. *Res. Vet. Sci.* **38**: 46–53.

MICHELL, A.R. and TAYLOR, E.A. 1982. The optimum pH of renal adenosine triphosphatase in rats; influence of vanadate, noradrenaline and potassium. *Enzyme* **28**: 309–316.

MICHELL, A.R., MOSS, P., VINCENT, I., HILL, R. and WILLIAMS, H. 1988. The effect of pregnancy and dietary sodium intake on water and electrolyte balance in sheep. *Br. Vet. J.* **144**: 147–157.

MICHELL, A.R., BYWATER, R., CLARKE, K.W., HALL, L.W. and WATERMAN, A. 1989. *Veterinary Fluid Therapy*. Blackwell, Oxford.

MICHELL, A.R., BROOKS, H.W., WAGSTAFF, A.J. and WHITE, D.G. 1992. The effectiveness of three commercial solutions in correcting fluid, electrolyte and acid–base disturbances caused by calf diarrhoea. *Br. Vet. J.* **148**: 507–522.

MICHELSON, O. 1982. The nutritional importance of sodium. In: *Sodium Intake—Dietary Concerns*, pp. 1–69. FREEMAN, T.M. and GREGY, O.W. (Eds). American Association of Cereal Chemists, St Paul, MN.

MICKELSON, O. 1982. In: *Sodium Intake, Dietary Concerns*, p. 1. FREEMAN, T.M. and GREGORY. O.W. (Eds). American Association of Cereal Chemists, St Paul, MN.

MIDDLETON, E. and WILLIAMS, P.C. 1974. Electrolyte and fluid appetite and behaviour in male guinea-pigs: effects of stilboestrol treatment and renal function tests. *J. Endocrinol.* **61**: 381–399.

MIDDLETON, P.G., GEDDES, D.M. and ALTON, E.W. 1993. Effect of amiloride and saline on nasal mucociliary clearance and potential difference in cystic fibrosis and normal subjects. *Thorax* **48**: 812–816.

MIDKIFF, E.E., FITTS, D.A., SIMPSON, J.B. and BERNSTEIN, I.C. 1987. Attenuated sodium appetite in response to sodium deficiency in Fischer-344 rats. *Am. J. Physiol.* **252**: R562–R566.

MIKAWA, K., UESHIMA, H., HASHIMOTO, T., FUJITA, Y., NARUSE, Y., NAKAGAWA, H., KASAMATSU, T. and KAGAMINORI, S. 1994. An Intersalt investigation: relationship between body mass index and blood pressure in the combined populations of three local centres in Japan. *J. Human Hyperten.* **8**: 101–105.

MILES, A.M. and FRIEDMAN, E.A. 1993. Strategies for slowing progression of diabetic nephropathy. In: *International Yearbook of Nephrology*, pp. 109–139. ANDREWS, V.E. and FINE, L.G. (Eds). Springer, London.

MILLIGAN, L.P. and MCBRIDE, B.W. 1985. Energy costs of ion–pumping by animal tissue. *J. Nutr.* **115**: 1374–1382.

MILLS, E.H., BUTKUS, A., COGHLAN, J.P., DENTON, D.A., REID, A.F., SPENCE, C.D., WHITWORTH, J.A. and SCOGGINS, B.A. 1988. The effect of potassium loading on sodium status and pressor responsiveness in the sheep. *Clin. Expl Hyperten.* **A10**: 433–446.

MITCHELL, K.D. and NAVAR, L.G. 1989. The renin–angiotensin–aldosterone system in volume control. *Bailliere's Clin. Endocrinol. Metab.* **3**: 393–430.

MIZELLE, H.L., MONTANI, J.P., HESTER, R.L., DIDLAKE, R.H. and HALL, J.E. 1993. Role of pressure natriuresis in long–term control of renal electrolyte excretion. *Hypertension* **22**: 102–110.

MODI, K., O'DONNELL, M.P. and KEANE, W.F. 1993. Dietary interventions for progressive renal disease in experimental animal models. In: *Prevention of Progressive Chronic Renal Failure*, pp. 117–172. EL-NAHAS, A. M., MALTICK, N.P. and ANDERSON, S. (Eds). Oxford University Press, Oxford.

MOE, K. 1986. The ontogeny of salt intake in rats. In: *The Physiology of Thirst and Sodium Appetite*, pp. 31–36. DE CARO, G., EPSTEIN, A.N. and MASSI, M. (Eds). Plenum Press, New York.

MOE, O.W., ALPERN, R.J. and HENRICH, W.L. 1993. Renal proximal tubule renin–angiotensin system. *Sem. Nephrol.* **13**: 552–557.

MOHRING, J. and MOHRING, B. 1972. Evaluation of sodium and potassium balance in rats. *J. Appl. Physiol.* **33**: 688–692.

MOHRING, J., PETRI, M., SZOKOL, M., HAACK, D. and MOHRING, B. 1976. Effects of saline drinking on malignant course of renal hypertension. *Am. J. Physiol.* **230**: 849–857.

MOLITORIS, B.A. 1992. The potential role of ischaemia in renal disease progression. *Kidney Int.* **41**: Suppl. 36, S21–S25.

MONCADA, S. and HIGGS, A. 1993. The L-arginine–nitric oxide pathway. *New Engl. J. Med.* **329**: 2002–2012.

MONTAIN, S.J. and COYLE, E.F. 1992. Fluid ingestion during exercise increases skin blood flow independent of increases in blood volume. *J. Appl. Physiol.* **73**: 903–910.

MOORE, K., WENDON, J., FRAZER, M., KARANI, J., WILLIAMS, R. and BADR, K. 1992. Plasma endothelin immunoreactivity in liver disease and the hepatorenal syndrome. *New Engl. J. Med.* **327**: 1774–1778.

MOORHEAD, J.F. 1991. Lipids in the pathogenesis of kidney disease. *Am. J. Kidney Dis.* Suppl. 1, 65–70.

MORGAN, T., MYERS, J. and FITZGIBBON, W. 1981. Sodium intake, blood pressure and red cell sodium efflux. *Clin. Expl Hyperten.* **A3:** 641–653.

MORIKOTI, H., YOSHIKAWA, T., HISAYAMA, T. and TAKEUCHI, S. 1992. Possible mechanisms of age-associated reduction of vascular relaxation caused by atrial natriuretic peptide. *Eur. J. Pharmacol.* **210:** 61–68.

MORITA, H., MATSUDA, T., FURUYA, F., KHANCHOWDHURY, M.R. and HOSOMI, H. 1993. Hepatorenal reflex plays an important role in natriuresis after high-NaCl food intake in conscious dogs. *Circ. Res.* **75:** 552–559.

MORRIS, J.G. and PETERSEN, R.G. 1975. Sodium requirements of lactating ewes. *J. Nutr.* **105:** 595–598.

MORRISON, G. and SINGER, I. 1994. Hyperosmolal states. In: *Maxwell and Kleeman's Clinical Disorders of Fluid and Electrolyte Metabolism*, 5th Edn, Chap. 21, pp. 617–658. NARINS, R.G. (Ed.). McGraw-Hill, New York.

MOSER, M. 1990. Antihypertensive medications: relative effectiveness and adverse reactions. *J. Hyperten.* **8:** Suppl. 2, S9–S16.

MOSS, N.G. 1989. Electrophysical characteristics of renal sensory receptors and afferent renal nerves. *Miner. Electrolyte Metab.* **15:** 59–65.

MOUW, D.R., ABRAHAM, S.F., BLAIR-WEST, J.R., COGHLAN, J.P., DENTON, D.A., McKENZIE, J.S., McKINLEY, M.J. and SCOGGINS, B.A. 1974. Brain receptors, renin secretion and renal sodium retention in conscious sheep. *Am. J. Physiol.* **226:** 56–67.

MUIRHEAD, E.E. 1983. Depressor functions of the kidney. *Sem. Nephrol.* **3:** 14–29.

MUIRHEAD, E.E. 1994. Renal medullary vasodepressor lipid: medullipin. In: *Textbook of Hypertension*, pp. 341–359. SWALES, J.D. (Ed.). Blackwell, Oxford.

MUJAIS, S.K. 1986. Transport and renal effects of general anaesthetics. *Sem. Nephrol.* **6:** 251–258.

MULTHAUF, R.P. 1978. *Neptune's Gift*. Johns Hopkins University Press, Baltimore, MD.

MUNTZEL, M. and DRUEKE, T.A. 1992. A comprehensive review of the salt and blood pressure relationship. *Am. J. Hyperten.* **5:** S1–S42.

MUNTZEL, M., POUZET, B., LACOUR, B., HANNEDOUCHE, T. and DRUEKE, T. 1991. Selective effects of sodium and chloride depletion on salt appetite in rats. *Am. J. Physiol.* **261:** R603–R608.

MURER, H., KRAPF, R. and HELMLE-KAB, C. 1994. Regulation of renal proximal tubular Na/H-exchange: a tissue culture approach. *Kidney Int.* **45:** Suppl. 44, S23–S31.

MURRAY, J.F. 1981. Respiration. In: *Pathophysiology*, pp. 921–1071. SMITH, L.H. and THIER, S.O. (Eds). Saunders, Philadelphia.

NACHMAN, M. 1963. Learned aversion of the taste of lithium chloride and generalisation to other salts. *J. Comp. Physiol. Psychol.* **56:** 343–349.

NACHMAN, M. and COLE, L.P. 1971. In: *Handbook of Sensory Physiology, Chemical Senses*, Vol. 2: Taste, p. 357. BEIDLER, L.M. (Ed). Springer, New York.

NADER, P.C., THOMPSON, J.R. and ALPERN, R.J. 1988. Complications of diuretic use. *Sem. Nephrol* **8:** 365–387.

NAKAMURA, T., ALBEROLA, A.M. and GRANGER, J.P. 1993. Role of renal interstitial pressure as a mediator of sodium retention during systemic blockade of nitric oxide. *Hypertension* **21:** 956–960.

NAKAYAMA, S., SIBLEY, L., GUNTHER, R.A., HOLCROFT, J.W. and KRAMER, G.C. 1984. Small volume resuscitation with hypertonic saline during haemorrhagic shock. *Circ. Shock* **13:** 149–159.

NAOMI, S., UMEDA, T., TWAOKA, T., YAMAUCHI, J., IDEGUCHI, Y., FUJIMOTO,

Y. *et al.* 1993. Endogenous erythropoietin and salt sensitivity of blood pressure in patients with essential hypertension. *Am. J. Hypertens.* **6**: 15–20.

NARINS, R.G., KRISHNA, G.G., YEE, J., IKEMIYASHIRO, D and SCHMIDT, R.J. 1994. The metabolic acidoses. In: *Maxwell and Kleeman's Clinical Disorders of Fluid and Electrolyte Metabolism*, 5th Edn, pp. 769–825. NARINS, R.G. (Ed.). McGraw-Hill, New York.

NAS 1980. *National Academy of Sciences. Recommended Dietary Allowances*, 9th Edn, pp. 1–15, 168–172. National Academy of Sciences, Washington, DC.

NAVAR, L.G., BELL, P.D. and EVAN, A.P. 1986. The regulation of glomerular filtration rate in mammalian kidneys. In: *Physiology of Membrane Disorders*, 2nd Edn, Chap. 36, pp. 637–667. ANDREOLI, T.E., HOFFMAN, J.F., FANESTIL, D.D. and SCHULTZ, S.G. (Eds). Plenum Press, New York.

NAYLOR, J.N. 1986. Alkalinising abilities of calf oral electrolyte solutions. *Proc. 14th World Buiatrics Congress*, Dublin. pp. 362–367.

NAYLOR, J. 1990. Oral fluid therapy in neonatal ruminants and swine. *Vet. Clin. N. Am. Food Anim. Prac.* **6**: 51–67.

NAZARIO, C.M., SZKLO, M., DIAMOND, E., ROMAN-FRANCO, A., CLIMENT, C., SUAREZ, E. and CONDE, J.G. 1993. Salt and gastric cancer: a case-control study in Puerto Rico. *Int. J. Epidem.* **22**: 790–797.

NELSON, W.J. 1993. Regulation of cell surface polarity in renal epithelia. *Pediatr. Nephrol.* **7**: 599–604.

NEUGEBAUER, E., DIETRICH, A., LECHLEUTHNER, A., BOUILLON, B. and EYPASCH, E. 1992. Pharmacotherapy in shock syndromes: the neglected field of pharmacokinetics and pharmacodynamics. *Circ. Shock* **36**: 312–320.

NEUMAN, W.F. 1969. The milieu interieur of the bone: Claude Bernard revisited. *Fed. Proc.* **28**: 1846–1850.

NEWMAN, J.C. and AMARASINGHAM, J.L. 1993. The pathogenesis of eclampsia: the 'magnesium ischaemia' hypothesis. *Med. Hypotheses* **40**: 250–256.

NICHOLLS, M.G. 1994. The natriuretic peptides in heart failure. *J. Intern. Med.* **235**: 515–526.

NICHOLS, W.M. and O'ROURKE, M.F. 1990. Measuring principles of arterial waves. In: *McDonald's Blood Flow in Arteries*, 3rd Edn, Chap. 6, pp. 143–269. Edward Arnold, London.

NICOLAIDIS, S., GALAVERNA, O. and MELTZER, C.G. 1990. Extracellular dehydration during pregnancy increases salt intake of offspring. *Am. J. Physiol.* **258** R281–R283.

NOAKES, T.D. 1992. The hyponatraemia of exercise. *Int. J. Sport Nutr.* **2**: 205–228.

NOLTEN, W.E. and EHRLICH, E.N. 1980. Sodium and mineralocorticoids in normal pregnancy. *Kidney Int.* **18**: 162–172.

NOLTEN, W.E. and EHRLICH, E.N. 1986. Sodium metabolism in normal pregnancy and in pre-eclampsia. In: *The Kidney in Pregnancy*, Chap. 6, pp. 81–94. ANDREUCCI, V.E. (Ed.). Martinus Nijhoff, Boston.

NOONE, P.G., OLIVIER, K.N. and KNOWLES, M.R. 1994. Modulation of the ionic milieu of the airway in health and disease. *A. Rev. Med.* **45**: 421–434.

NORTON, S.A. 1992. Salt consumption in ancient Polynesia. *Perspec. Biol. Med.* **35**: 160–181.

NRC 1974. National Research Council. Nutrient Requirements of Domestic Animals No. 8. In: *Nutrient Requirements of Dogs*, pp. 14, 35–36. National Academy of Sciences, Washington, DC.

NRC 1978. National Research Council. The laboratory rat. Nutrient Requirements of Domestic Animals No. 10. In: *Nutrient Requirements of Laboratory Animals*, p. 19. National Academy of Sciences, Washington, DC.

NRC 1985. National Research Council. *Nutrient Requirements of Dogs.* National Academy of Sciences, Washington, DC.

NRC 1989. National Research Council. *Nutrient Requirements of Horses*, pp. 12–13. National Academy of Sciences, Washington DC.

O'BRIEN, E.O. and FITZGERALD, D. 1991. The history of indirect blood pressure measurement. In: *Handbook of Hypertension*, Vol. 14: *Blood Pressure Measurement*, Chap. 1, pp. 1–54. O'BRIEN, E. and O'MALLEY, K. (Eds). Elsevier, Amsterdam.

O'CONNOR, W.J. and POTTS, D.J. 1988. Kidneys and drinking in dogs. In: *Renal Disease in Dogs and Cats: Comparative and Clinical Aspects*, pp. 30–47. MICHELL, A.R. (Ed.). Blackwell Scientific, Oxford.

ODDO, M., NEGRI, A.L., SANTOS, J.C., MARTIN, R.S. and ARRIZURIETA, E.E. 1993. Chronic indomethacin administration and its relation with the renal kallikrein kinin system during rat pregnancy. *Med. Buenos Aires* 53: 326–332.

OGUCHI, A., IKEDA, U., KANBE, T., TSURUYA, Y., YAMAMOTO, K., KAWAKAMI, K., MEDFORD, R.M. and SHIMADA, K. 1994. Regulation of Na–K ATPase gene expression by aldosterone in vascular smooth muscle cells. *Am. J. Physiol.* 265: H1167–H1172.

OH, M.S. and CARROLL, H.J. 1992. Disorders of sodium metabolism: hypernatraemia and hyponatraemia. *Crit. Care Med.* 20: 94–103.

OLIVER, W.J., NEEL, J.V., GREKIN, R.J. and COHEN, E.J. 1981. Hormonal adaptations to the stresses imposed by pregnancy and lactation in Yanomamo Indians, a culture without salt. *Circulation* 63: 110–116.

O'NEIL, R.G. 1990. Aldosterone regulation of sodium and potassium transport in the cortical collecting tubule. *Sem. Nephrol.* 10: 365–374.

ORFEUR, N.B. 1985. Urine drinking. *Vet. Rec.* 116: 527.

ORR, J.B. 1929. *Minerals in Pastures*, pp. 124–137. Lewis, London.

OSBORN, J.L. 1991. Relation between sodium intake, renal function and the regulation of arterial pressure. *Hypertension* 17: Suppl. 1, I91–I96.

OSBORN, J.L. and JOHNS, G.J. 1989. Renal neurogenic control of renin and prostaglandin release. *Miner. Electrolyte Metab.* 15: 51–58.

O'SHAUGHNESSY, W.B. 1832. *Report on the Chemical Pathology of Malignant Cholera.* S. Highley, London.

OSTER, J.R., PRESTON, R.A. and MATERSON, B.J. 1994a. Fluid and electrolyte disorders in congestive heart failure. *Sem. Nephrol.* 14: 485–505.

OSTER, J.R., KOPYT, N.P., KLEEMAN, C.R., DUNFEE, T.P., KREISBERG, R.A. and NARINS, R.G. 1994b. Diabetic acidosis and coma. In: *Maxwell and Kleeman's Clinical Disorders of Fluid and Electrolyte Metabolism*, 5th Edn, Chap. 25, pp. 769–826. NARINS, R.G. (Ed.). McGraw-Hill, New York.

OTT, C.E., WELCH, W.J. and LORENZ, J.N. 1989. Effect of salt deprivation on blood pressure in rats. *Am. J. Physiol.* 256: H1426–H1431.

PADFIELD, P.L. 1989. Disturbances in salt and water metabolism in hypertension. *Bailliere's Clin. Endocrinol. Metab.* 3: 531–557.

PAGE, L.B., DAMON, A. and MOELLERING, R.C. 1974. Antecedents of cardiovascular disease in six Solomon Islands societies. *Circulation* 49: 1132–1146.

PALLER, M.S. 1984. Mechanisms of decreased responsiveness to ANG II, NE and vasopressin in pregnant rats. *Am. J. Physiol.* 247: H100–H108.

PALLER, M.S. and FERRIS, T.F. 1983. Edema in pregnancy. *Sem. Nephrol.* 3: 241–248.

PALLER, M.S. and FERRIS, T.F. 1990. Fluid and electrolyte disorders of pregnancy. In *Fluids and Electrolytes*, 2nd Edn, Chap. 18, pp. 906–921. KOKKO, J.P. and TANNEN, R.L. (Eds). Saunders, Philadelphia.

PALLER, M.S. and FERRIS, T.F. 1994. Fluid and electrolyte metabolism during pregnancy. In: *Maxwell and Kleeman's Clinical Disorders of Fluid and Electrolyte Metabolism*, 5th Edn, Chap. 35, pp. 1121–1136. NARINS, R.G. (Ed.). McGraw-Hill, New York.

PALMER, B.F. and ALPERN, R.J. 1993. Integrated renal response to abnormalities in tonicity. In: *Clinical Disturbances of Water Metabolism*, Chap. 15, pp. 273–295. SELDIN, D.W. and GIEBISCH, G. (Eds). Raven Press, New York.

PALMER, L.G. 1989. The regulation of intracellular potassium. In: *The Regulation of Potassium Balance*, pp. 89–119. SELDIN, D.W. and GIEBISCH, G. (Eds). Raven Press, New York.

PANG, P.K.T. 1994. Parathormone/parathormone-like substances. In: *Textbook of Hypertension*, Chap. 13, pp. 298–302. SWALES, J.D. (Ed.). Blackwell, Oxford.

PANG, P.K.T., BENISHIN, C.G., SHAN, J. and LEWANCZUK, R.Z. 1994. PHF: the new parathyroid hypertensive factor. *Blood Pressure* **3**: 148–155.

PAPANEK, P.E., BOVEE, K.C., SKELTON, M.M. and COWLEY, A.W. 1993. Chronic pressure–natriuresis relationship in dogs with inherited essential hypertension. *Am. J. Hyperten.* **6**: 960–967.

PAPPONE, P.A. and CAHALAN, M.D. 1986. Ion permeation in cell membranes. In: *Physiology of Membrane Disorders*, 2nd Edn, Chap. 15, pp. 249–272. ANDREOLI, T.E., HOFFMANN, J.F., FANESTIL, D.D. and SCHULTZ, S.G. (Eds). Plenum Press, New York.

PAQUE, C. 1980. Saharan Bedouins and the salt water of the Sahara: a model for salt intake. In: *Biological and Behavioural Aspects of Salt Intake*, pp. 31–47. KARE, M.R., FREGLY, M.J. and BERNARD, M.A. (Eds). Academic Press, New York.

PARK, R.G., CLEVER, J., McKINLEY, M.J. and RUNDGREN, M. 1989. Renal denervation does not prevent dehydration-induced natriuresis in sheep. *Acta Physiol. Scand.* **137**: 199–206.

PARKER, J.C. and BERKOWITZ, L.R. 1986. Genetic variants affecting the structure and function of the human red cell membrane. In: *Physiology of Membrane Disorders*, 2nd Edn, Chap. 43, pp. 785–814. ANDREOLI, T.E., HOFFMAN, J.F., FANESTIL, D.D. and SCHULTZ, S.G. (Eds). Plenum Press, New York.

PERRIN, D.R. 1958. The chemical composition of the colostrum and milk of the ewe. *J. Dairy Res.* **25**: 70–74.

PERRONE, R.D. and ALEXANDER, E.A. 1994. Regulation of extrarenal potassium metabolism. In: *Maxwell and Kleeman's Clinical Disorders of Fluid and Electrolyte Metabolism*, 5th Edn, Chap. 6, pp. 129–145. NARINS, R.G. (Ed.). McGraw-Hill, New York.

PETERS, J.P. 1948. The role of sodium in the production of edema. *New Engl. J. Med.* **239**: 353–362.

PETERS, R.M. and HARGENS, A.R. 1981. Protein vs electrolytes and all of the Starling forces. *Arch. Surg.* **116**: 1293–1299.

PFAFFMAN, C. and BARE, J.K. 1950. Gustatory nerve discharges in normal and adrenalectomised rats. *J. Comp. Physiol. Psychol.* **43**: 320–324.

PHILLIPS, G.D. and SUNDARAM, S.K. 1966. Sodium depletion of pregnant ewes and its effects on foetuses and foetal fluids. *J. Physiol.* **184**: 889–898.

PHILLIPS, S.F. 1994. Small and large intestinal disorders: associated fluid and electrolyte complications. In: *Maxwell and Kleeman's Clinical Disorders of Fluid and Electrolyte Metabolism*, 5th Edn, Chap. 36, pp. 1137–1152. NARINS, R.G. (Ed.). McGraw-Hill, New York.

PICKAR, J.G., SPIER, S.J., HARROLD, D. and CARLSEN, R.C. 1993. Ouabain binding in skeletal muscle from horses with hyperkalaemic periodic paralysis. *Am. J. Vet. Res.* **54**: 783–787.

PICKERING, G.W. 1961. The nature of essential hypertension. In: *The Nature of Essential Hypertension*, Chap. 9, pp. 124–144. Churchill, London.

PICKERING, L.K. 1991. Therapy for acute infectious diarrhea in children. *J. Pediatr.* **118**: S118–S128.

PICKETT, E.E. and O'DELL, B.L. 1992. Evidence for dietary essentiality of lithium in the rat. *Biol. Trace Elem. Res.* **34**: 299–319.

PITTS, R.F. 1974. *Physiology of the Kidney and Body Fluids*, 3rd Edn, pp. 109, 249–250. Year Book, Chicago.

PLISHKER, G.A. and APPEL, S.H. 1986. Inherited membrane disorders of muscle: Duchenne muscular dystrophy and myotonic muscular dystrophy. In: *Physiology of Membrane Disorders*, 2nd Edn, Chap. 44, pp. 815–824. ANDREOLI, T.E., HOFFMAN, J.F., FANESTIL, D.D. and SCHULTZ, S.G. (Eds). Plenum Press, New York.

PODJARNY, E., MANDELBAUM, A. and BERNHEIM, J. 1994. Does nitric oxide play a role in normal pregnancy and pregnancy-induced hypertension? *Nephrol. Dial. Transplant.* **9**: 1527–1529.

POLLARD, H.B., ROJAS, E. and ARISPE, N. 1993. A new hypothesis for the mechanism of amyloid toxicity based on the calcium channel activity of amyloid beta protein (A beta P) in phospholipid bilayer membranes. *Ann. N.Y. Acad. Sci.* **695**: 165–168.

POOLE-WILSON, P.A. 1987. In: *Diuretics in Heart Failure*, pp. 75–80. Royal Society of Medicine, London.

POSNER, A.S. 1978. The chemistry of bone mineral. *Bull. Hosp. Joint Dis.* **39**: 126–144.

POSTON, L., WOOLFSON, R.G. and GRAVES, J.E. 1993. Effects of sodium–transport inhibition in human resistance arteries. *J. Cardiovasc. Pharmacol.* **22**: Suppl. 2, S1–S3.

POULTER, N.R. and SEVER, P.S. 1994. Low blood pressure populations and the impact of rural–urban migration. In: *Textbook of Hypertension*, Chap. 2, pp. 22–36. SWALES, J.D. (Ed.). Blackwell Scientific, Oxford.

POUYSSEGUR, J., FRANCHI, A., PARIS, S. and SARDET, C. 1988. Mechanisms of activation and molecular genetics of the mammalian Na^+/H^+ antiporter. In: *pH Homeostasis; Mechanisms and Control*, pp. 61–78. HAUSSINGER, D. (Ed.). Academic Press, London.

POWELL, D.W. 1986. Ion and water transport in the intestine. In: *Physiology of Membrane Disorders*, 2nd Edn, Chap. 33, pp. 559–596. ANDREOLI, T.E., HOFFMAN, J.F., FANESTIL, D.D. and SCHULTZ, S.G. (Eds). Plenum Press, New York.

POWELL-TUCK, J. 1993. Glutamine, parenteral feeding and intestinal nutrition. *Lancet* **342**: 451–452.

PRICE, L.H. and HENINGER, G.R. 1994. Lithium in the treatment of mood disorders. *New Engl. J. Med.* **331**: 591–596.

PRICHARD, B.N., SMITH, C.C., SEN, S. and BETTERIDGE, D.J. 1992. Hypertension and insulin resistance. *J. Cardiovasc. Pharmacol.* **20**: Suppl. 11, S77–S84.

PRIEN, T., THULIG, B., WUSTEN, R., SCHOOFS, J., WEYAND, M. and LAWIN, P. 1993. Hypertonic–hyperoncotic volume replacement in patients with coronary artery stenosis. *Zentralbl.-Chir.* **118**: 257–263.

PRINEAS, R.J. 1991. Measurement of blood pressure in the obese. *Ann. Epidemiol.* **1**: 321–336.

QI, L. and DONG, W. 1992. Protective action of phenytoin on cerebral ischemia in rats. *Chung Hua I Hsueh Tsa Chih Taipei* **72**: 420–423, 447. (Chinese language: abstract on Medline, February 1993.)

QUAMME, G.A. 1986. Loop diuretics. In: *Diuretics: Physiology, Pharmacology and Clinical Use*, Chap. 5, pp. 87–116. DIRKS, J.H. and SUTTON, R.A.L. (Eds). Saunders, Philadelphia.

QUAMME, G.A. and DIRKS, J.H. 1993. The physiology of renal magnesium handling. In: *Handbook of Physiology*, Section 8, *Renal Physiology*, Vol. II, Chap. 40, pp. 1917–1935. WINDHAGER, E.E. (Ed.). American Physiological Society/Oxford University Press, New York.

QUAMME, G.A. and DIRKS, J.H. 1994. Magnesium metabolism. In: *Maxwell and Kleeman's Clinical Disorders of Fluid and Electrolyte Metabolism*, 5th Edn, Chap. 13, pp. 373–397. NARINS, R.G. (Ed.). McGraw-Hill, New York.

QUILLEN, E.W. and NUWAYHID, B.S. 1992. Steady-state arterial pressure – urinary output relationships during ovine pregnancy. *Am. J. Physiol.* **263**: R1141–R1146.

RABINOVICI, R., KRAUSZ, M.M. and FEUERSTEIN, G. 1991. Control of bleeding is essential for a successful treatment of hemorrhagic shock with 7.5% sodium chloride solution. *Surg. Gynecol. Obstet.* **173**: 98–106.

RABINOWITZ, L. and AIZMAN, R.I. 1993. The central nervous system in potassium homeostasis. *Front. Neuroendocrinol.* **14**: 1–26.

RAFTERIE, E.B. 1991. Technical aspects of blood pressure measurement. In: *Handbook of Hypertension*, Vol. 14. *Blood Pressure Measurement*, pp. 55–71. Elsevier, Amsterdam.

RAKOVIC, M. and PILECKA, N. 1992. Two compartments of slowly exchangeable sodium in bone. *Phys. Med. Biol.* **37**: 1399–1402.

RALL, T.W. 1990. Drugs used in the treatment of asthma. In: *Goodman and Gilman's The Pharmacological Basis of Therapeutics*, 8th Edn, pp. 618–637. GILMAN, A.G., RALL, T.W., NIES, A.S. and TAYLOR, P. (Eds). Pergamon Press, Oxford.

RALL, T.W. and SCHLEIFER, L.S. 1990. Drugs effective in the therapy of epilepsies. In: *Goodman and Gilman's The Pharmacological Basis of Therapeutics*, 8th Edn, pp. 436–462. GILMAN, A.G., RALL, T.W., NIES, A.S. and TAYLOR, P. (Eds). Pergamon Press, Oxford.

RAM, C.V.S. 1988. Treatment of hypertension in renal failure. In: *Nephrology*, pp. 907–916. DAVISON, A.M. (Ed.). Bailliere Tindall, London.

RAMIRES, J.A.F. and PILEGGI, F. 1986. Diuretics in cardiac oedema. *Drugs* **31**: Suppl. 4, 68–75.

RAMSAY, D.J. 1989. The importance of thirst in maintenance of fluid balance. *Bailliere's Clin. Endocrinol. Metabol.* **3**: 371–391.

RAMSAY, D.J. and THRASHER, T.N. 1984. The defence of plasma osmolality. *J. Physiol. Paris* **79**: 416–420.

RAMSAY, D.J. and THRASHER, T.N. 1991. Regulation of fluid intake in dogs following water deprivation. *Brain Res. Bull.* **27**: 495–499.

RANG, H.P. and DALE, M.M. 1991. Local anaesthetics and other drugs that affect excitable membranes. In: *Pharmacology*, Chap. 31, pp. 746–759. Churchill Livingstone, Edinburgh.

RAO, R.K., RIVIERE, P.J., PASCAUD, X., JUNIEN, J.L. and PORRACA, F. 1994. Tonic regulation of mouse ileal ion transport by nitric oxide. *J. Pharmacol. Expl Ther.* **269**: 626–631.

RASKIND, M. and BARNES, R.F. 1989. Water metabolism in psychiatric disorders. *Sem. Nephrol.* **4**: 316–324.

RAUTANEN, T., SALO, E., VERASALO, M. and VESIKARI, T. 1994. Randomised double-blind trial of hypotonic oral rehydration solutions with and without citrate. *Arch. Dis. Child.* **70**: 44–46.

RAYMOND, K.H., HUNT, J.M. and STEIN, J.H. 1986. Acute and chronic renal failure. In: *Diuretics: Physiology, Pharmacology and Clinical Use*, Chap. 12, pp. 237–258. DIRKS, J.H. and SUTTON, R.A.L. (Eds). Saunders, Philadelphia.

RAYSON, B.M. and GILBERT, M.T. 1992. Regulation of Na–K ATPase in hypertension. *Sem. Nephrol.* **12**: 72–75.

REECE, W.O. 1993. The kidneys. In: *Duke's Physiology of Domestic Animals*, 11th Edn, Chap. 31, pp. 573–603. SWENSON, M.J. and REECE, W.O. (Eds). Cornell University Press, Ithaca.

REED, L.L., MANGLANO, R., MARTIN, M., HOCHMAN, M., KOCKA, F. and BARRETT, J. 1991. The effect of hypertonic saline resuscitation on bacterial translocation after hemorrhagic shock in rats. *Surgery* **110**: 685–690.

REED, W.E. and SABATINI, S. 1986. The use of drugs in renal failure. *Sem. Nephrol.* **6**: 259–295.

REEDERS, S.T. and GERMINO, G.G. 1989. The molecular genetics of autosomal dominant polycystic kidney disease. *Sem. Nephrol.* **9**: 122–134.

REEVES, W.B. and MOLONY, D.A. 1988. The physiology of loop diuretic action. *Sem. Nephrol.* **8**: 225–233.

REHRER N.J. 1994. The maintenance of fluid balance during exercise. *Int. J. Sports Med.* **15**: 122–125.

REID, A.F., COGHLAN, J.P., McDOUGALL, J.G., WHITWORTH, J.A. and SCOGGINS, B.A. 1988. The effect of potassium loading on bone sodium and total exchangeable sodium in sheep. *Clin. Expl Pharmacol. Physiol.* **15**: 781–788.

REIFART, N. and HAMPTON, J. 1987. In: *Diuretics in Heart Failure*, pp. 3–10, 11–14. POOLE-WILSON, P.A. (Ed.). Royal Society of Medicine, London.

RESNICK, L.M. 1992. Cellular ions in hypertension, insulin resistance, obesity and diabetes: a unifying theme. *J. Am. Soc. Nephrol.* **3**: S78–S85.

REYES, A.J. 1993. Renal excretory profiles of loop diuretics: consequences for therapeutic application. *J. Cardiovasc. Pharmacol.* **22**: Suppl. 3, S11–S23.

REYES, A.J. and LEARY, W.P. 1993. Clinicopharmacological reappraisal of the potency of diuretics. *Cardiovasc. Drug Ther.* **7**: 23–28.

REYNOLDS, P.E., RICHMOND, M.H. and WARING, M.J. 1981. Antibiotics affecting the function of the cytoplasmic membrane. In: *The Molecular Basis of Antibiotic Action*, pp. 226–245. GALE, E.F. and CUNDRIFFE, C. (Eds). John Wiley, London.

RHOADS, J.M., KEKU, E.O., BENNETT, L.E., QUINN, J. and LECCE, J.G. 1990. Development of L-glutamine-stimulated electroneutral sodium absorption in piglet jejunum. *Am. J. Physiol.* **259**: G99–G107.

RHOADS, J.M., KEKU, E.O., QUINN, J., WOOSELY, J. and LECCE, J.G. 1991. L-Glutamine stimulates jejunal sodium and chloride absorption in pig rotavirus enteritis. *Gastroenterology* **100**: 683–691.

RHOADS, J.M., KEKU, E.O., WOODWARD, J.P., BANGDIWALA, S., LECCE, J.E. and GATZY, J.T. 1992. L-Glutamine with D-glucose stimulates oxidative metabolism and NaCl absorption in piglet jejunum. *Am. J. Physiol.* **263**: G960–G966.

RIBEIRO, H. DA C. and LIFSHITZ, F. 1991. Alanine-based oral rehydration therapy for infants with acute diarrhoea. *J. Pediatr.* **118**: S86–S90.

RICHARDS, A.M. 1990. Is atrial natriuretic factor a physiological regulator of sodium excretion? A review of the evidence. *J. Cardiovasc. Pharmacol.* **16**: Suppl. 7, S39–S42.

RICHARDS, A.M. 1994. The natriuretic peptides and hypertension. *J. Intern. Med.* **235**: 543–560.

RICHTER, C.P. 1936. Increased salt appetite in adrenalectomised rats. *Am. J. Physiol.* **115**: 155–161.

RICHTER, C.P. 1943. Total self-regulatory function in animals and human beings. *Harvey Lect.* **38**: 63–103.

RIKER, W.F., OKAMOTO, M. and ARTUSIO, J.F. 1990. The interaction of ouabain with post-tetanic and facilitatory drug potentiations at cat soleus neuromuscular junctions *in vivo*. *Neurochem. Res.* **15**: 457–465.

RIVIN, B. and SANTOSHAM, M. 1993. Rehydration and nutritional management. *Bailliere's Clin. Gastroenterol.* **7**: 451–476.

ROBERTSON, G.L. 1993. Regulation of vasopressin secretion. In: *Clinical Disturbances of Water Metabolism*, Chap. 6, pp. 99–118. SELDIN, D.W. and GIEBISCH, G. (Eds). Raven Press, New York.

ROBERTSON, G.L. 1994. The use of vasopressin assays in physiology and pathophysiology. *Sem. Nephrol.* **14**: 368–383.

ROBERTSON, J.I.S. 1987. Salt, volume and hypertension: causation or correlation. *Kidney Int.* **32**: 590–602.

ROBERTSON, J.I.S. 1988. The role of the renin–angiotensin system in hypertension. *Sem. Nephrol.* **8**: 120–130.

ROCCHINI, A.P. 1994. The relationship of sodium sensitivity to insulin resistance. *Am. J. Med. Sci.* **307**: Suppl. 1, S75–S80.

RODGERS, W.L. 1967. Specificity of specific hungers. *J. Comp. Physiol. Psychol.* **59**: 98–101.

RODLAND, K.D. and DUNHAM, P.B. 1980. Kinetics of lithium efflux through the Na–K pump of human erythrocytes. *Biochim. Biophys. Acta* **602**: 376–388.

ROGOFF, J.M. and STEWART, G.N. 1927. Studies in adrenal insufficiency: 3. The influence of pregnancy upon the survival period of adrenalectomized dogs. *Am. J. Physiol.* **79**: 508–514.

ROJO-ORTEGA, J.M., QUERIOZ, F.P. and GENEST, T. 1979. Effects of sodium chloride on early and chronic phases of malignant hypertension in rats. *Am. J. Physiol.* **236**: H665–H671.

ROMERO, J.C. and KNOX F.G. 1988. Mechanisms underlying pressure–related natriuresis: the role of renin–angiotensin and prostaglandin systems. *Hypertension* **11**: 724–729.

ROMERO, J.C., LAHERA, V., SALOM, M.G. and BIONDI, M.L. 1992. Role of the endothelin-dependent relaxing factor nitric oxide on renal function. *J. Am. Soc. Nephrol.* **2**: 1371–1387.

ROOS, J.C., KOOMANS, H.A., MEES, E.J.D. and DELAWI, I.M.K. 1985. Renal sodium handling in normal humans subjected to low, normal and extremely high sodium supplies. *Am. J. Physiol.* **249F**: 941–947.

ROSE, A.M. and VALDES, R. 1994. Understanding the sodium pump and its relevance to disease. *Clin. Chem.* **40**: 1674–1685.

ROSE, B.D. 1986. New approach to disturbances in the plasma sodium concentration. *Am. J. Med.* **81**: 1033–1041.

ROSE, B.D. 1987. In: *Body Fluid Homeostasis*, pp. 409–438. BRENNER, B.M. and STEIN, R.J.H. (Eds). Churchill Livingstone, New York.

ROSE, R.J., EVANS, D.L., HENCKEL, P., KNIGHT, P.K., CLUER, D. and SALTIN, B. 1994. Metabolic responses to prolonged exercise in the racing camel. *Acta Physiol. Scand.* **150**: Suppl. 617, 49–60.

ROSENBERG, H.C., CHIU, T.H., PUTNEY, J.W. and ASKARI, A. 1986. Modification of membrane function by drugs. In: *Physiology of Membrane Disorders*, 2nd Edn, Chap. 23, pp. 369–382. ANDREOLI, T.E., HOFFMAN, J.F., FANESTIL, D.D. and SCHULTZ, S.G. (Eds). Plenum Press, New York.

ROSSAINT, R., KREBS, M., FORTHER, J., UNGER, V., FALKE, K. and KACZMARCZYK, G. 1993. Inferior vena caval pressure increase contributes to sodium and water retention during PEEP in awake dogs. *J. Appl. Physiol.* **75**: 2484–2492.

ROSSKOPF, D., FROMTER, E. and SIFFERT, W. 1993. Hypertensive sodium–proton exchanger phenotype persists in immortalized lymphoblasts from essential hypertensive patients. A cell culture model for human hypertension. *J. Clin. Invest.* **92**: 2553–2559.

ROUSE, D. and SUKI, W.N. 1994. Effects of neural and humoral agents on the renal tubules in congestive heart failure. *Sem. Nephrol.* **14**: 412–426.

ROUSE, I.L. and BEILIN, L.J. 1984. Vegetarian diet and blood pressure. *J. Hyperten.* **2**: 231–240.

RUDDY, M.C., ARORA, A., MALKA, E.S. and BIALY, G.B. 1993. Blood pressure variability and urinary electrolyte excretion in normotensive adults. *Am. J. Hyperten.* **6**: 480–486.

RUDOLPH, J.A., SPIER, S.J., BYRNS, G. and HOFFMAN, E.P. 1992. Linkage of hyperkalaemic periodic paralysis in Quarter horses to the horse adult skeletal muscle sodium channel gene. *Anim. Genet.* **23**: 241–250.

RUEGG, U.T. 1992. Ouabain – a link in the genesis of high blood pressure? *Experientia* **48**: 1102–1106.

RUFF, R.L. 1989. Periodic paralysis. In: *The Regulation of Potassium Balance*, pp. 303–323. SELDIN, D.W. and GIEBISCH, G. (Eds). Raven Press, New York.

RUILOPE, L.M., LAHERA, V., ARAQUE, A., SUAREZ, C., RODICIO, J.L. and ROMERO, J.C. 1994. Electrolyte excretion and sodium intake. *Am. J. Med. Sci.* **307**: Suppl. 1, S107–S111.

RUPPERT, M., OVERLACK, A., KOLLOCH, R., KRAFT, K., GOBEL, B. and STUMPFE K.O. 1993. Neurohormonal and metabolic effects of severe and moderate salt restriction in non-obese normotensive adults. *J. Hyperten.* **11**: 743–749.

RUPPERT, M., OVERLACK, A., KOLLOCH, R., KRAFT, K., LENNARZ, M. and STUMPE, K.O. 1994. Effects of severe and moderate salt restriction on serum lipids in nonobese normotensive adults. *Am. J. Med. Sci.* **307**: Suppl. 1, S87–S90.

RUSCH, N.J. and KOTCHEN, T.A. 1994. Vascular smooth muscle regulation by calcium, magnesium and potassium in hypertension. In: *Textbook of Hypertension*, pp. 188–199. SWALES, J.D. (Ed.). Blackwell, Oxford.

RUSSELL, F.C. and DUNCAN, D.C. 1956. *Minerals in Pasture*. Commonwealth Agricultural Bureaux, Farnham Royal.

RYAN, M.P. 1987. Diuretics and potassium/magnesium depletion. *Am. J. Med.* **82**: 38–47.

RYAN, M.P. 1993. Interrelationships of magnesium and potassium homeostasis. *Miner. Elect. Metab.* **19**: 290–295.

RYAN, M.P. and BRADY, H.R. 1984. The role of magnesium in the prevention and control of hypertension. *Am. J. Clin. Res.* **43**: Suppl. 82–88.

SABATINI, S. 1988. Analgesic-induced papillary necrosis. *Sem. Nephrol.* **8**: 41–54.

SABATINI, S. 1989. Pathophysiologic mechanisms of collecting duct function. *Sem. Nephrol.* **9**: 179–202.

SABATINI, S. and KURTZMAN, N.A. 1994. Metabolic alkalosis. In: *Maxwell and Kleeman's Clinical Disorders of Fluid and Electrolyte Metabolism*, 5th Edn, Chap. 29, pp. 933–956. NARINS, R.G. (Ed.). McGraw-Hill, New York.

SAFAR, M.E., ASMAR, R., BENETOS, A., LEVY, B.I. and LONDON, G.M. 1994. Sodium, large arteries and diuretic compounds in hypertension. *Am. J. Med. Sci.* **307**: Suppl. 1, S3–S8.

SAFWATE, A., DAVICCO, M.-J., BARLET, J.P. and DELOST, P. 1981. Sodium and potassium in blood and milk and plasma aldosterone in high-yield dairy cows. *Reprod. Nutr. Devl.* **21**: 601–610.

SAGIE, A., LARSON, M.G. and LEVY, D. 1993. The natural history of borderline isolated systolic hypertension. *New Engl. J. Med.* **329**: 1912–1917.

SAGNELLA, G.A. and MACGREGOR, G.A. 1994. Atrial natriuretic peptides. In: *Textbook of Hypertension*, Chap. 12, pp. 273–288. SWALES, J.D. (Ed.). Blackwell Scientific, Oxford.

SAKAI, R.R. 1986. The hormones of renal sodium conservation act synergistically to arouse a sodium appetite in the rat. In: *The Physiology of Thirst and Sodium Appetite*, pp. 425–430. DE CARO, G., EPSTEIN, A.N. and MASSI, M. (Eds). Plenum Press, New York.

SAKHAEE, K., HARVEY, J.A., PADALINO, P.K., WHITSON, P. and PAK, L.Y. 1993. The potential role of salt abuse on the risk for kidney stone formation. *J. Urol.* **150**: 310–312.

SALAKO, L.A. 1993. Hypertension in Africa and effectiveness of its management with various classes of antihypertensive drugs and in different socio–economic and cultural environments. *Clin. Expl Hyperten.* **15**: 997–1004.

SALLOUM, R.M., STEVENS, B.R., SCHULTZ, G.S. and SOUBA, W.W. 1993. Regulation of small intestinal glutamine transport by epidermal growth factor. *Surgery* **113**: 552–559.

SAMANI, N.J. 1994. Molecular genetics of susceptibility to the development of hypertension. *Br. Med. Bull.* **50**: 260–271.

SANCHEZ-CASTILLO, S.P., WARRENDER, S., WHITEHEAD, T.P. and JAMES, W.P.T. 1987. An assessment of the sources of dietary salt in a British population. *Clin. Sci.* **72**: 95–102.

SANCHO, J.M., GARCIA-METIN, E., GARCIA-ROBLES, R., SANTIRSO, R., VIKA, E., GURIERREZ-MERINO, C. and RICOTE, M. 1993. Properties of the purified hypothalamic pituitary Na/K ATP-ase inhibitor. *J. Cardiovasc. Pharmacol.* **22**: Suppl. 2, 32–34.

SANTOSHAM, M. and GREENOUGH, W.B. 1991. Oral rehydration therapy: a global perspective. *J. Pediatr.* **118**: S44–S52.

SCHADT, J.C. 1994. Experimental observations—animals. In: *Blood Loss and Shock*, pp. 11–23. SECHER, N.H., PAWELCZYK, J.A. and LUDBROOK, J. (Eds). Edward Arnold, London.

SCHARRER, E. and MEDL, S. 1982. The electrical potential difference, short circuit current and Na^+ and Cl^- transport across different segments of sheep colon. *Comp. Biochem. Physiol.* **73A:**, 413–416.

SCHECHTER, P.J., HORWITZ, D. and HENKIN, R.I. 1974. Salt preference in patients with untreated and treated essential hypertension. *Am. J. Med. Sci.* **267**: 320–326.

SCHEDL, H.P., MAUGHAN, R.J. and GISOLFI, C.V. 1993. Intestinal absorption during rest and exercise: implications for formulating an oral rehydration solution (ORS). *Med. Sci. Sports Exerc.* **26**: 267–280.

SCHEIBLER, T.H. and DANNER, K.G. 1978. The effect of sex hormones on the proximal tubules in the rat kidney. *Cell Tissue Res.* **192**: 527–549.

SCHELLING, J.R. and LINAS, S.L. 1990. Hepatorenal syndrome. *Sem. Nephrol.* **10**: 565–570.

SCHERTEL, E.R., VALENTINE, A.K., SCHMALL, L.M., ALLEN, D.A. and MUIR, W.W. 1991. Vagotomy alters the hemodynamic response of dogs in hemorrhagic shock. *Circ. Shock* **35**: 393–397.

SCHIELKE, G.P. and BETZ, A.L. 1992. Electrolyte transport. In: *Physiology and Pharmacology of the Blood Brain Barrier*, pp. 221–243. BRADBURY, M.W.B. (Ed.). Springer, Berlin.

SCHIFFMAN, S.S., LOCKHEAD, E. and MAES, F.W. 1987. Amiloride reduces the taste intensity of Na^+ and Li^+ salts and sweeteners. *Proc. Nat. Acad. Sci.* **80**: 6136–6140.

SCHLESSINGER, S.D., TANKERSLEY, M.R. and CURTIS, J.J. 1994. Clinical documentation of end–stage renal disease due to hypertension. *Am. J. Kid. Dis.* **23**: 655–660.

SCHMIDT-NIELSEN, K., SCHMIDT-NIELSEN, B., JARNUM, S.A. and HOUPT, T.R. 1957. Body temperature of the camel and its relation to water economy. *Am. J. Physiol.* **188**: 103–112.

SCHOTT, H.C., HODGSON, D.R., BAYLY, W.M. and GOLLNICK, P.D. 1991. Renal responses to high intensity exercise. *Equine Exercise Physiol.* **3**: 361–367.

SCHRADER, J., GRUWEC, H., DECKING, U. and ALVES, C. 1994. Why do endothelial cells require adenosine triphosphate? *Arzneimittelforschung* **44**: 436–438.

SCHRIER, R.W. 1976. *Renal and Electrolyte Disorders*, pp. 167–223, 263–288. Little Brown, Boston, MA.

SCHRIER, R.W. 1988. Effective blood volume revisited: pathogenesis of oedematous disorders. In: *Nephrology*, pp. 663–675. DAVISON, A.M. (Ed.). Bailliere Tindall, London.

SCHRIER, R.W. and BRINER, V.A. 1991. Peripheral arterial vasodilation hypothesis of sodium and water retention in pregnancy: implications for pathogenesis of pre–eclampsia– eclampsia. *Obstet. Gynecol.* **77**: 632–639.

SCHRIER, R.W. and DURR, J.A. 1987. Pregnancy: an overfill or underfill state. *Am. J. Kidney Dis.* **9**: 284–289.

SCHRIER, R.W., NIEDERBERGER, M., WEIGERT, A. and GINES, P. 1994a. Peripheral arterial vasodilation: determinant of functional spectrum of cirrhosis. *Sem. Liver Dis.* **14**: 14–22.

SCHRIER, R.W., SHAPIRO, J.I., CHAN, L. and HARRIS, D.C. 1994b. Increased nephron oxygen consumption: potential role in progression of chronic renal disease. *Am. J. Kidney Dis.* **23**: 176–182.

SCHRYVER, H.F., PARKER, M.T., DANILUK, P.D., PAGAN, K.I., WILLIAMS, J., SODERHOLM, L.V. and HINTZ, H.F. 1987. Salt consumption and the effect of salt on mineral metabolism in horses. *Cornell Vet.* **77**: 122–131.

SCHULKIN, J. 1991. *Sodium Hunger: the Search for a Salty Taste*. Cambridge University Press, Cambridge.

SCHULTZ, P.J. and TOLINS, J.P. 1993. Adaptation to increased dietary salt intake in the rat. Role of endogenous nitric oxide. *J. Clin. Invest.* **91**: 642–650.

SCHULTZ, S.G. 1981. Salt and water absorption by the mammalian small intestine. In: *Physiology of the Gastrointestinal Tract*, Vol. 2, pp. 991–1002. JOHNSON, L.R. (Ed.). Raven Press, New York.

SCHULTZ, S.G. and HUDSON, R.L. 1991. Biology of sodium–absorbing epithelial cells: dawning of a new era. In: *Handbook of Physiology*, Section **6**: *Gastrointestinal Tract*, Vol. IV, *Intestinal Secretion and Absorption*, Chap. 2, pp. 45–81. FIELD, M. and FRIZELL, R.A. (Eds). American Physiological Society, Bethesda, MD.

SCHUMER, W. 1986. Cellular metabolism in shock. *Klin. Wochenschr.* **64**: Suppl. 7, 7–13.

SCHUNKERT, H., HENSE, H.-W., HOLMES, S.R., STENDER, M., PERZ, S., KEIL, U., LORELL, B.H. and REIGGER, G.A.J. 1994. Association between a deletion polymorphism of the angiotensin–converting enzyme gene and left ventricular hypertrophy. *New Engl. J. Med.* **330**: 1634–1638.

SCHUSTER, V.L. 1989. Potassium deficiency: pathogenesis and treatment. In: *The Regulation of Potassium Balance*, Chap. 10, pp. 241–268. SELDIN, D.W. and GIEBISCH, G. (Eds). Raven Press, New York.

SCOTT, D.A. and EDELMAN, R. 1993. Treatment of gastrointestinal infections. *Baillieres* Clin. Gastroenterol. 7: 477–499.

SCOTT, J. 1987. How is peripheral input processed in the central nervous system? *Ann. N.Y. Acad. Sci.* **510**: 133–134.

SEBASTIAN, A., HARRIS, S.T., OTTAWAY, J.H., TODD, K.M. and MORRIS, R.C. 1994. Improved mineral balance and skeletal metabolism in postmenopausal women treated with sodium bicarbonate. *New Engl. J. Med.* **330**: 1821–1822.

SECHER, N.H. and FRIEDMAN, D.B. 1994. Treatment of hypovolaemic shock: pharmacological intervention. In: *Blood Loss and Shock*, pp. 173–178. SECHER, N.H., PAWELCZYK, J.A. and LUDBROOK, J. (Eds). Edward Arnold, London.

SECHER, N.H., PAWELCZYK, J.A. and LUDBROOK, J. 1994. *Blood Loss and Shock* Edward Arnold, London.

SEEGER, W. 1987. Clinical features and pathophysiology of lung failure in shock. *J. Clin. Chem. Clin. Biochem.* **25**: 209–211.

SEELEY, J.E. and LARAGH, J.H. 1990. The integrated regulation of electrolyte balance and blood pressure by the renin system. In: *The Regulation of Sodium and Chloride Balance* SELDIN, D.W. and GIEBISCH, G. (Eds). Raven Press, New York. pp. 133–193.

SELDIN, D.W. 1990. Sodium balance and fluid volume in normal and edematous states. In: *The Regulation of Sodium and Chloride Balance*, pp. 261–292. SELDIN, D.W. and GIEBISCH, G. (Eds). Raven Press, New York.

SELDIN, D.W. and GIEBISCH, G. 1989. *The Regulation of Potassium Balance*, pp. 241–268, 325–345. Raven Press, New York.

SELDIN, D.W. and GIEBISCH, G. 1990. *The Regulation of Sodium and Chloride Balance*. Raven Press, New York.

SELDIN, D.W., PREISIG, P.A. and ALPERN, R.J. 1991. Regulation of proximal reabsorption by effective arterial volume. *Sem. Nephrol.* **11**: 212–219.

SEMPLICINI, A., CEOLOTTO, G., MASSIMINO, M., VALLE, R., SERENA, L., DETONI, R., PESSINA, A.C. and DAL PALU, C. 1994. Interactions between insulin and sodium homeostasis in essential hypertension. *Am. J. Med. Sci.* **307**: Suppl. 1, S43–S46.

SEVER, P. 1993. Management guidelines in essential hypertension: report of the second working party of the British Hypertension Society. *Br. Med. J.* **306**: 983–987.

SEYMOUR, A.A., DAVIS, J.O., FREEMAN, R.H., DE FORREST, J.M., ROWE, B.P., STEVENS, G.A. and WILLIAMS, G.M. 1980. Hypertension produced by sodium depletion and unilateral nephrectomy: a new experimental model. *Hypertension* **2**: 125–129.

SFERRA, T.J. and COLLINS, F.S. 1993. The molecular biology of cystic fibrosis. *A. Rev. Med.* **44**: 133–144.

SHAN, J., PANG, P.K., LIN, H.C. and YANG, M.C. 1994. Cardiovascular effects of human parathyroid hormone and parathyroid hormone–related peptide. *J. Cardiovasc. Pharmacol.* **23**: Suppl. 2, 38–41.

SHANNON, R.P., MINAKER, K.L. and ROWE, J.W. 1984. Aging and water balance in humans. *Sem. Nephrol.* **4**: 346–353.

SHAPER, A.G. 1967. Blood pressure studies in East Africa. In: *Epidemiology of Hypertension*. STAMLER, J., STAMLER, R. and PULLMAN, T.N. (Eds). Grune and Stratton, New York.

SHAPIRO, J.T. and ANDERSON, R.J. 1987. Sodium depletion states. In: *Body Fluid Homeostasis*, pp. 245–276. STEIN, J.H. and BRENNER, B.M. (Eds). Churchill Livingstone, New York.

SHAPIRO, R.J. and DIRKS, J.H. 1990. Edema of congestive heart failure. In: *The Regulation of Sodium and Chloride Balance*, Chap. 9, pp. 293–320. SELDIN, D.W. and GIEBISCH, G. (Eds). Raven Press, New York.

SHARMA, A.M. and DISTLER, A. 1994. Acid–base abnormalities in hypertension. *Am. J. Med. Sci.* **307**: Suppl. 1, S112–S115.

SHAW, R.K. and PHILLIPS, P.H. 1953. The potassium and sodium requirements of certain mammals. *Lancet* **73**: 176–180.

SHIMAMOTO, K., HIRATA, A., KUKVOKA, M., HIGASHURA, K., MIYAZAKI, Y. and SHIIKI, M. 1994. Insulin sensitivity and the effect of insulin on renal sodium handling and pressure systems in essential hypertensive patients. *Hypertension* **23**: Suppl. I, 129–133.

SHIRAKAMI, G., MAGARIBUCHI, T., SHINGU, K., SUGA, S., TAMAI, S., NAKAO, K. and MORI, K. 1993. Positive end-expiratory pressure ventilation decreases plasma atrial and brain natriuretic peptide levels in humans. *Anaesth. Analy.* **77**: 1116–1121.

SHORTT, C. and FLYNN, A. 1990. Sodium–calcium inter-relationships with specific reference to osteoporosis. *Nutr. Res. Revs.* **3**: 101–115.

SHRAZI-BEECHEY, S.P., HARAYAMA, B.A., WANG, Y., SCOTT, D., SMITH, M.W. and WRIGHT, E.M. 1991. Ontogenic development of lamb intestinal sodium–glucose cotransporter is regulated by diet. *J. Physiol.* **437**: 699–708.

SHULTZ, P.J. and TOLINS, J.P. 1993. Adaptation to dietary salt intake in the rat: role of endogenous nitric oxide. *J. Clin. Invest.* **91**: 642–650.

SIMON, G. 1990. Increased vascular wall sodium in hypertension: where is it, how does it get there and what does it do there? *Clin. Sci.* **78**: 533–540.

SIMPSON, F.O. 1988. Sodium intake, body sodium and sodium excretion. *Lancet* **i**: 25–28.

SINCLAIR, K.B. and JONES, D.I.H. 1968. Comparison of the weight gain and composition of blood and saliva in sheep grazing timothy and ryegrass swards. *Br. J. Nutr.* **22**: 661–666.

SINGER, D.R., MARKANDU, N.D., BUCKLEY, M.G., MILLER, M.A., SAGNELLA, G.A., LACHNO, D.R., CAPPUCCIO, F.P., MURDAY, A., YACOUB, M.H. and MACGREGOR, G.A. 1994. Blood pressure and endocrine responses to changes in dietary sodium intake in cardiac transplant recipients. Implications for the control of sodium balance. *Circulation* **89**: 1153–1159.

SISCOVICK, D.S., RAGHUNATHAN, T.E., PSATY, B.M., KOEPSELL, T.D., WICKLUND, K.G., LIN, X. *et al.* 1994. Diuretic therapy for hypertension and the risk of cardiac arrest. *New Engl. J. Med.* **330**: 1852–1857.

SJOLIN, J. 1991. High-dose corticosteroid therapy in human septic shock: has the jury reached a correct verdict? *Circulatory shock* **35**: 139–151.

SLADEN, R.N. 1994. Renal physiology. In: *Anaesthesia*, 4th Edn, pp. 663–688. MILLER, R.D. (Ed.). Churchill Livingstone, Edinburgh.

SMITH, S., ANDERSON, S., BALLERMAN, B.J. *et al.* 1986. Atrial natriuretic peptide in adaptation of sodium excretion with reduced renal mass. *J. Clin. Invest.* **77**: 1395–1398.

SMITH, T.W. 1989. Therapy of heart failure: two centuries of progress. In: *Cardiac and Renal Failure: an Expanding Role for ACE Inhibitors*, 173–184. DOLLERY, C.T. and SHERWOOD, L.M. (Eds). Hanley and Belfus, Philadelphia.

SNOW, D.H., BILLAH, A. and RIDHA, A. 1988. Effects of maximal exercise on the blood composition of the racing camel. *Vet. Rec.* **123**: 311–312.

SOARES, A.M., FREIRE, J.E. and LIMA, A.A.M. 1991. Transport of glutamine, alanine and glucose by rabbit intestinal membrane. *Braz. J. Med. Biol. Res.* **24:** 111–113.

SOLOMON, R. 1987. The relationship between disorders of K$^+$ and Mg$^+$ homeostasis. *Sem. Nephrol.* **7:** 253–262.

SOMMARDAHL, C.S., ANDREWS, F.M., SAXTON, A.M., GEISER, D.R. and MAYKUTA, P.L. 1994. Alterations in blood viscosity in horses competing in cross-country jumping. *Am. J. Vet. Res.* **55:** 389–394.

SONGU-MIZE, E. and BEALER, S.L. 1993. Effect of hypothalamic lesions on the interaction of centrally administered ANF and the circulating sodium–pump inhibitor. *J. Cardiovasc. Pharmacol.* **22:** Suppl. 2, S4–S6.

SONNENBLICK, M., FRIEDLANDER, Y. and ROSIN, A.J. 1993. Diuretic-induced severe hyponatraemia. Review and analysis of 129 reported patients. *Chest* **103:** 601–606.

SOUBA, W.W., PAN, M. and STEVENS, B.R. 1992. Kinetics of the sodium-dependent glutamine transporter in human intestinal confluent monolayers. *Biochem. Biophys. Res. Commun.* **188:** 746–753.

SPANGLER, W.L., GRIBBLE, D.H. and WEISER, M.G. 1977. Canine hypertension: a review. *J. Am. Vet. Med. Assoc.* **170:** 995–998.

STAMLER, J. 1993. Dietary salt and blood pressure. *Ann. N.Y. Acad. Sci.* **676:** 122–156.

STANTON, B.A. 1987. Regulation of Na$^+$ and K$^+$ transport by mineralocorticoids. *Sem. Nephrol.* **7:** 82–90.

STAR, R.A. 1993. Clinical consequences of hypernatremia and its correction. In: *Clinical Disturbances of Water Metabolism*, Chap. 3, pp. 237–247. SELDIN, D.W. and GIEBISCH, G. (Eds). Raven Press, New York.

STEEFEY, E.P., GIRI, S.N., DUNLOP, C.I., CULLEN, L.K., HODGSON, D.S. and WILLITS, N. 1993. Biochemical and haematological changes following prolonged halothane anaesthesia in horses. *Res. Vet. Sci.* **55:** 338–345.

STEEGERS, E.A.P., LAKWIJK, H.P.J.M., JONGSMA, H.W., FAST, J.H., DE BOO, T., ESKES, T.K.A.B. and HEIN, P.R. 1991. Pathophysiological implications of chronic dietary sodium restriction during pregnancy; a longitudinal prospective randomized study. *Br. J. Obstet. Gynecol.* **98:** 980–987.

STEIN, J.H., LIFSHITZ, M.D. and BARNES, L.D. 1978. Pathophysiology of acute renal failure. *Am. J. Physiol.* **234F:,** 171–181.

STEIN, P. and BLACK, H.R. 1993. The role of diet in the genesis and treatment of hypertension. *Med. Clin. N. Am.* **77:** 831–847.

STEINHAUSEN, M., HOLZ, F.G. and PAREKH, N. 1988. Regulation of pre– and post-glomerular resistances visualised in the split hydronephrotic kidney. In: *Nephrology*, Vol. 1, pp. 37–45. DAVISON, A.M. (Ed.). Bailliere Tindall, London.

STELLAR, E. 1993. Salt appetite: its neurendocrine basis. *Acta Neurobiol. Expl Warsz.* **53:** 475–484.

STERN, N., PALANT, C., OZAKI, L. and TUCK, M.L. 1994. Dexamethasone enhances active cation transport in cultured aortic smooth muscle cells. *Am. J. Hyperten.* **7:** 146–150.

STERNS, R.H. 1990. The management of symptomatic hyponatraemia. *Sem. Nephrol.* **10:** 503–514.

STERNS, R.H. and SPITAL, A. 1987. Disorders of internal potassium balance. *Sem. Nephrol.* **7:** 206–222.

STERNS, R.H., CLARK, E.C. and SILVER, S.M. 1993. Clinical consequences of hyponatremia and its correction. In: *Clinical Disturbances of Water Metabolism*,

Chap. 12, pp. 225–236. SELDIN, D.W. and GIEBISCH, G. (Eds). Raven Press, New York.

STERNS, R.H., OCDAL, H., SCHRIER, R.W. and NARINS, R.G. 1994. Hyponatraemia: pathophysiology, diagnosis and therapy. In: *Maxwell and Kleeman's Clinical Disorders of Fluid and Electrolyte Metabolism*, 5th Edn, Chap. 20, pp. 583–616. NARINS, R.G. (Ed.). McGraw-Hill, New York.

STOKES, J.B. 1989. Potassium intoxication: pathogenesis and treatment. In: *The Regulation of Potassium Balance*, pp. 269–302. SELDIN, D.W. and GIEBISCH, G. (Eds). Raven Press, New York.

STOKES, J.B. 1994. Principles of epithelial transport. In: *Maxwell and Kleeman's Clinical Disorders of Fluid and Electrolyte Metabolism*, Chap. 2, pp. 21–44. NARINS, R.G. (Ed.). McGraw-Hill, New York.

STOKKE, E.S., NAESS, P.A., OSTENSEN, J., LANGBERG, H.C. and KIIL, F. 1993. Plasma potassium concentration as a determinant of proximal tubular NaCl and NaHCO$_3$ reabsorption in dog kidneys. *Acta Physiol. Scand.* 148: 45–54.

STONE, D.K., CRIDER, B.P. and XIE, X.-S. 1990. Aldosterone and urinary acidification. *Sem. Nephrol.* 10: 375–379.

STRANGE, K. 1993. Maintenance of cell volume in the central nervous system. *Pediatr. Nephrol.* 7: 689–697.

STRAUSS, M.B., LANDIN, E., PIERCE-SMITH, W. and BLEIFER, D.J. 1958. Surfeit and deficit of sodium, a kinetic concept of sodium excretion. *Arch. Intern. Med.* 102: 527–536.

STRICKER, E.M., HOSUTT, J.A. and VERBALIS, J.G. 1987. Neurohypophyseal secretion in hypovolaemic rats: inverse relation to sodium appetite. *Am. J. Physiol.* 252: R889–R896.

STROHMAN, R.C. 1993. Ancient genomes, wise bodies, unhealthy people: limits of a genetic paradigm in biology and medicine. *Perspec. Biol. Med.* 37: 112–145.

SULLIVAN, J.M. and JOHNSON, J.G. 1983. The management of hypertension in patients with renal insufficiency. *Sem. Nephrol.* 3: 40–51.

SULLIVAN, S.K. and FIELD, M. 1991. Ion transport across mammalian small intestine. In: *Handbook of Physiology*, Section 6: *Gastrointestinal Secretion and Absorption*, pp. 320–327. FIELD, M. and FRIZELL. R.A. (Eds). American Physiological Society, Bethesda, MD.

SUTTERS, M., CARMICHAEL, J.S., LIGHTMAN, S.L. and PEART, W.S. 1993. Effect of chronic low-dose arginine vasopressin infusion on body fluid homeostasis during adaptation from a high- to a low-sodium diet in normal men. *Clin. Sci.* 85: 465–470.

SUTTON, R.A.L. 1986. Calcium disorders. In: *Diuretics: Physiology, Pharmacology and Clinical Use*, Chap. 13, pp. 259–272. DIRKS. J.H. and SUTTON. R.A.L. (Eds). Saunders, Philadelphia.

SUTTON, R.A.L. and DRANCE, S.M. 1986. Other uses of diuretics. In: *Diuretics: Physiology, Pharmacology and Clinical Use*, Chap. 14, pp. 273–286. DIRKS, J.H. and SUTTON, R.A.L. (Eds). Saunders, Philadelphia.

SWALES, D. 1990. Studies of salt intake in hypertension: what can epidemiology teach us? *Am. J. Hyperten.* 3: 645–649.

SWALES, J.D. 1975. *Sodium Metabolism in Disease*, p. 1. Lloyd–Luke, London.

SWALES, J.D. 1988. Salt saga continued. *Br. Med. J.* 297: 307–308.

SWALES, J.D. 1991. Dietary salt and blood pressure: the role of meta-analyses. *J. Hyperten.* 9: Suppl. 6, S42–S46.

SWALES, J.D. 1994. Diet and blood pressure. In: *Consensus in Clinical Nutrition*, pp. 444–459. HEATLEY, R.V., GREEN, J.H. and LOSOWKY, M.S. (Eds). Cambridge University Press, Cambridge.

SWARTZ, R.D. 1990. Fluid, electrolyte and acid–base changes during renal failure. In: *Fluids and Electrolytes*, 2nd Edn, Chap. 12, pp. 689–780. KOKKO, J.P. and TANNEN, R.L. (Eds). Saunders, Philadelphia.

SWENSON, M.J. 1993. Physiological properties and cellular and chemical constituents of blood. In: *Dukes' Physiology of Domestic Animals*, 11th Edn, Chap. 3, pp. 22–48. SWENSON, M.J. and REECE, W.O. (Eds). Comstock Publishing, Cornell University Press., Ithaca, NY.

SYKES, M.K., VICKERS, M.D. and HULL, C.J. 1991. Direct measurement of intravascular pressure. In: *Principles of Measurement and Monitoring in Anaesthesia and Intensive Care*, Chap. 13, pp. 163–174. Blackwell Scientific, Oxford.

SZILAGYI, M., ANKE, M., BALOGH, I., REGIUS-MOLSENYI, A. and SURI, A. 1989. Lithium status and animal metabolism. 6th *Int. Trace Element Symp.*, Vol. 4, 1249–1261.

TAKAHASHI, M., NISHIKAWA, A., FURUKAWA, F., ENAMI, T., HASEGAWA, T. and HAYASHI, Y. 1994a. Dose-dependent promoting effects of sodium chloride (NaCl) on rat glandular stomach carcinogenesis initiated with *N*-methyl-*N*-nitro-*N*-nitrosoguanidine. *Carcinogenesis* **15**: 1429–1432.

TAKAHASHI, M., TOTSUNE, K. and MOURI, T. 1994b. Endothelin in chronic renal failure. *Nephron* **66**: 373–379.

TAKAMATA, A., MACK, G.W., GILLEN, C.M. and NADEL, E.R. 1994. Sodium appetite, thirst and body fluid regulation in humans during rehydration without sodium replacement. *Am. J. Physiol.* **266**: R1493–R1502.

TAMURA, R. and NARGREN, R. 1993. Acute sodium depletion enhances taste responses in the nucleus of the solitary tract of rats. *Proc. 11th Int. Conf. Physiology of Food and Fluid Intake*, Oxford. p.48.

TAN, A.C., RUSSEL, F.G., THEIN, T. and BENRAAD, T.J. 1993. Atrial natriuretic peptide. An overview of clinical pharmacology and pharmacokinetics. *Clin. Pharmacokinet.* **24**: 28–45.

TAUFIELD, P.A., ALES, K.L., RESNICK, L.M., DRUZIN, M.L., GERTNER, J.M. and LARAGH, J.H. 1987. Hypocalciuria in pre–eclampsia. *New Engl. J. Med.* **316**: 715–718.

TAYLOR, A.E. and RIPPE, B. 1986. Pulmonary edema. In: *Physiology of Membrane Disorders*, 2nd Edn, Chap. 56, pp. 1025–1039. ANDREOLI, T.E., HOFFMAN, J.F., FANESTIL, D.D., SCHULTZ, S.G. (Eds). Plenum Press, New York.

TAYLOR, D.J. 1983. *Pig Diseases*, 3rd Edn, pp. 226–227. Burlington Press, Cambridge.

TAYLOR, R.E., GLASS, J.T., RADKE, K.J. and SCHNEIDER, E.G. 1987. Specificity of effect of osmolality on aldosterone secretion. *Am. J. Physiol.* **252**: E118–123.

TAYLOR, S.H. 1987. In: *Diuretics in Heart Failure*, pp. 15–36. POOLE-WILSON, P.A. (Ed.). Royal Society of Medicine, London.

TEITELBAUM, I., KLEEMAN, C.R. and BERL, T. 1994. The physiology of the renal concentrating and diluting mechanisms. In: *Maxwell and Kleeman's Clinical Disorders of Fluid and Electrolyte Metabolism*, 5th Edn, pp. 101–128. McGraw-Hill, New York.

TERRAGNO, N.A. and TERRAGNO, A. 1988. Mechanisms of hypertension in pregnancy. *Sem. Nephrol.* **8**: 138–146.

TESTA, M.A., ANDERSON, R.B., NACKLEY, J.F. and HOLLENBERG, N.K. 1993. Quality of life and antihypertensive therapy in men. *New Engl. J. Med.* **328**: 907–913.

THATCHER, C.D. and KEITH, J.C. 1988. Pregnancy-induced hypertension: development of a model in the pregnant sheep. *Am. J. Obstet. Gynecol.* **155**: 201–207.

THIER, S.O. 1981. The kidney. In: *Pathophysiology*, Chap. 9, pp. 799–920. SMITH, L.H. and THIER, S.O. (Eds). Saunders, Philadelphia.

THILLAINAYAGAM, A.V., DIAS, J.A., SALIM, A.F., MOURAD, F.H., CLARK, M.L. and FARTHING, M.J. 1994. Glucose polymer in the fluid therapy of acute diarrhoea: studies in a model of rotavirus infection in neonatal rats. *Clin. Sci.* **86**: 469–477.

THOMSEN, J.K., FOGH-ANDERSEN, N. and JASZCZAK, P. 1993. Atrial natriuretic peptide decrease during normal pregnancy as related to hemodynamic changes and volume regulation. *Acta Obstet. Gynecol. Scand.* **72**: 103–110.

THOMSEN, J.K., FOGH-ANDERSEN, N. and JASZCZAK, P. 1994. Atrial natriuretic peptide, blood volume, aldosterone and sodium excretion during twin pregnancy. *Acta Obstet. Gynecol. Scand.* **73**: 14–20.

THORN, G.W., NELSON, K.R. and THORN, D.W. 1938. A study of the mechanism of edema associated with menstruation. *Endocrinology* **22**: 155–163.

THORNBOROUGH, J.R. and PASSO, S.S. 1975. The effects of estrogens on sodium and potassium metabolism in rats. *Endocrinology* **97**: 1528–1536.

THRASHER, T.N. and RAMSAY, D.J. 1993. Interactions between vasopressin and atrial natriuretic peptides. *Ann. N. Y. Acad. Sci.* **689**: 426–437.

THURAU, K. 1989. The formation of a glomerular filtrate. In: *Cardiac and Renal Failure: an Expanding Role for ACE-Inhibitors*, pp. 230–238. DOLLERY, C.T. and SHERWOOD, L.M. (Eds).

TOAL, C.B. and LEENAN, F.H.H. 1983. Dietary sodium restriction and development of hypertension in spontaneously hypertensive rats. *Am. J. Physiol.* **245**: H1081–H1084.

TOBIAN, L. 1979. Dietary salt and hypertension. *Am. J. Clin. Nutr.* **32**: 2659–2662, 2740–2748.

TOBIAN, L. and BINION, J.T. 1952. Tissue cations and water in arterial hypertension. *Circulation* **5**: 754–760.

TOBIAS, T.A., SCHERTEL, E.R., SCHMALL, L.M., WILBUR, N. and MUIR, W.M. 1993. Comparative effects of 7.5% NaCl in 6% dextran 70 and 0.9% saline on cardiorespiratory parameters after cardiac output-controlled resuscitation from canine haemorrhagic shock. *Circ. Shock* **39**: 139–146.

TOLINS, J.P. and RAIJ, L. 1991. Antihypertensive therapy and the progression of chronic renal disease. Are there renoprotective drugs? *Sem. Nephrol.* **11**: 538–548.

TOMKIEWICZ, R.P., APP, E.M., ZAYAS, J.G., RAMIREZ, O., CHURCH, N., BOUCHER, R.C., KNOWLES, M.R. and KING,M. 1993. Amiloride inhalation therapy in cystic fibrosis. Influence on ion content, hydration and rheology of sputum. *A. Rev. Respir. Dis.* **148**: 1002–1007.

TOMLINSON, R.W.S. 1971. The action of progesterone derivatives and other steroids on the sodium transport of isolated frog skin. *Acta Physiol. Scand.* **83**: 407–411.

TORDOFF, M.G., SCHULKIN, J. and FRIEDMAN, M.I. 1987. Further evidence for hepatic control of salt appetite in rats. *Am. J. Physiol.* **253**: R1095–R1102.

TORDOFF, M.G., HUGHES, R.L. and PILCHAK, D.M. 1993. Independence of salt intake from the hormones regulating calcium homeostasis. *Am. J. Physiol.* **264R**: 500–512.

TOSTESON, D.C., DE FRIEZ, A.K., ADAMS, M., GOTTSCHALK, C.W. and LANDIS, E.M. 1951. Effects of adrenalectomy, desoxycorticosterone acetate and increased fluid intake on intake of sodium chloride and bicarbonate by hypertensive and normal rats. *Am. J. Physiol.* **164**: 369–379.

TSUGANE, S., TEI, Y., TAKAHASHI, T., WATANABE, S. and SUGANO, K. 1994. Salty food intake and risk of *Helicobacter pylori* infection. *Jpn J. Cancer Res.* **85**: 474–478.

TUCK, M.L. 1992. Obesity, the sympathetic nervous system and essential hypertension. *Hypertension* **19**: Suppl. 1, I67–I77.

TUCK, M.L. and CORRY, D.R. 1994. Electrolytes in the pathogenesis of hypertension: Part II—cation transport. In: *Maxwell and Kleeman's Clinical Disorders of Fluid and Electrolyte Metabolism*, 5th Edn, Chap. 48, pp. 1591–1618. NARINS, R.G. (Ed.). McGraw-Hill, New York.

TUSO, P.J., NISSENSON, A.R. and DANOVITCH, G.M. 1994. Electrolyte disorders in chronic renal failure. In: *Maxwell and Kleeman's Clinical Disorders of Fluid and Electrolyte Metabolism*, 5th Edn, Chap. 39, pp. 1195–1211. NARINS, R.G. (Ed.). McGraw-Hill, New York.

TUTTLE, K.R., DE FRONZO, R.A. and STEIN, J.H.S. 1991. Treatment of diabetic nephropathy: a rational approach based on its pathophysiology. *Sem. Nephrol.* **11**: 220–235.

TYLER, R.D., QUALLS, C.W., HEALD, R.D., COWELL, R.L. and CLINKENBEARD, K.D. 1987. Renal function in hyponatraemic dogs. *J. Am. Vet. Med. Assoc.* **191**: 1095–1099.

UNWIN, R.J. 1989. Neuropeptides and renal function: the potential for interaction between the gastrointestinal tract and kidney. In: *Concepts in Hypertension*, pp. 5–16. MATHIAS, C.J. and SEVER, P.S. (Eds). Springer, New York.

UNWIN, R.J., MOSS, S., PEART, W.S. and WADSWORTH, J. 1985. Renal adaptation and gut hormone release during sodium restriction in ileostomized man. *Clin. Sci.* **69**: 299–308.

USABA, A., MOTOKI, R., MIYAUCHI, Y., SUZUKI, K. and TAKAHASHI, A. 1991. Effect of neo red cells on the canine hemorrhagic shock model. *Int. J. Artif. Organs* **14**: 739–744.

VAGELLI, G., CALABRESE, G., MAZOTTA, A., PRATESI, G. and GONELLA, M. 1994. Arterial pressure in idiopathic calcium nephrolithiasis. *Minerva Urol. Nefrol.* **46**: 69–71.

VALENTIN, J.-P., QIU, C.Q., MULDOWNEY, W.P., YING, W.Z., GARDNER, D.G. and HUMPHREYS, M.H. 1992. Cellular basis for blunted volume expansion natriuresis in experimental nephrotic syndrome. *J. Clin. Invest.* **90**: 1302–1312.

VALENTIN, J.-P., WIEDEMANN, E. and HUMPHREYS, M.H. 1993. Natriuretic properties of melanocyte-stimulating hormones. *J. Cardiovasc. Pharmacol.* **22**: Suppl. 2, S114–S118.

VAN DE KERK, P. 1968. Sodium supply for cattle on farms in the Vorden-Hengelo area. Tijdschr. *Diergeneesk.* **93**: 55–65.

VAN DER HULST, V.N., VAN KREEL, B.K., MEYENFELDT, M.F., BRUMMER, R.J.M., ARENDS, J.W., DEUTZ, N.E.P. and SOETERS, P.B. 1993. Glutamine and the preservation of gut integrity. *Lancet* **341**: 1363–1365.

VEITH, I. 1966. *The Yellow Emperor's Classic in Internal Medicine*. Berkeley University Press, Berkeley, CA.

VERREY, F. 1990. Regulation of gene expression by aldosterone in tight epithelia. *Sem. Nephrol.* **10**: 410–420.

VIJANDE, M., COSTALES, M. and FITZSIMONS, J.T. 1986. Increased sodium appetite and polydipsia in Goldblatt hypertension. In: *The Physiology of Thirst and Sodium Appetite*, pp. 413–418. DE CARO, G., EPSTEIN, A.N. and MASSI, M. (Eds). Plenum Press, New York.

VINCENT, I.C. and MICHELL, A.R. 1992. Comparison of cortisol concentration in saliva and plasma of dogs. *Res. Vet. Sci.* **53**: 342–345.

VINCENT, I.C., MICHELL, A.R. and LEAHY, R. 1993. Non-invasive measurement of arterial blood pressure in dogs: a potential indicator for identification of stress. *Res. Vet. Sci.* **54**: 196–202.

VIVAS, L. and CHIARAVIGLIO, G. 1987. Effects of agents which alter sodium transport or sodium appetite in rats. *Brain Res. Bull.* **19**: 679–685.

VOELKER, J.R. and BRATER, D.C. 1990. Diuretics: applied pharmacokinetics and drug resistance. In: *The Regulation of Sodium and Chloride Balance*, Chap. 18, pp. 503–524. SELDIN, D.W. and GIEBISCH, G. (Eds). Raven Press, New York.

VOLPE, M. 1992. Atrial natriuretic peptide and the baroreflex control of circulation. *Am. J. Hyperten.* **5**: 488–493.

VON BUNGE, G. 1902. *Textbook of Physiological and Pathological Chemistry*, 2nd English Edn, from 4th German Edn, Lecture 7. STARLING, F.A. and STARLING, E.H. (Eds). Blakiston, Philadelphia.

VON HATTINGBERG, H.M. 1992. Water; mechanisms of oral rehydration, water deficiency=deficiency in salt. *Meth. Findings Clin. Pharmacol.* **14**: 289–295.

WADE, C.E., HANNON, J.P., BOSSONE, C.A., HUNT, M.M., LOVEDAY, J.A., COPPES, R.I. and GILDENGORIN, V.L. 1991. Neuroendocrine responses to hypertonic saline/dextran resuscitation following hemorrhage. *Circ. Shock* **35**: 37–43.

WADE, C.E., TILLMAN, F.J., LOVEDAY, J.A., BLACKMON, A., POTANKO, E., HUNT, E.M. and HANNON, J.P. 1992. Effect of dehydration on cardiovascular responses and electrolytes after hypertonic saline/dextran treatment for moderate hemorrhage. *Ann. Emerg. Med.* **21**: 113–119.

WALKER, B.R. and PADFIELD, P.L. 1994. Subgroups of hypertension. In: *Textbook of Hypertension*, Chap. 39, pp. 735–739. SWALES, J.D. (Ed.). Blackwell Scientific, Oxford.

WALLACE, A.G. 1981. Pathophysiology of cardiovascular disease. In *Pathophysiology*, Chap. 11, pp. 1072–1264. SMITH, L.H. and THIER, S.O. (Eds). Saunders, Philadelphia.

WALSER, M. 1985a. Phenomenological analysis of renal regulation of sodium and potassium balance. *Kidney Int.* **27**: 837–841.

WALSER, M. 1985b. Phenomenological analysis of electrolyte and water homeostasis. In: *The Kidney: Physiology and Pathophysiology*, pp. 3–13. SELDIN, D.W. and GIEBISCH, G. (Eds). Raven Press, New York.

WANG, G.K., SCHMEID, R., EBNER, R. and KORTH, M. 1993. Intracellular sodium activity and its regulation in guinea-pig atrial myocardium. *J. Physiol.* **465**: 73–84.

WARNER, L., SKORECKI, K., BLENDIS, L.M. and EPSTEIN, M. 1993. Atrial natriuretic factor and liver disease. *Hepatology* **17**: 500–511.

WAR OFFICE 1908. Animal Management. Prepared in the Veterinary Department for General Staff, War Office. Harrison, London, p.276.

WATT, J.G. 1965. The use of fluid replacement in the treatment of neonatal diseases in calves. *Vet. Rec.* **77**: 1474–1486.

WEBB, G.D., ASHMEAD, G.G., AL-MAHDI, S., AULETTA, F.J. and MCLAUGHLIN, M.K. 1993. Changes in sodium transport during the human menstrual cycle and pregnancy. *Clin. Sci.* **84**: 401–405.

WEDER, A.B. 1991. Membrane sodium transport and salt sensitivity of blood pressure. *Hypertension* **17**: Suppl. 1, I75–I80.

WEDER, A.B. 1994. Sodium metabolism, hypertension and diabetes. *Am. J. Med. Sci.* **307**: Suppl. 1, S53–S59.

WEINBERG, M.S. 1993. Renal effects of angiotensin converting enzyme inhibitors in heart failure: a clinician's guide to minimizing azotemia and diuretic-induced electrolyte imbalances. *Clin. Ther.* **15**: 3–17.

WEINBERGER, M.H. 1987. Sodium chloride and blood pressure. *New Engl. J. Med.* **317**: 1084–1085.

WEISINGER, R.S., DENTON, D.A., MCKINLEY, M.J., MISEUS, R.R., PARK, R.G. and SIMPSON, J.B. 1993. Forebrain lesions that disrupt water homeostasis do not eliminate the sodium appetite of sodium deficiency in sheep. *Brain Res.* **628**: 166–178.

WEISS, M.L. 1986. Sodium appetite induced by sodium depletion is suppressed by cerebroventricular captopril. In: *The Physiology of Thirst and Sodium Appetite*, pp. 405–411. DE CARO, G., EPSTEIN, A.N. and MASSI, M. (Eds). Plenum Press, New York.

WEISS, N.M. and ROBERTSON, G.L. 1984. Water metabolism in endocrine disorders. *Sem. Nephrol.* **4**: 303–315.

WEISSTUCH, J.M. and DWARKIN, L.D. 1992. Does essential hypertension cause end–stage renal disease? *Kidney Int.* **41**: Suppl. 36., S33–S37.

WELBOURNE, T.C. 1990. Glucocorticoid control of ammoniogenesis in the proximal tubule. *Sem. Nephrol.* **10**: 339–349.

WELLARD, R.M. and ADAM, W.R. 1987. Water depletion, not oral sodium loading, increases levels of sodium, potassium-dependent adenosine triphosphatase inhibitors in rat plasma. *Clin. Sci.* **73**: 87–92.

WELLS, T.G. 1990. The pharmacology and therapeutics of diuretics in the pediatric patient. *Pediat. Clin. N. Am.* **37**.2: 463–503.

WELLS, T., FORSLING, M.L. and WINDLE, R.J. 1990. The vasopressin response to centrally administered hypertonic solutions in the conscious rat. *J. Physiol.* **427**: 483–493.

WESTENFELDER, C., BRUCE, H. and BARANOWSKI, R.L. 1993. Uremia alters rat brain expression of atrial, brain and C–type natriuretic peptides, growth factors and proto–oncogenes. *J. Am. Soc. Nephrol.* **4**: 787.

WHANG, R. 1987. Magnesium deficiency. *Am. J. Med.* **82**: 24–29.

WHELTON, P.K. and KALG, M.J. 1989. Magnesium and blood pressure: review of the epidemiologic and clinical trial experience. *Am. J. Cardiol.* **63**: 26G–30G.

WHITE, P.C. 1994. Disorders of aldosterone biosynthesis and action. *New Engl. J. Med.* **331**: 250–258.

WHO/UNICEF 1985. *The Management of Diarrhoea and Use of Oral Rehydration Therapy*, 2nd Edn. WHO, Geneva.

WHITLOCK, R.H., KESSLER, M.J. and TASKER, J.B. 1975. Salt (sodium). deficiency in dairy cattle: polyuria and polydipsia as prominent clinical features. *Cornell Vet.* **65**: 512–526.

WHITWORTH, J.A. and BROWN, M.A. 1993. Hypertension and the kidney. *Kidney Int.* **42**: S52–S58.

WIDMAN, L., WESTER, P., STEGMAYR, B. and WIRELL, M. 1993. The dose-dependent reduction in blood pressure through administration of magnesium. *Am. J. Hyperten.* **6**: 41–45.

WILCOX, C.S. 1989. Diuretics and potassium. In: *The Regulation of Potassium Balance*, Chap. 13, pp. 325–346. SELDIN, D.W. and GIEBISCH, G. (Eds). Raven Press, New York.

WILKINSON, R. 1994. Renal and renovascular hypertension. In: *Textbook of Hypertension*, Chap. 45, pp. 831–857. SWALES, J.D. (Ed.). Blackwell Scientific, Oxford.

WILKINSON, S.P. 1981. The kidney and liver diseases. *J. Clin. Path.* **34:** 1241–1244.

WILKINSON, S.P. 1984. Ascites, diuretics and surgery. In: *Liver,* pp. 308–330. WILLIAMS, R. and MADDREY, W.C. (Eds). Butterworths, London.

WILLIAMS, G.H. and DLUHY, R.G. 1994. Hypertensive states: associated fluid and electrolyte disturbances. In: *Maxwell and Kleeman's Clinical Disorders of Fluid and Electrolyte Metabolism,* 5th Edn, pp. 1619–1648. NARINS, R.G. (Ed.). McGraw-Hill, New York.

WILLIAMS, M.E. and EPSTEIN, F.H. 1989. Internal exchanges of potassium. In: *The Regulation of Potassium Balance,* pp. 3–30. SELDIN, D.W. and GIEBISCH, G. (Eds). Raven Press, New York.

WILLIAMS, G.H. and HOLLENBERG, N.K. 1991. Non-modulating hypertension: a subset of sodium-sensitive hypertension. *Hypertension* **17:** Suppl. 1, I81–I85.

WILLIAMS, R.R., HASSTEDT, S.J., HUNT, S.C., WU, L.L., HOPKINS, P.N., BERRY, T.D. *et al.* 1991. Genetic traits related to hypertension and electrolyte metabolism. *Hypertension* **17:** Suppl. 1, I69–I73.

WILLS, M.R. 1986. Magnesium and potassium interrelationships in cardiac disorders. *Drugs* **31:** Suppl. 4, 121–131.

WILSON, D.R. 1980. Pathophysiology of obstructive nephropathy. *Kidney Int.* **18:** 281–292.

WILSON, P.D. 1991. Cell biology of human autosomal dominant polycystic kidney disease. *Sem. Nephrol.* **11:** 607–616.

WINTERS, C.J., REEVES, W.B. and ANDREOLI, T.E. 1991. A survey of transport properties of the thick ascending limb. *Sem. Nephrol.* **11:** 236–247.

WITTEMAN, J.C.M., WILLETT, W.C. and STAMPFER, M.C. 1989. A prospective study of nutritional factors and hypertension among U.S. women. *Circulation* **80:** 1320–1327.

WOLF, G. 1965. The effect of DOCA on salt appetite in intact and adrenalectomised rats. *Am. J. Physiol.* **208:** 1281–1285.

WOLF, G. 1969. Innate mechanisms for regulation of sodium intake. In: *Olfaction and Taste,* pp. 548–553. PFAFFMANN, C. (Ed.). Rockefeller University Press, New York.

WOLF, G. and NEILSON, E.G. 1993. Angiotensin II as a hypertrophogenic cytokine for proximal tubular cells. *Kidney Int.* **43:** Suppl. 39, S100–107.

WOLF, G., DAHL, L.K. and MILLER, N.E. 1965. Voluntary sodium intake in two strains of rat with opposite genetic susceptibility to experimental hypertension. *Proc. Soc. Expl Biol. Med.* **120:** 301–305.

WOLF, G., McGOVERN, J.F. and DI CARA, L.V. 1974. Sodium appetite: some conceptual and methodologic aspects of a model drive system. *Behav. Biol.* **10:** 27–42.

WONG, F. and BLENDIS, L. 1994. Pathophysiology of sodium retention and ascites formation in cirrhosis: role of atrial natriuretic factor. *Sem. Liver Dis.* **14:** 59–70.

WONG, K.C., SCHAFER, P.G. and SCHULTZ, J.R. 1993. Hypokalemia and anesthetic implications. *Anaesth. Anal.* **77:** 1238–1260.

WONG, K.S., WILLIAMSON, P.M., BROWN, M.A., ZAMMIT, V.C., DENTON, D.A. and WHITWORTH, J.A. 1993. Effects of cortisol on blood pressure and salt preference in normal humans. *Clin. Expl Pharmacol. Physiol.* **20:** 121–126.

WOO, J., LAU, E., CHAN, A., COCKRAM, C. and SWAMINATHAN, R. 1992. Blood pressure and urinary cations in a Chinese population. *J. Human Hyperten.* **6:** 299–304.

WOODBURY, D.M. and KEMP, J.W. 1971. Pharmacology and mechanisms of action of diphenylhydantoin. *Psychiat. Neurol. Neurochir.* **74:** 91–115.

WOODS, J.E. 1983. Renal function in essential hypertension. *Sem. Nephrol.* **3**: 30–39.

WOOLFSON, R.G., POSTON, L. and DE WARDENER, H.E. 1994. Digoxin-like inhibitors of active sodium transport and blood pressure: the current status. *Kidney Int.* **46**: 297–309.

WRIGHT, F.S. 1982. Effects of urinary tract obstruction on glomerular filtration rate and renal blood flow. *Sem. Nephrol.* **2**: 5–16.

WRONG, O.M., EDMONDS, C.J. and CHADWICK, V.S. 1981. *The Large Intestine: its Role in Mammalian Nutrition and Homeostasis*. M.T.P., Lancaster.

YAGINUMA, T. and O'ROURKE, M. 1993. Modification of wave travel and reflection by vasodilator therapy. In: *Arterial Vasodilation: Mechanisms and Therapy*, pp. 50–61. O'ROURKE, M.F., SAFAR, M.E. and DZAI, V.J. (Eds). Edward Arnold, London.

YANOVER, M.J., BICHET, D.G. and ANDERSON, R.J. 1986. Cirrhosis, ascites and the nephrotic syndrome. In: *Diuretics; Physiology, Pharmacology and Clinical Use*, pp. 374–382. DIRKS, J.H. and SUTTON, R.A.L. (Eds). Saunders, Philadelphia.

YARED, A. and YOSHIOKA, T. 1989. Autoregulation of glomerular filtration in the young. *Sem. Nephrol.* **9**: 94–97.

YARGER, W.E. and BUERKERT, J. 1982. Effect of urinary tract obstruction on renal tubular function. *Sem. Nephrol.* **2**: 17–31.

YOKOTA, N., DE BOLD, M.L.K. and DE BOLD, A.J. 1993. Effects of HS-142-1 (HS) an antagonist of natriuretic peptide NP. bioactive receptors, on DOCA escape. *J. Am. Soc. Nephrol.* **4**: 450.

YOUNG, J.B. 1994. Do digitalis glycosides still have a role in congestive heart failure? *Cardiol. Clinics* **12.1**: 51–61.

YOUNG, J.B. and PRATT, C.M. 1994. Hemodynamic and hormonal alterations in patients with heart failure: toward a contemporary definition of heart failure. *Sem. Nephrol.* **14**: 427–440.

YOUNG, J.B. and ROBERTS, R. 1986. In: *Body Fluid Homeostasis*, pp. 151–169. BRENNER, B.M. and STEIN, R.J.H. (Eds). Churchill Livingstone, New York.

YOUNG, P.T. 1966. Hedonic organisation and regulation of behaviour. *Psychol. Rev.* **72**: 59–86.

YUAN, C., MANUNTA, P., CHEN, S., HAMLYN, J.M., HADDY, F.J. and PAMNANI, M.B. 1993. Role of ouabain–like factors in hypertension: effects of ouabain and certain endogenous ouabain-like factors in hypertension. *J. Cardiovasc. Pharmacol.* **22**: Suppl. 2, S10–S12.

ZACHARIAH, P.K. and SUMNER, W.E. 1993. The clinical utility of blood pressure load in hypertension. *Am. J. Hyperten.* **6**: S194–S197.

ZAMBRASKI, E.J. 1989. Renal nerves in renal sodium–retaining states: cirrhotic ascites, congestive heart failure, nephrotic syndrome. *Miner. Electrolyte Metab.* **15**: 88–96.

ZAMBRASKI, E.J. and DUNN, M.J. 1993. Effects of non–steroidal anti–inflammatory drugs on renal function. In: *Analgesic and NSAID-Induced Kidney Disease*, Chap. 13, pp. 147–173. STEWART, J.H. (Ed.). Oxford University Press, Oxford.

ZARDETTO-SMITH, A.M., THUNHORST, R.L., CICHA, M.Z. and JOHNSON, A.K. 1993. Afferent signaling and forebrain mechanisms in the behavioural control of extracellular fluid volume. *Ann. N. Y. Acad. Sci.* **689**: 161–176.

ZEIDEL, M.L. and BRENNER, B.M. 1987. Actions of atrial natriuretic peptides on the kidney. *Sem. Nephrol.* **7**: 91–97.

ZEILMANN, M. 1993. Hyperkalaemic periodic paralysis in horses. *Tierarztl. Prax.* **21**: 524–527.

ZEMEL, M.B., KRANIAK, J., STANDLEY, P.R. and SOWERS, J.R. 1988. Erythrocyte cation metabolism in salt-sensitive hypertensive blacks as affected by dietary sodium and calcium. *Am. J. Hyperten.* **1**: 386–392.

ZERBE, R.L. and ROBERTSON, G.L. 1994. Osmotic and non–osmotic regulation of thirst and vasopressin secretion. In: *Maxwell and Kleeman's Clinical Disorders of Fluid and Electrolyte Metabolism*, 5th Edn, Chap. 4, pp. 81–100. McGraw-Hill, New York.

ZHANG, C.L. and POPP, F.A. 1994. Log–normal distribution of physiological parameters and the coherence of biological systems. *Med. Hypotheses* **43**: 11–16.

ZHANG, W., FRANKEL, W.L., SINGH, A., LAITIN, E., KLURFELD, D. and ROMBEAU, J.L. 1993. Improvement of structure and function in orthotopic small bowel transplantation in the rat by glutamine. *Transplantation* **56**: 512–517.

ZHOU, J. and HOFFMAN, E.P. 1994. Pathophysiology of sodium channelopathies. *J. Biol. Chem.* **269**: 18563–18571.

ZONGAZO, M.A., CARAYON, A., MASSON, F., ISNARD, R., EURIN, J., MAISTRE, G. *et al.* 1992. Atrial natriuretic peptide during water deprivation or hemorrhage in rats. Relationship with arginine, vasopressin and osmolality. *J. Physiol. Paris* **86**: 167–175.

Abbreviations

Na	Sodium
Cl	Chloride
K	Potassium
HCO_3^-	Bicarbonate, Hydrogen carbonate
Ca	Calcium
H^+	Hydrogen ion: proton
Mg	Magnesium
Li	Lithium
A-II (& III)	Angiotensin II (& III)
ACE-inhibitor	Angiotensin converting enzyme inhibitor
ACTH	Adrenocorticotrophic hormone
ADH	Antidiuretic hormone, vasopressin
ANP	Atrial natriuretic peptide
ARC	Agricultural Research Council (U.K.)
ARF	Acute renal failure
ASTI/EDLI	Active sodium transport inhibitor/Endogenous digitalis-like inhibitor (of sodium transport).
ATN	Acute tubular necrosis
ATP	Adenosine triphosphate (the currency of energy metabolism)
AV3V	Anteroventral 3rd ventricle (of the brain)
BBB	Blood–brain barrier
BNP	Brain natriuretic peptide
cAMP, cGMP	Cyclic adenine or guanosine monophosphate
CD	Collecting duct
CF	Cystic fibrosis
CNS	Central nervous system
CNT	Connecting tubule
CRF	Chronic renal failure
CSF	Cerebrospinal fluid
DCT	Distal convoluted tubule

DHCC	Dihydroxycholecalciferol, the fully activated form in which vitamin D becomes a renal hormone
DOC(A)	Deoxycorticosterone (acetate), an adrenal mineralocorticoid
DPH	Diphenylhydantoin
ECF	Extracellular fluid
EDRF	Endothelium derived relaxing factor, probably nitric oxide
EGF	Epidermal growth factor
EPO	Erythropoietin, the renal hormone responsible for normal maturation of RBC
ET	Endothelin
FWC	Free water clearance
GABA	γ-Amino butyric acid
GFR	Glomerular filtration rate
GIP	Gastric inhibitory polypeptide (= glucose-dependent insulinotropic peptide)
G6PD	Glucose-6 phosphate dehydrogenase
5-HT	5-Hydroxytryptamine
HWI	Head-out water immersion
ICF	Intracellular fluid
ICT	Initial collecting tubule
IDDM	Insulin-dependent diabetes mellitus
ISF	Interstitial fluid
K_f	Filtration coefficient
LK	Low potassium
L-NMMA	Inhibitor of nitric oxide synthesis (L-N monomethyl arginine)
mmol	millimole: molecular (formula) weight in milligrams
mOsm	milli-osmole
MSH	Melanocyte stimulating hormone
Na–K ATPase	Sodium, potassium adenosine triphosphatase, the enzyme system underlying the 'sodium' pump.
NAS	National Academy of Sciences (U.S.A.)
NIDDM	Non-insulin-dependent diabetes mellitus
NSAI	Non-steroidal anti-inflammatory drug
ORS	Oral rehydration solution
ORT	Oral rehydration therapy
PAF	Platelet activating factor
pAH	*para*-Amino-hippuric acid; used to measure RBF
PCO_2	Partial pressure of carbon dioxide
PCV	Packed cell volume, haematocrit
PD	Potential difference
PG	Prostaglandin

PHF	Parathyroid hypertensive factor
PTH	Parathyroid hormone, parathormone
RAS	Renin–angiotensin system
RBC	Red blood cell, erythrocyte
RBF	Renal blood flow
RPF	Renal plasma flow
SIADH	Syndrome of inappropriate ADH secretion
SNGFR	Single nephron GFR
TAL	Thick ascending limb (of loop of Henlé)
VIP	Vasoactive intestinal peptide
WBC	White blood cell
WHO	World Health Organisation

Index

Page numbers in *italic* type indicate an illustration or table.